LOGIC AS A TOOL

LOGIC AS A TOOL
A GUIDE TO FORMAL LOGICAL REASONING

Valentin Goranko

Stockholm University, Sweden

Library of Congress Cataloging-in-Publication Data

Names: Goranko, Valentin, author.
Title: Logic as a tool : a guide to formal logical reasoning / Valentin
   Goranko.
Description: Chichester, UK ; Hoboken, NJ : John Wiley & Sons, 2016. |
   Includes bibliographical references and index.
Identifiers: LCCN 2016010458 (print) | LCCN 2016014532 (ebook) | ISBN 9781118880005 (cloth) |
ISBN 9781118880050 (pdf) | ISBN 9781118880043 (epub)
Subjects: LCSH: Logic–Textbooks.
Classification: LCC BC71 .G67 2016 (print) | LCC BC71 (ebook) | DDC
   511.3–dc23
LC record available at http://lccn.loc.gov/2016010458

A catalogue record for this book is available from the British Library.

Set in 10.5/12pt, TimesLTStd by SPi Global, Chennai, India.
Printed and bound in Malaysia by Vivar Printing Sdn Bhd

1   2016

*This book is dedicated to those from whom I have learned
and to those who will learn from it.*

Contents

Preface

Unlike most books and textbooks on logic, this one purports to teach logic not so much as a *subject to study*, but rather as a *tool to master and use* for performing and structuring correct reasoning. It introduces classical logic rather informally, with very few theorems and proofs (which are mainly located in the supplementary sections). Nevertheless, the exposition is systematic and precise, without compromising on the essential technical and conceptual issues and subtle points inherent in logic.

Aims

This textbook covers only the core of *classical logic*, which itself is just the heart of the vast and growing body of modern logic. The main aims of the book are:

1. to explain the language, grammar, meaning, and formal semantics of logical formulae, to help the reader understand the use of classical logical languages and be able both to formalize natural language statements in them and translate back from logical formulae to natural language;

2. to present, explain, and illustrate with examples the use of the most popular deductive systems (namely, axiomatic systems, Semantic Tableaux, Natural Deduction, and Resolution with the only notable exclusion being Sequent Calculus, which is essentially inter-reducible with Natural Deduction) for mechanizing and "computing" logical reasoning both on propositional and on first-order level, and to provide the reader with the necessary technical skills for practical derivations in them; and

3. to offer systematic advice and guidelines on how to organize and perform a logically correct and well-structured reasoning using these deductive systems and the reasoning techniques that they provide.

Summary of the content and main features

The structure of the book reflects the two levels of expression and reasoning in classical logic: propositional and first-order.

The first two chapters are devoted to propositional logic. In Chapter 1 I explain how to understand propositions and compute their truth values. I then introduce propositional

formulae and their truth tables and then discuss logical validity of propositional arguments. The fundamental notion here is that of *propositional logical consequence*. Then,
in Chapter 2, I present several *deductive systems* used for deriving logical consequences
in propositional logic and show how they can be used for checking the logical correctness of propositional arguments and reasoning. In a supplementary section at the end of
the chapter I sketch generic proofs of soundness and completeness of the propositional
deductive systems.

The exposition of propositional logic is uplifted to first-order logic in the following two
chapters. In Chapter 3 I present first-order structures and languages and then the syntax
and semantics (first informally, and then more rigorously) of first-order logic. Then I focus
on using first-order languages and translations between them and natural languages. In
the last section of this chapter I present and discuss the fundamental semantic concepts
of logical validity, consequence, and equivalence in first-order logic. Deductive systems
for first-order logic are introduced in Chapter 4 by extending the respective propositional
deductive systems with additional rules for the quantifiers. Derivations in each of these
are illustrated with several examples. Again in a supplementary section, I sketch generic
proofs of soundness and completeness of the deductive systems for first-order logic.

Chapter 5 contains some applications of classical logic to mathematical reasoning and
proofs, first in general and then specifically, for sets functions, relations, and arithmetic.
It consists of concise presentations of the basic theories of these, where the proofs are left
as exercises. The chapter ends with applications of classical logic to automated reasoning
and theorem proving, as well as to logic programming, illustrated briefly with Prolog.

The book ends with a comprehensive set of detailed solutions or answers to many of
the exercises.

The special features of this book include:

- concise exposition, with semi-formal but rigorous treatment of the minimum necessary
  theory;

- emphasis both on conceptual understanding by providing many examples, and on developing technical skills and building experience by providing numerous exercises, most
  of them standard, yet non-trivial, as well as full solutions or answers for many of them;

- solid and balanced coverage of semantic, syntactic, and deductive aspects of logic;

- some refreshing extras, such as a few logic-related cartoons scattered around, as well
  as many biographical boxes at the end of each section with photos and short texts on
  distinguished logicians, providing some background to their lives and contributions;

- selected references to other books on logic, listed at the end of each section, which are
  suitable for further reading on the topics covered in the section; and

- a supplementary website with slides, additional exercises, more solutions, and errata,
  which can be viewed at https://logicasatool.wordpress.com

For the instructor

The textbook is intended for introductory and intermediate courses in classical logic,
mainly for students in both mathematics and computer science, but is also suitable and
useful for more technically oriented courses for students in philosophy and social sciences.

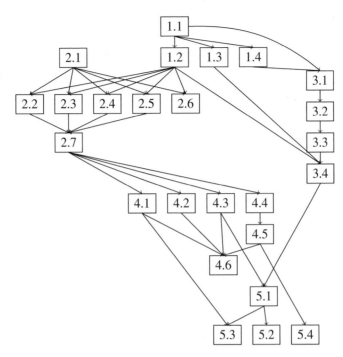

Dependancy chart

Some parts of the text and some exercises are much more relevant to only one of the main target audiences, and I have indicated them by using *Mathematics Track* 𝛑 and *Computer Science Track* ⊟ markers in the text. Everything else which is not explicitly included in either of these tracks should be suitable for both groups. Likewise, some specific topics and exercises are somewhat more advanced and are indicated with an *Advanced Track* marker ⚙. These are, of course, only indications.

The whole book can be covered in one or two semester courses, depending on the background and technical level of the audience. It assumes almost no specific prior knowledge, except some general background in college maths for specific topics and examples, usually indicated in *Mathematics* or *Advanced* tracks. A dependency chart of the different sections is provided in the figure above.

Acknowledgements

This textbook grew out of lecture notes that I have compiled for introductory logic courses for students in mathematics and computer science since the late 1990s. Many people have contributed in various ways to this book over these years. I am grateful to former colleagues who have helped with valuable comments, technical or editorial corrections, and some solutions to exercises, including Thomas Bolander, Willem Conradie, Ruaan Kellerman and Claudette Robinson, as well as to many students at the University of Johannesburg, the University of the Witwatersrand, and the Technical University of Denmark, who have sent me useful feedback and have noticed some errors in the lecture notes from which this book has evolved.

I also thank Christopher Burke, Mike Cavers, and Andreï Kostyrka for generously allowing me to include some of their cartoons in the book.

I gratefully acknowledge Wikipedia, the Stanford Encyclopedia of Philosophy, and the MacTutor History of Mathematics archive of the School of Mathematics and Statistics University of St Andrews as the main sources of the historical and biographical information provided.

The core content of the present book is a substantially extended version of the two chapters in logic in the recently published *Logic and Discrete Mathematics: A Concise Introduction* (written by Willem Conradie and myself, published by Wiley). I am grateful to Wiley for the continued collaboration. In particular, I thank everyone employed or contracted by Wiley who took part in the different stages of the technical preparation of this book.

Lastly, I owe very special thanks to my life partner Nina for her personal support and understanding during the work on this book, as well as for her invaluable technical help with producing many figures and diagrams and collecting photos and references for many of the historical boxes and the cover of the book.

Attributions for the photos used in the book

Most of the photos are public domain or under Creative Commons licence with free permissions. The photos of Cook, Fitting, Prawitz and Hodges are obtained from their owners or used with their permission. Attributions for the rest are listed below.

Davis
Author of photo: David Monniaux

Gentzen
Author of photo: "Eckart Menzler-Trott", permission obtained from the MFO "Archives of the Mathematisches Forschungsinstitut Oberwolfach"

Henkin
Author of photo: George M. Bergman

Herbrand
Author of photo: Journal des Sciences

Jaśkowski
Author of photo: Alojzy Czarnecki

Kleene
Author of photo: Konrad Jacobs, Erlangen, permission obtained from the MFO "Archives of the Mathematisches Forschungsinstitut Oberwolfach"

Levin
Author of photo: Wikipedia user Sergio01

Maltsev
Author of photo: Shiltsev

Robinson
Author of photo: David Monniaux

Putnam
Author of photo: Hilary Putnam

Skolem
Author of photo: Oslo Museum: image no. OB.F06426c (Byhistorisk samling/City historic collection), via oslobilder.no

Tarski
Author of photo: George M. Bergman, permission obtained from the MFO "Archives of the Mathematisches Forschungsinstitut Oberwolfach"

Wittgenstein
Author of photo: Moritz Nähr

Zeno
Author of photo: Wikipedia user Shakko

Introduction

What is logic about? What does it study and what does it offer? A usual definition found in the encyclopedia is that it is the branch of philosophy that studies the laws and rules of human reasoning. Little of this is actually correct. First, logic left the cradle of philosophy long ago and is now a truly interdisciplinary area, related and relevant not only to philosophy but also to mathematics, computer science, artificial intelligence, linguistics, social sciences, and even economics. Second, is logic really about how we reason? If that were the case, as a professional logician for many years I should already know quite well how exactly humans reason. Alas, the more experience I gain in life, the less I understand that. One thing is certain: most people use in their everyday reasoning emotions, analogies, clichés, ungrounded beliefs and superstitions, that is, everything but logic. But then, maybe logic studies the reasoning of the *rational* human, for whom reasoning is a purely rational brain activity? Well, while many (but far from all) of us humans reason with their brains, this is not sufficient to understand how we do it. As the American scientist Emerson M. Pugh brilliantly put it: "If the human brain were so simple that we could understand it, we would be so simple that we couldn't."

What does logic tell us, after all, if not how we reason? A better answer is: it tells us how we *can* – and ideally *should* – reason in a systematic and well-structured way that would *guarantee* that we always derive true and correct conclusions, providing we *only* use true assumptions and *only* apply logically correct rules of reasoning. Logic is therefore not just concerned with what is true and what is false, but rather with the *correctness* of our argumentation of what implies what and with the *validity* of our reasoning. What exactly does all that mean? This book aims to answer this question, beginning with some food for thought here.

The famous Greek philosopher Aristotle (384–322 BC), regarded as the founding father of formal logic, was the first who systematically studied and classified logically correct and incorrect forms and rules of reasoning. Aristotle studied specific patterns of arguments called *syllogisms*. Here is a typical example (mine, *not* Aristotle's) of a syllogism:

> All logicians are clever.
> All clever people are rich.
> ———————————————
> All logicians are rich.

The way we actually read this is as follows.

If all logicians are clever **and** all clever people are rich, **then** all logicians are rich.

This sounds intuitively like a correct piece of reasoning, and it *is*, but it does not mean that the conclusion is necessarily *true*. (In fact, unfortunately, it is not.) What then makes it correct?

Here is another example:

> All natural numbers are integers.
> Some integers are odd numbers.
> ———————————————
> Some natural numbers are odd numbers.

Note that all statements above are true. So, is this a correct argument? If you think so, then how about taking the same argument and replacing the words "natural numbers" by "mice," "integers" by "animals," and "odd numbers" by "elephants." This will not change the logical shape of the argument and, therefore, should not change its logical correctness. The result speaks for itself, however.

> All mice are animals.
> Some animals are elephants.
> ———————————————
> Some mice are elephants.

So what makes an argument logically correct? You will also find answers to this question in this book.

Let me say a few more concrete words about the main aspects and issues of classical logic treated in this book. There are two levels of logical discourse and reasoning in classical logic. The lower level is *propositional logic*, introduced and discussed in the first two chapters of this book, and the higher level is *first-order logic*, also known as *predicate logic*, treated in the rest of the book.

Propositional logic is about reasoning with *propositions*, sentences that can be assigned a *truth value* of either *true* or *false*. They are built from simple, *atomic propositions* by using *propositional logical connectives*. The truth values propagate over all propositions through *truth tables* for the propositional connectives.

Propositional logic can only formalize simple logical reasoning that can be expressed in terms of propositions and their truth values, but it is quite insufficient for practical knowledge representation and reasoning. For that, it needs to be extended with several additional features, including *constants (names)* and *variables* for objects of any nature (numbers, sets, points, human beings, etc.), *functions and predicates* over objects, as well as *quantifiers* such as "for all objects $x(\ldots x \ldots)$," and "there exists an object x such that $(\ldots x \ldots)$." These lead to *first-order languages*, which (in many-sorted versions) are essentially sufficient to formalize most common logical reasoning. Designing appropriately expressive logical languages and using them to capture fragments of natural languages and reasoning is one of the main tasks of modern logic.

There are three major aspects of a logical system: *semantic*; *syntactic*; and *deductive*. The former deals mostly with the semantic notions of *truth, validity and logical consequence*, whereas the latter two deal respectively with the *syntax and grammar of logical languages* and with systems for *logical deduction* and *derivations and deductive consequences*. Deductive systems are purely mechanical procedures designed to *derive (deduce)* logical validities and consequences by means of formal *rules of inference* and possibly some postulated derived formulae called *axioms*. Thus, a deductive system does not refer explicitly to the meaning (semantics) of the formulae but only treats them as special strings of symbols and acts on their shape (syntax). In principle, a deductive system can be used successfully without any understanding of what formulae mean, and derivations in a deductive system can be performed not only by humans but also by artificial "agents" or computers. However, deductive systems are always meant to capture (or even determine) logical consequence so, ideally, semantic logical consequence and deductive consequence should precisely match each other. If that is the case, we say that the deductive system is *sound and complete*, or just *adequate*. Design and study of adequate and practically useful deductive systems is another major logical task.

The main syntactic, semantic, and deductive aspects of classical logic are discussed in detail in the book; there is much more that is not treated here however, both inside and outside of classical logic. In particular, logic is deeply related to: the foundations of mathematics, via *axiomatic theories of sets*; mathematics itself via *model theory*; the important notions of *algorithmic decidability and computability* via *recursion theory*; and the fundamentals and limitations of the deductive approach via *proof theory*. All of these are major branches of logic that I will only mention briefly in the text, but much more can be seen in the references. Furthermore, there is a rich variety of other, more specialized *non-classical* logical languages and systems that are better suited for specific modes and aspects of reasoning, such as *intuitionistic, modal, temporal, epistemic, deontic, and non-monotonic logics* that will not (except briefly intuitionistic logic) be discussed at all in this book. References to relevant publications covering these topics are provided throughout.

Finally, a few final words on the role of logic in the modern world. As I mentioned earlier, contemporary logic has become a highly interdisciplinary area with fundamental applications to a wide variety of scientific fields including mathematics, philosophy, computer science, artificial intelligence, and linguistics. Today logic not only provides methodology for correct human reasoning, but also techniques and tools for automated reasoning of intelligent agents. It also provides theoretical foundations for basic concepts in computer science such as *computation and computability*, *algorithms and complexity*, and *semantics of programming languages*, as well as practical tools for formal specification, synthesis, analysis, and verification of software and hardware, development and management of intelligent databases, and logic programming. The impact of logic on computer science nowadays is often compared to the impact of differential and integral calculus on natural sciences and engineering from the 17th century.

I end this introduction with a humble hope that this book will help the reader understand and master the use of this great intellectual tool called Logic. Enjoy it!

Valentin Goranko
Stockholm, November 2015

An Appetizer: Logical Paradoxes and Self-Reference

Dear reader,

The sentence that you are reading now is not true.

Is this claim true or false? If true, then it truly claims that it is not true, so it can't be true. But then, it is *not true* that it is not true, it *must* be true! Or . . . ?

This is a version of probably the oldest known *logical paradox* since antiquity, also known as the **liar's paradox** which refers to the quote "I am lying now."

What is a logical paradox? It is a statement or an argument that presents an apparent logical contradiction, either with well-known and accepted truths or simply with itself. Unlike a *fallacy*, a paradox is not due to an incorrect reasoning, but it could be based on wordplay or on a subtle ambiguity in the assumptions or concepts involved. Most commonly however, logical paradoxes arise when using *self-reference*, such as in the opening sentence above. Logicians love playing with self-reference. For instance, I have added this sentence in order to make a reference to itself. And, this one, which *does not* make a reference to itself. (Or, does it . . . ?)

A variation of the liar's paradox is **Jourdain's card paradox**, which does not rely on immediate self-reference but on a circular reference. Here is a simple version:

The next sentence is true. The previous sentence is false.

I end this appetizer two more paradoxes which are not exactly logical but *semantic*, again a self-referential play but now with natural language.

The first is known as **Berry's paradox.** Clearly every natural number can be defined in English with sufficiently many words. However, if we bound the number of words to be used, then only finitely many natural numbers can be defined. Then, there will be numbers that cannot be defined with that many words. Hence, there must be a *least* so undefinable natural number. Now, consider the following sentence "*The least natural number that is not definable in English with less than twenty words.*" There is a uniquely determined natural number that satisfies this description, so it *is a definition* in English, right? Well, count how many words it uses.

The second is the **Grelling–Nelson paradox**. Divide all adjectives into two groups: **autological**, if and only if it describes itself, such as "English," "short," and "fourteen-letter;" and **heterological**, if and only if it does not describes itself, such as "wet," "white," and "long." Now, is the adjective "heterological" autological or heterological?

By Andreï Victorovitch Kostyrka, http://kostyrka.ru/blog

1

Understanding Propositional Logic

Propositional logic is about reasoning with propositions. These are sentences that can be assigned a truth value: *true* or *false*. They are built from primitive statements, called *atomic propositions*, by using *propositional logical connectives*. The truth values propagate over all propositions through *truth tables* for the propositional connectives. In this chapter I explain how to understand propositions and compute their truth values, and how to reason using schemes of propositions called *propositional formulae*. I will formally capture the concept of *logically correct propositional reasoning* by means of the fundamental notion of *propositional logical consequence*.

1.1 Propositions and logical connectives: truth tables and tautologies

1.1.1 Propositions

The basic concept of propositional logic is **proposition**. A proposition is a sentence that can be assigned a unique **truth value**: `true` or `false`.

Some simple examples of propositions include:

- The Sun is hot.
- The Earth is made of cheese.
- 2 plus 2 equals 22.
- The 1000th decimal digit of the number π is 9.
  (You probably don't know whether the latter is true or false, but it is surely *either true or false*.)

The following are not propositions (why?):

- Are you bored?
- Please, don't go away!
- She loves me.
- x is an integer.
- This sentence is false.

Logic as a Tool: A Guide to Formal Logical Reasoning, First Edition. Valentin Goranko.
© 2016 John Wiley & Sons, Ltd. Published 2016 by John Wiley & Sons, Ltd.

Here is why. The first sentence above is a question, and it does not make sense to declare it true or false. Likewise for the imperative second sentence. The truth of the third sentence depends on who "she" is and who utters the sentence. Likewise, the truth of the fourth sentence is not determined as long as the variable x is not assigned a value, integer or not. As for the last sentence, the reason is trickier: assuming that it is true it truly claims that it is false – a contradiction; assuming that it is false, it falsely claims that it is false, hence it is not false – a contradiction again. Therefore, no truth value can be consistently ascribed to it. Such sentences are known as *self-referential* and are the main source of various *logical paradoxes* (see the appetizer and Russell's paradox in Section 5.2.1).

1.1.2 Propositional logical connectives

The propositions above are very simple. They have no logical structure, so we call them **primitive** or **atomic** propositions. From primitive propositions one can construct **compound** propositions by using special words called **logical connectives**. The most commonly used connectives are:

- not, called **negation**, denoted $\neg$;
- and, called **conjunction**, denoted $\wedge$ (or sometimes $\&$);
- or, called **disjunction**, denoted $\vee$;
- if . . . then . . . , called **implication**, or **conditional**, denoted $\rightarrow$;
- . . . if and only if . . . , called **biconditional**, denoted $\leftrightarrow$.

Remark 1 *It is often not grammatically correct to read compound propositions by simply inserting the names of the logical connectives in between the atomic components. A typical problem arises with the negation: one does not say "Not the Earth is square." A uniform way to get around that difficulty and negate a proposition P is to say "It is not the case that P."*

In natural language grammar the binary propositional connectives, plus others like *but, because, unless, although, so, yet*, etc. are all called "conjunctions" because they "conjoin", that is, connect, sentences. In logic we use the propositional connectives to connect propositions. For instance, given the propositions

"Two plus two equals five" and "The Sun is hot"

we can form the propositions

- "It is **not** the case that two plus two equals five. "
- "Two plus two equals five **and** the Sun is hot."
- "Two plus two equals five **or** the Sun is hot."
- "**If** two plus two equals five **then** the Sun is hot."
- "Two plus two equals five **if and only if** the Sun is hot."

For a more involved example, from the propositions (we assume we have already decided the truth value of each)

> "Logic is fun", "Logic is easy", and "Logic is boring"

we can compose a proposition

> "Logic is not easy or if logic is fun then logic is easy and logic is not boring."

It sounds better smoothed out a bit:

> "Logic is not easy or if logic is fun then it is easy and not boring."

1.1.3   Truth tables

How about the truth value of a compound proposition? It can be *computed* from the truth values of the components[1] by following the rules of 'propositional arithmetic':

- *The proposition $\neg A$ is true if and only if*
  the proposition A is false.
- *The proposition $A \wedge B$ is true if and only if*
  both A and B are true.
- *The proposition $A \vee B$ is true if and only if*
  either of A or B (possibly both) is true.
- *The proposition $A \rightarrow B$ is true if and only if*
  A is false or B is true, that is, if the truth of A implies the truth of B.
- *The proposition $A \leftrightarrow B$ is true if and only if*
  A and B have the same truth values.

We can systematize these rules in something similar to multiplication tables. For that purpose, and to make it easier for symbolic (i.e., mathematical) manipulations, we introduce a special notation for the two truth values by denoting the value true by T and the value false by F. Another common notation, particularly in computer science, is to denote true by **1** and false by **0**.

The rules of the "propositional arithmetic" can be summarized by means of the following **truth tables** (p and q below represent arbitrary propositions):

p	$\neg p$		p	q	$p \wedge q$	$p \vee q$	$p \rightarrow q$	$p \leftrightarrow q$
T	F		T	T	T	T	T	T
F	T		T	F	F	T	F	F
			F	T	F	T	T	F
			F	F	F	F	T	T

---

[1] Much in the same way as we can compute the value of the algebraic expression $a \times (b - c) + b/a$ as soon as we know the values of a, b, c.

1.1.4   The meaning of the connectives in natural language and in logic

The use and meaning of the logical connectives in natural language does not always match their formal logical meaning. For instance, quite often the conjunction is loaded with a temporal succession and causal relationship that makes the common sense meanings of the sentences "The kid threw the stone and the window broke" and "The window broke and the kid threw the stone" quite different, while they have the same truth value by the truth table of the conjunction. Conjunction in natural language is therefore often non-commutative, while the logical conjunction is commutative. The conjunction is also often used to connect not entire sentences but only parts, in order to avoid repetition. For instance "The little princess is clever and beautiful" logically means "The little princess is clever and the little princess is beautiful." Several other conjunctive words in natural language, such as *but, yet, although, whereas, while* etc., translate into propositional logic as logical conjunction.

The disjunction in natural language also has its peculiarities. As for the conjunction, it is often used in a form which does not match the logical syntax, as in "The old stranger looked drunk, insane, or completely lost". Moreover, it is also used in an *exclusive* sense, for example in "I shall win or I shall die", while in formal logic we use it by convention in an *inclusive* sense, so "You will win or I will win" will be true if we both win. However, "exclusive or", abbreviated *Xor*, is sometimes used, especially in computer science. A few other conjunctive words in natural language, such as *unless*, can translate into propositional logic as logical disjunction, for instance "I will win, unless I die." However, it can also equivalently translate as an implication: "I will win, if I do not die."

Among all logical connectives, however, the implication seems to be the most debatable. Indeed, it is not so easy to accept that a proposition such as "If 2+2=5, then the Moon is made of cheese", if it makes any sense at all, should be assumed true. Even more questionable seems the truth of the proposition "If the Moon is made of chocolate then the Moon is made of cheese." The leading motivation to define the truth behavior of the implication is, of course, the logical meaning we assign to it. The proposition $A \to B$ means:

$$\textit{\textbf{If } A \textit{ is true, } \textbf{then } B \textit{ must be true,}}$$

Note that if A is not true, then the (truth of the) implication $A \to B$ requires *nothing* regarding the truth of B. There is therefore only one case where that proposition should be regarded as false, namely when A is true, and yet B is not true. In all other cases we have no reason to consider it false. For it to be a proposition, it must be regarded true. This argument justifies the truth table of the implication. It is very important to understand the idea behind that truth table, because the implication is the logical connective which is most closely related to the concepts of logical reasoning and deduction.

Remark 2 *It helps to think of an implication as a promise. For instance, Johnnie's father tells him: "If you pass your logic exam, then I'll buy you a motorbike." Then consider the four possible situations: Johnnie passes or fails his exam and his father buys or does not buy him a motorbike. Now, see in which of them the promise is kept (the implication is true) and in which it is broken (the implication is false).*

Some terminology: the proposition A in the implication $A \to B$ is called the **antecedent** and the proposition B is the **consequent** of the implication.

The implication $A \to B$ can be expressed in many different but "logically equivalent" (to be defined later) ways, which one should be able to recognize:

- A implies B.
- B follows from A.
- If A, B.
- B if A.
- A only if B.
- B whenever A.
- A is sufficient for B.
  (*Meaning: The truth of A is sufficient for the truth of B.*)
- B is necessary for A.
  (*Meaning: The truth of B is necessary for A to be true.*)

1.1.5   Computing truth values of propositions

It can be seen from the truth tables that the truth value of a compound proposition does not depend on the meaning of the component propositions, but only on their truth values. To check the truth of such a proposition, we merely need to replace all component propositions by their respective truth values and then "compute" the truth of the whole proposition using the truth tables of the logical connectives. It therefore follows that

- "It is not the case that two plus two equals five" is true;
- "Two plus two equals five and the Sun is hot" is false;
- "Two plus two equals five or the Sun is hot" is true; and
- "If two plus two equals five, then the Sun is hot" is true (even though it does not make good sense).

For the other example, suppose we agree that

"Logic is fun" is true,
"Logic is boring" is false,
"Logic is easy" is true.

Then the truth value of the compound proposition

"Logic is not easy or if logic is fun then it is easy and not boring."

can be determined just as easily. However, in order to do so, we first have to analyze the *syntactic structure* of the proposition, that is, to determine how it has been composed, in other words in what order the logical connectives occurring therein have been applied. With algebraic expressions such as $a \times (b - c) + b/c$ that analysis is a little easier, thanks to the use of parentheses and the established priority order among the arithmetic operations. We also make use of parentheses and rewrite the sentence in the way (presumably) we all understand it:

"(Logic is not easy) or ((if logic is fun) then ((logic is easy) and (logic is not boring)))."

The structure of the sentence should be clear now. We can however go one step further and make it look exactly like an algebraic expression by using letters to denote the occurring primitive propositions. For example, let us denote

"Logic is fun" A,
"Logic is boring" B, and
"Logic is easy" C.

Now our compound proposition can be neatly rewritten as

$$(\neg C) \lor (A \rightarrow (C \land \neg B)).$$

In our rather informal exposition we will not use parentheses very systematically, but only whenever necessary to avoid ambiguity. For that purpose we will, like in arithmetic, impose a priority order among the logical connectives, namely:

- the negation has the strongest binding power, that is, the highest priority;
- then come the conjunction and disjunction;
- then the implication; and
- the biconditional has the lowest priority.

Example 3 *The proposition* $\neg A \lor C \rightarrow A \land \neg B$ *is a simplified version of* $((\neg A) \lor C) \rightarrow (A \land \neg B)$.

The last step is to compute the truth value. Recall that is not the actual meaning of the component propositions that matters but *only their truth values*, so we can simply replace the atomic propositions A, B, and C by their truth values and perform the formal computation following the truth tables step-by-step:

$$(\neg T) \lor (T \rightarrow (T \land \neg F)) = F \lor (T \rightarrow (T \land T)) = F \lor (T \rightarrow T) = F \lor T = T.$$

So, logic *is* easy after all! (At least, so far.)

1.1.6   Propositional formulae and their truth tables

If we only discuss particular propositions our study of logic would be no more useful than a study of algebra based on particular equalities such as $2 + 3 = 5$ or $12345679 \times 9 = 111111111$. Instead, we should look at *schemes of propositions* and their properties, just like we study algebraic formulae and equations and their properties. We call such schemes of propositions **propositional formulae**.

1.1.6.1   Propositional formulae: basics

I first define a **formal language** in which propositional formulae, meant to be templates for composite propositions, will be special words. That language involves:

- **propositional constants**: special fixed propositions $\top$, that always takes a truth value `true`, and $\bot$, that always takes a value `false`;

- **propositional variables** $p, q, r \ldots$, possibly indexed, to denote unspecified propositions in the same way as we use algebraic variables to denote unspecified or unknown numbers;
- the **logical connectives** that we already know; and
- **auxiliary symbols**: parentheses (and) are used to indicate the order of application of logical connectives and make the formulae unambiguous.

Using these symbols we can construct propositional formulae in the same way in which we construct algebraic expressions from variables and arithmetic operations. Here are a few examples of propositional formulae:

$$\top, p, \neg\bot, \neg\neg p, p \vee \neg q, \quad p_1 \wedge \neg(p_2 \rightarrow (\neg p_1 \wedge \bot))$$

There are infinitely many possible propositional formulae so we cannot list them all here. However, there is a simple and elegant way to give a precise definition of propositional formulae, namely the so-called **inductive definition** (or **recursive definition**). It consists of the following clauses or **formation rules**:

1. Every propositional constant or variable is a propositional formula.
2. If A is a propositional formula then $\neg A$ is a propositional formula.
3. If A, B are propositional formulae then each of $(A \vee B), (A \wedge B), (A \rightarrow B)$, and $(A \leftrightarrow B)$ is a propositional formula.

We say that a propositional formula is any string of symbols that can be constructed by applying – in some order and possibly repeatedly – the rules above, and only objects that can be constructed in such a way are propositional formulae.

Note that the notion of propositional formula that we define above is used in its own definition; this is the idea of *structural induction*. The definition works as follows: the first rule above gives us some initial stock of propositional formulae; as we keep applying the other rules, we construct more and more formulae and use them further in the definition. Eventually, every propositional formula can be obtained in several (finitely many!) steps of applying these rules. We can therefore think of the definition above as a construction manual prescribing how new objects (here, propositional formulae) can be built from already constructed objects. I discuss inductive definitions in more detail in Section 1.4.5.

From this point, I omit the unnecessary pairs of parentheses according to our earlier convention whenever that would not lead to syntactic ambiguity.

The formulae that are used in the process of the construction of a formula A are called **subformulae of** A. The last propositional connective introduced in the construction of A is called the **main connective of** A and the formula(e) to which it is applied is/are the **main subformula(e) of** A. I make all these more precise in what follows.

Example 4 (Construction sequence, subformulae and main connectives) *One construction sequence for the formula*

$$(p \vee \neg(q \wedge \neg r)) \rightarrow \neg\neg r$$

is

$$p, q, r, \neg r, \neg\neg r, q \wedge \neg r, \neg(q \wedge \neg r), p \vee \neg(q \wedge \neg r), (p \vee \neg(q \wedge \neg r)) \rightarrow \neg\neg r$$

For instance, the subformula $(q \wedge \neg r)$ has main connective (the only occurrence of) $\wedge$ in it, and its main subformulae are q land $\neg r$; the first occurrence of $\neg$ is the main connective of $\neg(q \wedge \neg r)$ and its only main subformula is $(q \wedge \neg r)$; and the only occurrence of $\rightarrow$ is the main connective of the whole formula, the main subformulae of which are $(p \vee \neg(q \wedge \neg r))$ and $\neg\neg r$.

1.1.6.2 Construction tree and parsing tree of a formula

A sequence of formulae constructed in the process of applying the definition and ending with A is called a **construction sequence of a formula** A. A formula has many construction sequences and a construction sequence may contain many redundant formulae. A better notion for capturing the construction of a formula is the **construction tree** of that formula. A construction tree is a tree-like directed graph with nodes labeled with propositional constants, variables, and propositional connectives, such that:

1. Every leaf is labeled by a propositional constant or variable.

2. Propositional constants and variables label only leaves.

3. Every node labeled with $\neg$ has exactly one successor node.

4. Every node labeled with any of $\wedge, \vee, \rightarrow$ or $\leftrightarrow$ has exactly two successor nodes: *left* and *right* successor.

A construction tree therefore looks like:

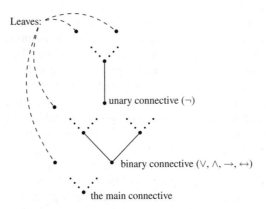

Every construction tree defines a formula C, built starting from the leaves and going towards the root, by applying at every node the formula construction rule corresponding to the label at that node. The formulae constructed in the process are precisely the subformulae of C, and the propositional connective labeling the root of the construction tree of a formula C is the main connective of C.

Example 5 (Construction tree) *The formula $(p \vee \neg(q \wedge \neg r)) \rightarrow \neg\neg r$ has the following construction tree:*

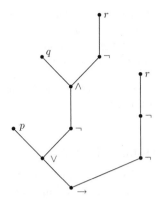

The **parsing tree** of a formula looks the same as its construction tree but is produced in inverse order, starting from the main connective (if any), drawing edges to the main components, and then recursively producing the parsing trees for each of them.

1.1.6.3 Truth assignments: satisfaction of propositional formulae

A propositional formula is a scheme that becomes a proposition whenever we substitute propositions for all occurring propositional variables. I, of course, mean **uniform substitutions**, that is, the same variables are replaced by the same propositions throughout the formula.

We cannot attribute a truth value to a propositional formula before we assign concrete propositions to all occurring propositional variables, for the same reason that we cannot evaluate $x(y+z)$ before we have assigned values to x, y, z. However, remember that in order to evaluate the truth value of a compound proposition, we only need to know the *truth values* of the occurring atomic propositions and not the propositions themselves.

For instance, if we substitute the propositions "0.5 is an integer" for p, "2 is less than 3" for q, and "the Moon is larger than the Sun" for r in the propositional formula

$$(p \lor \neg(q \land \neg r)) \to \neg\neg r$$

we find that the resulting proposition is true:

$$(F \lor \neg(T \land \neg F)) \to \neg\neg F = (F \lor \neg(T \land T)) \to \neg T = (F \lor \neg T) \to$$
$$F = (F \lor F) \to F = F \to F = T.$$

If, however, we substitute *any* true propositions for p and q and a false proposition for r, then the resulting proposition will be false:

$$(T \lor \neg(T \land \neg F)) \to \neg\neg F = (T \lor \neg(T \land T)) \to \neg T = (T \lor \neg T) \to$$
$$F = (T \lor F) \to F = T \to F = F.$$

Definition 6 *A function that assigns truth values to propositional variables in a given set is called **truth assignment** for that set of variables. If a truth assignment τ renders a formula A true, we say that τ **satisfies** A, denoted $\tau \models A$. A propositional formula is **satisfiable** if it is satisfied by some truth assignment.*

For instance, the formula $p \wedge \neg q \wedge r$ is satisfiable by the assignment $p : \mathrm{T}, q : \mathrm{F}, r : \mathrm{T}$, while the formula $p \wedge \neg p$ is not satisfiable.

1.1.6.4   Truth tables of propositional formulae

Clearly, the truth of a given propositional formula only depends on the truth values assigned to the variables occurring in that formula. We can therefore think of propositional formulae as functions from truth assignments to truth values. We can tabulate the "behavior" of any propositional formula in a truth table where we list all possible truth assignments of the occurring variables, and for each of them we compute the corresponding truth value of the formula. We can do that by successively computing the truth values of all occurring subformulae, as we did just now. For example, the truth table for the above formula is compiled as follows:

p	q	r	$\neg r$	$\neg\neg r$	$q \wedge \neg r$	$\neg(q \wedge \neg r)$	$p \vee \neg(q \wedge \neg r)$	$(p \vee \neg(q \wedge \neg r)) \rightarrow \neg\neg r$
T	T	T	F	T	F	T	T	T
T	T	F	T	F	T	F	T	F
T	F	T	F	T	F	T	T	T
T	F	F	T	F	F	T	T	F
F	T	T	F	T	F	T	T	T
F	T	F	T	F	T	F	F	T
F	F	T						
F	F	F						

Exercise 7 *Complete the last two rows.*

Truth tables can be somewhat simplified if we notice that every occurrence of a logical connective in a propositional formula determines a *unique subformula* where that occurrence is the main connective of that subformula. We can now simplify the truth table of the formula by listing only the occurring variables and the whole formula, and then computing the truth table of every subformula in the column below the corresponding main connective:

p	q	r	$(p$	$\vee$	$\neg$	$(q$	$\wedge$	$\neg$	$r))$	$\rightarrow$	$\neg$	$\neg$	r
T	T	T	T	T	T	T	F	F	T	**T**	T	F	T
T	T	F	T	T	F	T	T	T	F	**F**	F	T	F
T	F	T	T	T	T	F	F	F	T	**T**	T	F	T
T	F	F	T	T	T	F	F	T	F	**F**	F	T	F
F	T	T	F	T	T	T	F	F	T	**T**	T	F	T
F	T	F	F	F	F	T	T	T	F	**T**	F	T	F
F	F	T								$\cdots$			
F	F	F								$\cdots$			

We therefore see that a propositional formula can represent a true or a false proposition, depending of the choice of the propositions substituted for the occurring propositional variables, or rather on their truth values.

1.1.7 Tautologies

Definition 8 *A propositional formula A is a* **tautology** *if it obtains a truth value T for any assignment of truth values to the variables occurring in A.*
   The claim that A is a tautology is denoted

$$\models A.$$

*Tautologies are also called **(logically) valid formulae**.*

Thus, tautology always renders a true proposition; it represents a **logical law** in the same way as the identity $x + y = y + x$ represents a law of arithmetic and, therefore, holds no matter what values we assign to x and y.
   Here are a few simple tautologies that represent some important features of propositional logic:

- $(p \vee \neg p)$: the **law of excluded middle**, which states that every proposition p is either true or false (and, in the latter case, $\neg p$ must be true);

- $\neg(p \wedge \neg p)$: the **law of non-contradiction**, which states that it cannot be the case that both p and $\neg p$ are true;

- $(p \wedge q) \rightarrow p$: this is always true by the very meaning (and truth table) of $\wedge$;

- likewise, $p \rightarrow (p \vee q)$ is always true by the very meaning and truth table of $\vee$; and

- $((p \wedge (p \rightarrow q)) \rightarrow q)$: if p is true and it is true that p implies q, then q is true. This reflects the meaning of the implication.

1.1.7.1 Checking tautologies with truth tables

How can we determine if a propositional formula A is a tautology? Quite easily: complete the truth table of A and check if it always takes a truth value `true`. Let us check some of the examples mentioned above:

p	$\neg p$	$(p \vee \neg p)$	$(p \wedge \neg p)$	$\neg(p \wedge \neg p)$
T	F	T	F	T
F	T	T	F	T

p	q	$(p \rightarrow q)$	$(p \wedge (p \rightarrow q))$	$((p \wedge (p \rightarrow q)) \rightarrow q)$
T	T	T	T	T
T	F	F	F	T
F	T	T	F	T
F	F	T	F	T

The opposite concept of a tautology is a **contradictory formula**, that is, a formula that always takes a truth value `false`. For example, $(p \land \neg p)$ is a contradictory formula. A formula that is not contradictory is called **falsifiable**.

How are the concepts of tautology and contradictory formula, and the concepts of satisfiable and falsifiable formula related?

1. A formula A is a tautology precisely when its negation $\neg A$ is a contradictory formula, and A is contradictory precisely when its negation $\neg A$ is a tautology.

2. A formula A is satisfiable precisely when its negation $\neg A$ is falsifiable, and A is falsifiable precisely when its negation $\neg A$ is satisfiable.

1.1.7.2 Checking tautologies by searching for falsifying assignments

Checking tautologies with truth tables is straightforward but rather laborious and – let's admit – not very exciting. There is a somewhat streamlined and more intelligent method whereby we attempt to show that the formula *is not* a tautology by searching for an appropriate **falsifying truth assignment**, that is, a combination of truth values of the propositional variables that renders it false. If we succeed in finding such an assignment, then the formula is indeed not a tautology. If, however, when exploring a possible case we reach a state where some variable is required to be both true and false, that is clearly a *contradiction*, meaning that the case we are exploring is impossible and we must abandon that case. If *all* possible attempts to produce a falsifying assignment end up with a contradiction, then we have actually proved that the formula cannot be falsified; it must therefore be a tautology.[2]

The systematic search for a falsifying assignment is based on a step-by-step decomposition of the formula by using the truth tables of the propositional connectives occurring in it. Let us see how this works on some examples.

To make it more succinct I will use **signed formulae**, that is, expressions of the type $A : \mathrm{T}$, meaning "*A must be true*", and $A : \mathrm{F}$, meaning "*A must be false*".

1. Consider the formula $\neg(p \to \neg q) \to (p \lor \neg r)$.
   To falsify it, it must be the case that $\neg(p \to \neg q)$:T and $p \lor \neg r$: F.
   For the former, it must be the case that $p \to \neg q$: F, hence p : T and $\neg q$: F.
   For the latter, it must be the case that p : F and $\neg r$: F.
   This implies that p must be both true and false, which is impossible. Our attempt to falsify the formula has failed, and it is therefore a tautology.

2. Consider now $\neg p \to \neg(p \lor \neg q)$.
   To falsify it, it must be the case that $\neg p$: T while $\neg(p \lor \neg q)$: F.
   Therefore p : F and $p \lor \neg q$: T.
   For the latter, p : T or $\neg q$: T. Let us consider both cases:
   Case 1: p : T. This contradicts p : F.
   Case 2: $\neg q$: T. Then q : F. In this case we have not reached any contradiction and there is nothing more we can do in order to obtain one. Indeed, we can check that p : F and q : F renders the formula false.

3. A contradiction in the truth values can be reached on any *subformula*, not necessarily a variable. For instance, take $(p \lor \neg q) \to (\neg p \to (p \lor \neg q))$. For it to be false, $(p \lor \neg q)$: T

---

[2] This method of proof is called a *proof by contradiction* and is discussed in more detail in Section 5.1.

and $(\neg p \to (p \lor \neg q))$: F must be the case. From the latter, $\neg p$: T and $(p \lor \neg q)$: F, which contradicts the former.

4. Finally, take $((p \land \neg q) \to \neg r) \leftrightarrow ((p \land r) \to q)$. For it to be false, there are two possible cases:

Case 1: $((p \land \neg q) \to \neg r)$: T and $((p \land r) \to q)$: F. The latter implies $(p \land r)$: T and q : F, hence p : T, q : F, r : T. For the former, two sub-cases are possible:
Case 1a: $\neg r$: T. Then r : F, which is a contradiction with r : T.
Case 1b: $(p \land \neg q)$: F. Then:
Case 1bi: p : F, a contradiction with p : T.
Case 1bii: $\neg q$: F and q : T, again a contradiction but now with q : F.
Note that the consideration of these sub-cases could have been avoided, if we had noticed that the assignment p : T, q : F, r : T renders $((p \land \neg q) \to \neg r)$: F.

Case 2: $(((p \land r) \to q)$: T and $(p \land \neg q) \to \neg r)$: F. The former implies $(p \land \neg q)$: T and $\neg r$: F, that is, p : T, q : F, r : T. Again, we can either notice that this assignment renders $((p \land r) \to q)$: F, or consider the cases for $((p \land r) \to q)$: T and see that all lead to a contradiction.
The formula cannot be falsified in either case, so it is a tautology.

This method can be formalized and mechanized completely into a kind of *deductive system*, called Semantic Tableaux, which is presented in Section 2.3.

References for further reading
For helpful and accessible introductions to propositional logic and discussions of the issues covered here plus more, see Tarski (1965), Gamut (1991), Jeffrey (1994), Barwise and Echemedy (1999), Nederpelt and Kamareddine (2004), Boole (2005), Bornat (2005), Hodges (2005), Chiswell and Hodges (2007), Makinson (2008), Ben-Ari (2012), and van Benthem *et al.* (2014).

Some suitable books on philosophical logic, old and new, include Carroll (1897), Kalish and Montague (1980), Smith (2003), Copi *et al.* (2010), and Halbach (2010).

For more technical books on mathematical logic, see Shoenfield (1967), van Dalen (1983), Hamilton (1988), Mendelson (1997), Enderton (2001), and Hedman (2004),

For books on computational and applied logic the reader is referred to Gallier (1986), Nerode and Shore (1993), and Fitting (1996).

Last but not least, some fun logic books include Carroll (1886) and Smullyan (1998, 2009a, b, 2011, 2013, 2014).

Exercises

1.1.1 Which of the following are propositions? (Assume that John, Mary and Eva are concrete individuals.)

(a) $2^3 + 3^2 = 19$ (c) $2^3 + 3^2 = x$
(b) $2^3 + 3^2 = 91$ (d) Will you marry me?

(e) John married on 1 January 1999.
(f) John must marry Mary!
(g) I told her about John.
(h) Mary is not happy if John married Eva.
(i) Who is Eva?
(j) Mary is not happy if $2^3 + 3^2 = 19$.
(k) This sentence refers to itself.
(l) This sentence is true.
(m) If you are reading this sentence now, then it is not true.

1.1.2 If A and B are true propositions and C and D are false propositions, determine the truth values of the following compound propositions without using truth tables.

(a) $A \wedge (B \vee C)$
(b) $(C \to A) \to D$
(c) $C \to (A \to D)$
(d) $\neg(\neg A \vee C) \wedge B$
(e) $\neg(\neg(\neg D \wedge (B \to \neg A)))$
(f) $\neg(C \to A) \vee (C \to D)$
(g) $(C \leftrightarrow \neg B) \vee (A \to \neg A)$
(h) $(A \leftrightarrow \neg B) \leftrightarrow (C \leftrightarrow \neg D)$

1.1.3 Determine the antecedent and the consequent in each of the following implications.
(a) Whenever John talks everyone else listens.
(b) Everyone else listens if John talks.
(c) John talks only if everyone else listens.
(d) If everyone else listens, John talks.
(e) An integer is positive if its cube is positive.
     (Hint: To make it easier to reason, introduce a name for the object in question, for example "An integer n is positive if the cube of n is positive.")
(f) An integer is positive only if its cube is positive.
(g) A function is continuous whenever it is differentiable.
(h) The continuity of a function is necessary for it to be differentiable.
(i) The continuity of a function is sufficient for it to be differentiable.

1.1.4 A positive integer n is called **prime** if $n > 1$ and n is divisible only by 1 and by itself. Which of the following conditions are sufficient and which are necessary for the truth of "n is not prime", where n is some (given) positive integer?

(a) n is divisible by 3.
(b) n is even.
(c) n is divisible by 6.
(d) n has at least two different factors.
(e) n has more than two different factors.
(f) $n = 15$.
(g) n has a factor different from n.
(h) n has a prime factor.
(i) n has a prime factor different from n.

1.1.5 Write each of the following composite propositions in a symbolic form by identifying its atomic propositions and logical structure. Then determine its truth value.
(a) The Earth rotates around itself and, if the Moon rotates around the Earth, then the Sun rotates around the Moon.
(b) If the Sun rotates around the Earth or the Earth rotates around the Moon then the Sun rotates around the Moon.
(c) The Moon does not rotate around the Earth if the Sun does not rotate around the Earth and the Earth does not rotate around the Moon.

    (d) The Earth rotates around itself only if the Sun rotates around the Earth or the Moon does not rotate around the Earth.

    (e) The Earth rotates around itself if and only if the Moon does not rotate around the Earth or the Earth does not rotate around the Moon.

1.1.6 Determine the truth value of the proposition A in each of the following cases, without using truth tables. (Hint: if necessary, consider the possible cases.)

    (a) B and $B \rightarrow A$ are true.

    (b) $A \rightarrow B$ is true and B is false.

    (c) $\neg B$ and $A \vee B$ are true.

    (d) Each of $B \rightarrow \neg A$, $\neg B \rightarrow \neg C$, and C is true.

    (e) Each of $\neg C \wedge B$, $C \rightarrow (A \vee B)$, and $\neg(A \vee C) \rightarrow C$ is true.

1.1.7 Let P, Q, R, and S be propositions. Show that:

    (a) If the propositions P and $P \rightarrow Q$ are true, then Q is true.

    (b) If the propositions $(P \vee Q) \rightarrow R$ and $P \vee R$ are true, then R is true.

    (c) If $P \rightarrow Q$, $Q \rightarrow R$, and $P \wedge S$ are true, then $R \wedge S$ is true.

    (d) If $\neg P \rightarrow \neg Q$, $\neg(P \wedge \neg R)$, $\neg R$ are true, then Q is false.

    (e) If $P \rightarrow Q$, $R \vee (S \wedge \neg Q)$, and $\neg R$ are true, then P is false.

    (f) If $Q \rightarrow (R \wedge S)$ is true and $Q \wedge S$ is false, then $R \wedge Q$ is false.

    (g) If $P \rightarrow Q$ and $Q \rightarrow (R \vee S)$ are true and $P \rightarrow R$ is false, then $\neg R \rightarrow S$ is true.

    (h) If $\neg P \rightarrow (\neg Q \vee \neg R)$ and $Q \wedge (P \vee R)$ are true, then P is true.

    (i) If $P \rightarrow Q$ and $Q \rightarrow (R \vee S)$ are true and $P \rightarrow R$ is false, then S is true.

    (j) If $Q \rightarrow (R \wedge S)$ is true and $Q \wedge S$ is false, then Q is false.

1.1.8 Construct the truth tables of the following propositional formulae, and determine which (if any) of them are tautologies and which are contradictory formulae.

    (a) $\neg(p \rightarrow \neg p)$

    (b) $p \vee (p \rightarrow \neg p)$

    (c) $p \wedge (q \vee \neg q)$

    (d) $(p \wedge \neg p) \rightarrow q$

    (e) $((p \rightarrow q) \rightarrow p) \rightarrow p$

    (f) $\neg p \wedge \neg(p \rightarrow q)$

    (g) $(p \vee \neg q) \rightarrow \neg(q \wedge \neg p)$

    (h) $(p \rightarrow q) \wedge (q \rightarrow r) \wedge \neg(\neg p \vee r)$

    (i) $\neg(\neg p \leftrightarrow q) \wedge (r \vee \neg q)$

    (j) $\neg((p \wedge \neg q) \rightarrow r) \leftrightarrow (\neg(q \vee r) \rightarrow \neg p)$

1.1.9 Determine which (if any) of the following propositional formulae are tautologies by searching for falsifying truth assignments.

    (a) $q \rightarrow (q \rightarrow p)$

    (b) $p \rightarrow (q \rightarrow p)$

    (c) $((p \rightarrow q) \wedge (p \rightarrow \neg q)) \rightarrow \neg p$

    (d) $((p \rightarrow q) \vee (p \rightarrow \neg q)) \rightarrow \neg p$

    (e) $(p \vee \neg q) \wedge (q \rightarrow \neg(q \wedge \neg p))$

    (f) $((p \rightarrow q) \wedge (q \rightarrow r)) \rightarrow (r \rightarrow p)$

    (g) $((p \rightarrow q) \wedge (q \rightarrow r)) \rightarrow (p \rightarrow r)$

    (h) $((p \rightarrow q) \wedge (p \rightarrow r)) \rightarrow (p \rightarrow (q \wedge r))$

    (i) $((p \wedge q) \rightarrow r) \rightarrow ((p \rightarrow r) \wedge (q \rightarrow r))$

    (j) $((p \vee q) \rightarrow r) \rightarrow ((p \rightarrow r) \wedge (q \rightarrow r))$

(k) $((\neg p \wedge q) \rightarrow \neg r) \rightarrow (\neg q \rightarrow \neg(p \wedge \neg r))$
(l) $((p \rightarrow r) \vee (q \rightarrow r)) \rightarrow ((p \vee q) \rightarrow r)$
(m) $((p \rightarrow r) \wedge (q \rightarrow r)) \rightarrow ((p \vee q) \rightarrow r)$
(n) $p \rightarrow ((q \rightarrow r) \rightarrow ((p \rightarrow q) \rightarrow (p \rightarrow r)))$

1.1.10 Lastly, some logical puzzles[3]. On the remote planet Nologic there are two types of intelligent creatures:

- *truth-tellers*, who always tell the truth, and
- *liars*, who (you guessed it) always lie.

It is not possible to distinguish them by appearance, but only by the truth or falsity of the statements they make.

(a) A space traveler visited Nologic and met two inhabitants, P and Q. He asked them: "Is any of you a liar?" "At least one of us is a liar", replied P. Can you find out what P and Q are?

(b) Next, the stranger met two other inhabitants and asked one of them "Is any of you a liar?". He got a "yes" or "no" answer and from that answer was able to determine for each of them whether he is a liar or not. What was the answer and what was the stranger's conclusion?

(c) Walking about Nologic, the stranger met two other inhabitants A and B and asked A, "Is any of you a truth-teller?" "If B is a liar, then I am a liar too", replied A. What are A and B?

(d) The stranger went on and met three locals X, Y, and Z and asked X: "Are you a liar?" X answered something which the stranger did not hear, so he asked Y: "What did X say?" "He said that he is a liar", replied Y. Then Z added "Don't believe Y, he is a liar". Can you identify all liars?

(e∗) The stranger went on. In the evening, he began to look for a shelter for the night, but was very cautious because he knew that some of the inhabitants were man-eaters and it was not possible to recognize them by appearance. He met three inhabitants, C, D, and E. He asked C, "How many of you are truth-tellers?" "Flam flim" answered C in her language. "What did she say?" asked the stranger D. "Just one", replied D. "Do not trust D, he is a liar. Come with me, I'm not a man-eater" said E. "No, come with me, I'm not a man-eater" countered D.
What should the stranger do?

(f∗) The stranger decided to go back to the port where his spaceship was, but he got lost. After a long walk he got to a fork of the road. He knew that one of the two roads would take him to the spaceship port, but did not know which one. Luckily, he saw two of the inhabitants, one on each road. He had met them before so he knew that one of them was a liar and the other a truth-teller, but could not remember who was who.

Can the stranger ask just one question to either of these inhabitants in order to find out which is the correct road to the spaceship port? If so, what question should he ask?

---

[3] For many more puzzles of this type, I warmly recommend to the reader the marvellous logical puzzle books by Raymond Smullyan (1998, 2009a, b, 2011, 2013, 2014).

1.1.11

THE HARDEST LOLGIC PUZZLE EVAR!

THREE LOLCATS A, B, AND C ARE CALLED, IN NO PARTICULAR ORDER, TRUE, FALSE, AND RANDOM. TRUE ALWAYS MEOWS TRULY, FALSE ALWAYS MEOWS FALSELY, BUT WHETHER RANDOM MEOWS TRULY OR FALSELY IS A COMPLETELY RANDOM MATTER. YOUR TASK IS TO DETERMINE THE IDENTITIES OF A, B, AND C BY ASKING THREE YES-NO QUESTIONS; EACH QUESTION MUST BE PUT TO EXACTLY ONE LOLCAT. THE LOLCATS UNDERSTAND ENGRISH, BUT WILL ANSWER ALL QUESTIONS IN THEIR OWN LANGUAGE, IN WHICH THE WORDS FOR YES AND NO ARE DA AND JA, IN SOME ORDER. YOU DO NOT KNOW WHICH WORD MEANS WHICH.

spikedmath.com © 2012

The early origins of propositional logic

The Megarian school of philosophy

The **Megarian school** of philosophy was founded by **Euclid of Megara** (c. 430–360 BC). (This is not the famous geometer Euclid of Alexandria.) He was a disciple of Socrates and, following on his ideas, claimed that there is one universal Good in the world, sometimes also called Wisdom, God, or Reason, and that nothing that is not Good exists. Euclid used logic, in a dialogue form, to defend his ideas and win arguments. He applied extensively **reductio ad absurdum** (see Section 2.4) in his argumentation.

Eubulides (4th century BC) was a pupil of Euclid of Megara and a strong opponent of Aristotle. He was most famous for inventing several **paradoxes**, still boggling the mind today. The most popular of them is the *Liar's paradox*, also attributed to **Epimenides** (6th century BC), a Cretan, who is claimed to have said that "*All Cretans are liars*" (which is not a paradox yet, just a necessarily false statement).

Diodorus Cronus (?–c. 284 BC) was another prominent philosopher from the Megarian **Dialectical school**. He made important early contributions to logic, especially on the **theory of conditionals** and the concepts of "possible" and "necessary", thus following Aristotle in laying the foundations of **modal logic**. He is most famous for his **Master argument** in response to Aristotle's discussion of **future contingents**, such as "*There will be a sea battle tomorrow*". Diodorus' argument implied that whatever is possible is actually necessary, hence there are no contingencies.

Philo of Megara (c. 400 BC) was a disciple of Diodorus Cronus. He was also his most famous opponent in their disputes concerning the modal notions of "possible" and "necessary" and on the criteria for truth of conditional statements. Notably, Philo regarded a conditional as false only if it has both a true antecedent and a false consequent, essentially inventing the **truth-functional implication** which we now use in classical propositional logic, also known as **material implication**.

The Stoic school of philosophy

Greek philosopher **Zeno of Citium** (c. 335–265 BC), a pupil of Diodorus Cronus, founded the **Stoic school** in the early 3rd century BC. Zeno and his school had an elaborated theory of philosophy as a way of life, and also made influential contributions to physics, cosmology, epistemology and ethics.

Zeno taught that the *Universal Reason* (*Logos*, from which the word *Logic* originated) was the greatest good in life and living in accordance with it was the purpose of human life. The Stoic school was engaged in logical argumentation and essentially laid the foundations of propositional logic as an alternative to the Aristotelian logic of Syllogisms (see Section 3.5).

Chrysippus (c. 280–207 BC) was a philosopher and logician from the Stoic School. He wrote over 300 books (very few of survived to be studied) on many fundamental topics of logic, including propositions and propositional connectives (he introduced the implication, conjunction and exclusive disjunction), logical consequence, valid arguments, logical deduction, causation, and logical paradoxes, and on the most popular **non-classical logics**, including modal, tense, and epistemic logics. Chrysippus is often regarded as the founder of propositional logic, and is one of the most important early formal logicians along with Aristotle.

1.2   Propositional logical consequence: logically correct inferences

The central problem of logic is the study of *correct argumentation and reasoning*. In this section I define and discuss what it means for an argument to be logically correct or not by formalizing the fundamental logical concept of *logical consequence*. This is done here just for a simple type of logical arguments that only involve propositional reasoning, called *propositional arguments*.

1.2.1   Propositional logical consequence

The intuition behind logically correct reasoning is simple: starting from true premises should always lead to true conclusions. Let us first make this intuition precise.

Definition 9 *A propositional formula B is a **logical consequence** of the propositional formulae $A_1, \ldots, A_n$, denoted*[4]

$$A_1, \ldots, A_n \models B$$

---

[4] Note that I use here the same symbol we used to indicate tautologies. This will be justified soon.

if B is true whenever all $A_1, \ldots, A_n$ are true. That means: if every truth assignment to the variables occurring in $A_1, \ldots, A_n, B$ for which the formulae $A_1, \ldots, A_n$ is true, then the formula B is also true.

When $A_1, \ldots, A_n \models B$, we also say that B **follows logically from** $A_1, \ldots, A_n$, or that $A_1, \ldots, A_n$ **imply logically** B.

In the context of $A_1, \ldots, A_n \models B$, the formulae $A_1, \ldots, A_n$ are called **assumptions** while the formula B is called a **conclusion**.

When $A_1, \ldots, A_n \models B$ is not the case, we write $A_1, \ldots, A_n \not\models B$.

If $A_1, \ldots, A_n \models B$ then every substitution of propositions for the variables occurring in $A_1, \ldots, A_n, B$ which turns the formulae $A_1, \ldots, A_n$ into true propositions also turns the formula B into a true proposition.

In order to check whether $A_1, \ldots, A_n \models B$ we can simply complete the truth tables of $A_1, \ldots, A_n, B$ and check, row by row, if the following holds: *whenever all formulae $A_1, \ldots, A_n$ have a truth value T in that row, B must also have a truth value T.* (Of course, it is possible for B to be true without any of $A_1, \ldots, A_n$ being true.) If that holds in *every row* in the table, then B *does* follow logically from $A_1, \ldots, A_n$; if that fails *in at least one row*, then B does *not* follow logically from $A_1, \ldots, A_n$.

Thus, B *does not* follow logically from $A_1, \ldots, A_n$, just in case there is a truth assignment which renders all formulae $A_1, \ldots, A_n$ true and B false.

Example 10 (Some simple cases of logical consequences)

1. *Any formula B follows logically from any set of formulae that contains B. (Why?)*

2. *Any tautology follows logically from any set of formulae,* even from the empty set!

3. *For any formulae P and Q we claim that $P, P \to Q \models Q$.*
   Note first that, whatever the formulae P and Q, any truth assignment eventually renders each of them true or false and all combinations of these truth values can be possible, so we can treat P and Q as propositional variables and consider the truth tables for P, Q and $P \to Q$:

P	Q	P	$P \to Q$	Q
T	T	T	T	T
T	F	T	F	F
F	T	F	T	T
F	F	F	T	F

   Indeed, in every row where the 3rd and 4th entries are T, the 5th entry is also T.

4. $P \to R, Q \to R \models (P \vee Q) \to R$ *for any formulae P, Q, R.*
   Likewise, it suffices to show that $p \to r, q \to r \models (p \vee q) \to r$, for propositional variables p, q, r.
   Indeed, in every row of the truth table where the truth values of $p \to r$ and $q \to r$ are T, the truth value of $(p \vee q) \to r$ is also T.

p	q	r	$p \to r$	$q \to r$	$p \vee q$	$(p \vee q) \to r$
T	T	T	T	T	T	T
T	T	F	F	F	T	F
T	F	T	T	T	T	T
T	F	F	F	T	T	F
F	T	T	T	T	T	T
F	T	F	T	F	T	F
F	F	T	T	T	F	T
F	F	F	T	T	F	T

5. *Is it true that $p \vee q, q \to p \models p \wedge q$? We check the truth table:*

p	q	$p \vee q$	$q \to p$	$p \wedge q$
T	T	T	T	T
T	F	T	T	F
F	T	$\ldots$	$\ldots$	$\ldots$
F	F	$\ldots$	$\ldots$	$\ldots$

and see that in the 2nd row both assumptions are true while the conclusion is false. This suffices to conclude that $p \vee q, q \to p \not\models p \wedge q$, so there is no need to fill in the truth table any further.

Recall that $\models A$ means that A is a tautology. Tautologies and logical consequences are closely related. First, note that a formula A is a tautology if and only if (iff) it follows logically from the empty set of formulae. Indeed, we have already noted that if $\models A$ then $\emptyset \models A$. Now, to see that if $\emptyset \models A$ then $\models A$, suppose $\emptyset \models A$ and take *any* truth assignment. Note that it satisfies *every formula* from $\emptyset$. Why? Well, there can be *no* formula in $\emptyset$ which is not satisfied, because there are no formulae in $\emptyset$ at all! Since $\emptyset \models A$, that truth assignment must also satisfy A. Thus, A is satisfied by every truth assignment[5].

In general, we have the following equivalences.

Proposition 11 *For any propositional formulae $A_1, \ldots, A_n, B$, the following are equivalent:*

1. $A_1, \ldots, A_n \models B$
2. $A_1 \wedge \cdots \wedge A_n \models B$
3. $\models (A_1 \wedge \cdots \wedge A_n) \to B$
4. $\models A_1 \to (\cdots \to (A_n \to B) \cdots)$

I leave the proofs of these equivalences as easy exercises.

Checking logical consequences can be streamlined, in the same way as checking tautologies, by organizing a systematic search for a **falsifying assignment**. In order to check

---

[5] Here we did some logical reasoning based on the very same concepts of truth and logical consequence that we are discussing. When reasoning about logic, this kind of bootstrapping reasoning is inevitable!

if B follows logically from $A_1, \ldots, A_n$ we look for a truth assignment to the variables occurring in $A_1, \ldots, A_n, B$ that renders all $A_1, \ldots, A_n$ true and B false. If we succeed, then we have proved that B *does not* follow logically from $A_1, \ldots, A_n$; otherwise we want to prove that no such assignment is possible by showing that the assumption that it exists leads to a contradiction.

For example, let us check again that $p \to r, q \to r \models (p \lor q) \to r$. Suppose that for some assignment $(p \to r) : \text{T}$, $(q \to r) : \text{T}$ and $((p \lor q) \to r) : \text{F}$. Then $(p \lor q) : \text{T}$ and $r : \text{F}$, hence $p : \text{T}$ or $q : \text{T}$.

Case 1: $p : \text{T}$. Then $(p \to r) : \text{F}$, that is, a contradiction.
Case 2: $q : \text{T}$. Then $(q \to r) : \text{F}$, again, a contradiction.

Thus, there is no assignment that falsifies the logical consequence above.

1.2.2   Logically sound rules of propositional inference and logically correct propositional arguments

We now apply the notion of logical consequence to define and check whether a given propositional argument is logically correct. Let us first introduce some terminology.

Definition 12 *A **rule of propositional inference** (**inference rule**, for short) is a scheme:*

$$\frac{P_1, \ldots, P_n}{C}$$

where $P_1, \ldots, P_n, C$ are propositional formulae. The formulae $P_1, \ldots, P_n$ are called **premises** *of the inference rule, and C is its* **conclusion**.

*An **instance** of an inference rule is obtained by uniform substitution of concrete propositions for the variables occurring in all formulae of the rule. Every such instance is called a **propositional inference**, or a **propositional argument** based on that rule.*

Definition 13 *An inference rule is **(logically) sound** if its conclusion follows logically from the premises. A propositional argument is **logically correct** if it is an instance of a logically sound inference rule.*

Example 14

1. *The following inference rule*

$$\frac{p, p \to q}{q}$$

    *is sound, as we have already seen. This rule is known as the **Detachment rule** or **Modus Ponens**, and is very important in the logical deductive systems called* axiomatic systems *which we will study in Chapter 2.*
    The inference

    > Alexis is singing.
    > If Alexis is singing, then Alexis is happy.
    > ————————————————————
    > Alexis is happy.

    is therefore logically correct, being an instance of that inference rule.

2. *The inference rule*

$$\frac{p, q \rightarrow p}{q}$$

is not sound: if p is true and q is false, then both premises are true while the conclusion is false.
Therefore the inference

> 5 is greater than 2.
> 5 is greater than 2 if 5 is greater than 3.
> ───
> 5 is greater than 3.

is not logically correct, despite the truth of both premises and the conclusion, as it is an instance of an unsound rule.

Some remarks are in order here.

- It is very important to realize that the logical correctness of an inference *does not always guarantee the truth of the conclusion, but only when all premises are true.* In other words, if at least one premise of a logically correct inference is false, then the conclusion may also be false. For instance, the inference

> 5 divides 6.
> If 5 divides 6, then 5 divides 11.
> ──────────────────────────────────
> 5 divides 11.

is logically correct (being an instance of the rule Modus Ponens) in spite of the falsity of the conclusion. This does not contradict the idea of logical correctness of an inference, because the first premise is false. (What about the second premise?)

- Conversely, if the conclusion of an inference happens to be true, this does not necessarily mean that the inference is logically correct as in the second example above.

- Moreover, it may happen that the truth of the premises of an inference *does imply* the truth of the conclusion, and yet the inference is not *logically* correct. For instance, the correctness of the inferences

> Today is Monday.
> ─────────────────────────
> Tomorrow will be Tuesday.

or

$$\frac{a = 2, a + b = 5}{b = 3}$$

is based *not* on logical consequence but, in the first case, on the commonly known fact that Tuesday always follows after Monday, and in the second case on some laws of arithmetic. Indeed, the first inference is based on the rule

$$\frac{p}{q}$$

and the second inference (although the statements are not really propositions) on

$$\frac{p, q}{r},$$

both of which are clearly unsound.

To summarize, the meaning and importance of logically correct inferences is that only such inferences guarantee that *if* all premises are true, *then* the conclusions will also be true. That is why only logically correct inferences are safe to be employed in our reasoning.

Let us now look at a few more examples.

1. The inference rule

$$\frac{q, p \vee \neg q}{p}$$

is logically sound. You can check this in two ways: by applying the definition or by showing that the corresponding formula

$$(q \wedge (p \vee \neg q)) \rightarrow p$$

is a tautology.

Consequently, the inference

> Olivia is crying or Olivia is not awake.
> Olivia is awake.
> ___
> Olivia is crying.

is logically correct, being an instance of the rule above.

2. Now, take the argument

> If a divides b or a divides c, then a divides bc.
> a divides bc.
> a does not divide b.
> ___
> Therefore a divides c.

where a, b, c are certain integers. This argument is an instance of the following inference rule:

$$\frac{(p \vee q) \rightarrow r, r, \neg p}{q}.$$

Let us see if we can invalidate this rule. For that we need an assignment such that $((p \vee q) \rightarrow r) : \mathrm{T}$, $r : \mathrm{T}$, and $\neg p : \mathrm{T}$, hence $p : \mathrm{F}$ and $q : \mathrm{F}$. Indeed, the assignment $p : \mathrm{F}$, $q : \mathrm{F}$, and $r : \mathrm{T}$ renders all premises true and the conclusion false. (Check this.)

The rule is therefore not logically sound, hence the argument above is not logically correct.

I develop the method behind the last argument above in the next chapter.

1.2.3   Fallacies of the implication

As an application let us analyze some very common forms of correct and incorrect reasoning related to implications. Given the implication

$$A \rightarrow B$$

we can form the so-called **derivative implications**:

- the **converse** of $A \rightarrow B$ is $B \rightarrow A$;
- the **inverse** of $A \rightarrow B$ is $\neg A \rightarrow \neg B$; and
- the **contrapositive** of $A \rightarrow B$ is $\neg B \rightarrow \neg A$.

Now, suppose we know that $A \rightarrow B$ is true. What can we say about the truth of its derivatives? To answer that question, look at each of the inferences:

$$\frac{A \rightarrow B}{B \rightarrow A}, \qquad \frac{A \rightarrow B}{\neg A \rightarrow \neg B}, \qquad \frac{A \rightarrow B}{\neg B \rightarrow \neg A}$$

Exercise 15 *Show that the first two of these inferences are incorrect, while the third inference is correct.*

The truth of an implication $A \rightarrow B$ therefore *only implies the truth of its contrapositive*, but not the truth of the converse or inverse. These are mistakes that people often make, respectively called **the fallacy of the converse implication** and **the fallacy of the inverse implication**. For example, the truth of the implication "If it has just rained, then the tennis court is wet" does not imply that either of "If the tennis court is wet, then it has just rained" and "If it has not just rained, then the tennis court is not wet" is true – someone may have just watered the court on a clear sunny day – but it certainly implies that "If the court is not wet, then it has not just rained." In fact, it can easily be checked that

$$\frac{\neg B \rightarrow \neg A}{A \rightarrow B}$$

is also logically correct. This is the basis of the method of *proof by contraposition*, which is discussed further in Section 2.5.

References for further reading
Propositional logical consequence, as well as propositional arguments, inference rules and their logical correctness, are treated in more details in Carroll (1897), Tarski (1965), Kalish and Montague (1980), Gamut (1991), Nerode and Shore (1993), Jeffrey (1994), Barwise and Echemendy (1999), Smith (2003), Boole (2005), Bornat (2005), Chiswell and Hodges (2007), Copi *et al.* (2010), Halbach (2010), Ben-Ari (2012), and van Benthem *et al.* (2014).

Exercises

1.2.1 Prove Proposition 11. (Hint: you do not have to prove all pairs of equivalences. It is sufficient to show, for instance, that claim 1 implies 2, which implies 3, which implies 4, which implies 1.)

1.2.2 Show that the first two of the following inference rules, corresponding to the derivative implications, are logically unsound while the third rule is sound.

(a)
$$\frac{p \rightarrow q}{q \rightarrow p}$$

(b)
$$\frac{p \rightarrow q}{\neg p \rightarrow \neg q}$$

(c)
$$\frac{p \rightarrow q}{\neg q \rightarrow \neg p}$$

1.2.3 Using truth tables, check if the following inference rules are sound.

(a)
$$\frac{p \rightarrow q, \neg q \vee r, \neg r}{\neg p}$$

(c)
$$\frac{p \rightarrow q, p \vee \neg r, \neg r}{\neg q \vee r}$$

(b)
$$\frac{\neg p \rightarrow \neg q, q, \neg(p \wedge \neg r)}{r}$$

(d)
$$\frac{((p \wedge q) \rightarrow r), \neg(p \rightarrow r)}{q \rightarrow r}$$

1.2.4 Write down the inference rules on which the following arguments are based and check their logical soundness, using truth tables.

(a) In the following argument, X is a certain number.

$$\frac{X \text{ is greater than 3.}}{X \text{ is greater than or equal to 3.}}$$

(b) In the following argument, Y is a certain number.

If Y is greater than –1, then Y is greater than –2.
$$\frac{Y \text{ is not greater than –2.}}{Y \text{ is not greater than –1.}}$$

(c)

If the triangle ABC has a right angle, then it is not equilateral.
$$\frac{\text{The triangle ABC does not have a right angle.}}{\text{Therefore, the triangle ABC is equilateral.}}$$

(d)

If Victor is good at logic, then he is clever.
$$\frac{\text{If Victor is clever, then he is rich.}}{\text{Therefore, if Victor is good at logic, then he is rich.}}$$

(e) In the following argument n is a certain integer.

If n is divisible by 2 and n is divisible by 3, then n is divisible by 6.
If n is divisible by 6, then n is divisible by 2.
$$\frac{n \text{ is not divisible by 3.}}{\text{Therefore, } n \text{ is not divisible by 6.}}$$

1.2.5 For each of the following implications construct the converse, inverse and the contrapositive, phrased in the same way.
(a) If a is greater than –1, then a is greater than –2.
(b) x is not prime if x is divisible by 6.
(c) x is positive only if its square is positive.

(d) The triangle ABC is equilateral whenever its medicentre and orthocentre coincide.

(e) For the function f to be continuous, it is sufficient that it is differentiable.

(f) For a function not to be differentiable, it is sufficient that it is discontinuous.

(g) For the integer n to be prime, it is necessary that it is not divisible by 10.

George Boole (2.11.1815–8.12.1864) was an English mathematician who first proposed and developed an algebraic approach to the study of logical reasoning. Boole's first contribution to logic was a pamphlet called *Mathematical Analysis of Logic*, written in 1847. He published his main work on logic, *An Investigation of the Laws of Thought, on which are Founded the Mathematical Theories of Logic and Probabilities*, in 1854. In it Boole developed a general mathematical method of logical inference, laying the foundations of modern mathematical logic. His system proposed a formal algebraic treatment of propositions by processing only their two possible truth values: yes–no, true–false, zero–one. In Boole's system, if x stands for "white things" then $1 - x$ stands for "non-white things;" if y stands for "sheep", then xy stands for "white sheep", etc. $x(1 - x)$ denotes things that are both white and non-white, which is impossible. A proposition of the shape $x(1 - x)$ is therefore always false, that is, has a truth value 0. The algebraic law $x(1 - x) = 0$ therefore emerges. The resulting algebraic system is known today as (the simplest) **Boolean algebra**.

Boole also argued that symbolic logic is needed in other mathematical disciplines, especially in probability theory. He wrote: "$\cdots$ no general method for the solution of questions in the theory of probabilities can be established which does not explicitly recognise those universal laws of thought which are the basis of all reasoning $\cdots$"

Propositional logic today is somewhat different from Boole's system of logic 150 years ago, but the basic ideas are the same. That is why propositional logic is also often called **Boolean logic** in recognition of Boole's ground-breaking contribution. It is not only a fundamental system of formal logical reasoning, but it also provides the mathematical basis of the **logical circuits** underlying the architecture of modern digital computers.

William Stanley Jevons (1.09.1835–13.08.1882) was an English economist and logician known for his pioneering works on political and mathematical economics, including the *theory of utility*. As well as contributing to the early development of modern logic, in 1869 he designed one of the first mechanical computers which he called the *logic piano*.

Jevons studied natural sciences and moral philosophy at the University College of London and, in 1866, was appointed Professor of Logic, Mental and Moral Philosophy and Professor of Political Economy at Owens College. His book *A General Mathematical Theory of Political Economy* (1862) is one of the first works on mathematical methods in economics which, being concerned with quantities, he regarded as an essentially mathematical science.

Jevons' most important work on scientific methods is his *Principles of Science* (1874). In 1870 he published *Elementary Lessons on Logic*, which soon became the most widely read elementary textbook on logic in the English language, later supplemented by his *Studies in Deductive Logic*.

Jevons developed a general theory of induction, which he regarded as an inverse method of deduction; he also developed and published his own treatments of Boole's approach to logic and on the general theory of probability, and studied the relation between probability and induction.

1.3   Logical equivalence: negation normal form of propositional formulae

1.3.1   Logically equivalent propositional formulae

Definition 16 *The propositional formulae A and B are **logically equivalent**, denoted $A \equiv B$, if for every assignment of truth values to the variables occurring in them they obtain the same truth values.*

Being a little imprecise (you'll see why), we can say that A and B are logically equivalent if they have the same truth tables.

- Every tautology is equivalent to $\top$. For example, $p \vee \neg p \equiv \top$.
- Every contradiction is equivalent to $\bot$. For example, $p \wedge \neg p \equiv \bot$.
- $\neg\neg p \equiv p$: a double negation of a proposition is equivalent to the proposition itself.
- $\neg(p \wedge q) \equiv (\neg p \vee \neg q)$ and $\neg(p \vee q) \equiv (\neg p \wedge \neg q)$. These are known as **De Morgan's laws**. Let us check the first, using simplified truth tables:

p	q	$\neg$	$(p$	$\wedge$	$q)$	$(\neg$	p	$\vee$	$\neg$	$q)$
T	T	F	T	T	T	F	T	F	F	T
T	F	T	T	F	F	F	T	T	T	F
F	T	T	F	F	T	T	F	T	F	T
F	F	T	F	F	F	T	F	T	T	F

- $(p \wedge (p \vee q)) \equiv (p \wedge p)$. There is a small problem here: formally, the truth tables of these formulae are *not* the same, as the first contains two variables (p and q) while the second contains only p. However, we can always consider that q occurs *vacuously* in the second formula and include it in its truth table:

p	q	$(p$	$\wedge$	$(p$	$\vee$	$q))$	$(p$	$\wedge$	$p)$
T	T	T	T	T	T	T	T	T	T
T	F	T	T	T	T	F	T	T	T
F	T	F	F	F	T	T	F	F	F
F	F	F	F	F	F	F	F	F	F

Checking logical equivalence can be streamlined, just like checking logical validity and consequence, by systematic search for a falsifying assignment, as follows. In order to check if $A \equiv B$ we try to construct a truth assignment to the variables occurring in A and B which renders one of them true while the other false. If such an assignment exists, the formulae are *not* logically equivalent; otherwise, they are. For example, let us check the second De Morgan's law: $\neg(p \vee q) \equiv (\neg p \wedge \neg q)$. There are two possibilities for an assignment to falsify that equivalence:

(i) $\neg(p \vee q) : \text{T}$ and $(\neg p \wedge \neg q) : \text{F}$. Then $p \vee q : \text{F}$ hence $p : \text{F}$, and $q : \text{F}$, but then $(\neg p \wedge \neg q) : \text{T}$: a contradiction.

(ii) $\neg(p \lor q)$: F, and $(\neg p \land \neg q)$: T. Then $\neg p$: T and $\neg q$: T, hence p : F, and q : F. But then $\neg(p \lor q)$: T: a contradiction again.

There is therefore no falsifying assignment and the equivalence holds.

1.3.2   Basic properties of logical equivalence

Caution: do not confuse the propositional connective $\leftrightarrow$ and logical equivalence between formulae. These are different things: the former is a logical connective, a symbol in our *object language*, whereas the latter is a relation between formulae, that is, a statement in our *metalanguage*. However, as I show below, there is a simple relation between them.

1. Logical equivalence is reducible to logical validity:

$$A \equiv B \quad \text{iff} \quad \models A \leftrightarrow B.$$

  Indeed, they mean the same: that A and B always take the same truth values.

2. Logical equivalence is likewise reducible to logical consequence:

$$A \equiv B \quad \text{iff} \quad A \models B \quad \text{and} \quad B \models A$$

3. The relation $\equiv$ is an **equivalence relation**, that is, for every formulae A, B, C it is:
   (a) **reflexive**: $A \equiv A$;
   (b) **symmetric**: if $A \equiv B$ then $B \equiv A$;
   (c) **transitive**: if $A \equiv B$ and $B \equiv C$ then $A \equiv C$.

4. Moreover, $\equiv$ is a **congruence** with respect to the propositional connectives, that is:
   (a) if $A \equiv B$ then $\neg A \equiv \neg B$;
   (b) if $A_1 \equiv B_1$ and $A_2 \equiv B_2$ then $(A_1 \bullet A_2) \equiv (B_1 \bullet B_2)$, for $\bullet \in \{\land, \lor, \rightarrow, \leftrightarrow\}$.

5. The following property of **equivalent replacement** holds. For any propositional formulae A, B, C and a propositional variable p, presumably occurring in C, if $A \equiv B$ then $C(A/p) \equiv C(B/p)$, where $C(X/p)$ is the result of simultaneous substitution of all occurrences of p by X.

Logical equivalence between propositional formulae can therefore be treated just like equality between algebraic expressions.

1.3.3   Some important logical equivalences

I now briefly present and discuss some important and useful logical equivalences. Verifying all of these are easy, but useful, exercises.

1.3.3.1   Algebraic laws for the logical connectives

I begin with some important logical equivalences which are used, for example, for equivalent transformations of propositional formulae to so-called *conjunctive and disjunctive normal forms* that I introduce later.

- **Idempotency**: $p \wedge p \equiv p$; $p \vee p \equiv p$.
- **Commutativity**: $p \wedge q \equiv q \wedge p$; $p \vee q \equiv q \vee p$.
- **Associativity**: $(p \wedge (q \wedge r)) \equiv ((p \wedge q) \wedge r)$; $(p \vee (q \vee r)) \equiv ((p \vee q) \vee r)$. This law allows parentheses to be omitted in multiple conjunctions and disjunctions.
- **Absorption**: $p \wedge (p \vee q) \equiv p$; $p \vee (p \wedge q) \equiv p$.
- **Distributivity**: $p \wedge (q \vee r) \equiv (p \wedge q) \vee (p \wedge r)$; $p \vee (q \wedge r) \equiv (p \vee q) \wedge (p \vee r)$.

1.3.3.2 Equivalences mutually expressing logical connectives

The following logical equivalences, the proofs of which are left as easy exercises, can be used to define logical connectives in terms of others:

- $\neg A \equiv A \rightarrow \bot$. We sometimes use this equivalence in colloquial expressions, such as "If this is true then I can fly", when we mean "This cannot be true."
- $A \leftrightarrow B \equiv (A \rightarrow B) \wedge (B \rightarrow A)$. This equivalence allows us to consider the biconditional as definable connective, which we will often do.
- $A \vee B \equiv \neg(\neg A \wedge \neg B)$.
- $A \wedge B \equiv \neg(\neg A \vee \neg B)$.
- $A \rightarrow B \equiv \neg A \vee B$.
- $A \rightarrow B \equiv \neg(A \wedge \neg B)$.
- $A \vee B \equiv \neg A \rightarrow B$.
- $A \wedge B \equiv \neg(A \rightarrow \neg B)$.

We therefore see that each of $\wedge$, $\vee$, and $\rightarrow$ can be expressed by means of any other of these using negation.

1.3.3.3 Some simplifying equivalences

Other useful logical equivalences can be used to simplify formulae:

- $A \vee \neg A \equiv \top$, $A \wedge \neg A \equiv \bot$
- $A \wedge \top \equiv A$, $A \wedge \bot \equiv \bot$
- $A \vee \top \equiv \top$, $A \vee \bot \equiv A$
- $A \rightarrow \top \equiv \top$, $A \rightarrow \bot \equiv \neg A$
- $\top \rightarrow A \equiv A$, $\bot \rightarrow A \equiv \top$.
- $\neg A \rightarrow \neg B \equiv B \rightarrow A$ (Every implication is equivalent to its contrapositive.)

1.3.3.4 Negating propositional formulae: negation normal form

In mathematical and other arguments we sometimes have to negate a formalized statement and then use the result in further reasoning. For that, it is useful to transform, up to logical equivalence, the formula formalizing the statement in **negation normal form**, where negation may only occur in front of propositional variables. Such transformation can be done by step-by-step importing of all occurrences of negations inside the other logical connectives using the following equivalences, some of which we already know:

- $\neg\neg A \equiv A$
- $\neg(A \wedge B) \equiv \neg A \vee \neg B$
- $\neg(A \vee B) \equiv \neg A \wedge \neg B$
- $\neg(A \rightarrow B) \equiv A \wedge \neg B$
- $\neg(A \leftrightarrow B) \equiv (A \wedge \neg B) \vee (B \wedge \neg A)$.

Example 17 *Equivalent transformation to negation normal form:*

$$\neg((A \vee \neg B) \rightarrow (\neg C \wedge D))$$
$$\equiv (A \vee \neg B) \wedge \neg(\neg C \wedge D)$$
$$\equiv (A \vee \neg B) \wedge (\neg\neg C \vee \neg D)$$
$$\equiv (A \vee \neg B) \wedge (C \vee \neg D).$$

References for further reading
To read more on propositional equivalence and negation normal form of propositional formulae see Nerode and Shore (1993), Jeffrey (1994), Barwise and Echemendy (1999), Hedman (2004), Nederpelt and Kamareddine (2004), Boole (2005), Chiswell and Hodges (2007), and Ben-Ari (2012).

Exercises

1.3.1 Verify the following logical laws:
    (a) *Idempotency*: $p \wedge p \equiv p$, $p \vee p \equiv p$
    (b) *Commutativity*: $p \wedge q \equiv q \wedge p$, $p \vee q \equiv q \vee p$
    (c) *Associativity*: $(p \wedge (q \wedge r)) \equiv ((p \wedge q) \wedge r)$, $(p \vee (q \vee r)) \equiv ((p \vee q) \vee r)$
    (d) *Absorption*: $p \wedge (p \vee q) \equiv p$, $p \vee (p \wedge q) \equiv p$
    (e) *Distributivity*: $p \wedge (q \vee r) \equiv (p \wedge q) \vee (p \wedge r)$, $p \vee (q \wedge r) \equiv (p \vee q) \wedge (p \vee r)$
    (f) *De Morgan's laws* $\neg(p \wedge q) \equiv (\neg p \vee \neg q)$, $\neg(p \vee q) \equiv (\neg p \wedge \neg q)$

1.3.2 Prove the following logical equivalences:
    (a) $\neg(p \leftrightarrow q) \equiv (p \wedge \neg q) \vee (q \wedge \neg p)$ (f) $p \rightarrow (q \rightarrow r) \equiv (p \wedge q) \rightarrow r$
    (b) $(p \rightarrow q) \wedge (p \rightarrow r) \equiv p \rightarrow (q \wedge r)$ (g) $p \rightarrow (q \rightarrow r) \equiv q \rightarrow (p \rightarrow r)$
    (c) $(p \rightarrow q) \vee (p \rightarrow r) \equiv p \rightarrow (q \vee r)$ (h) $(p \rightarrow r) \wedge (q \rightarrow r) \equiv (p \vee q) \rightarrow r$
    (d) $\neg(p \leftrightarrow q) \equiv (p \wedge \neg q) \vee (q \wedge \neg p)$ (i) $p \leftrightarrow q \equiv q \leftrightarrow p$
    (e) $\neg(p \leftrightarrow q) \equiv \neg p \leftrightarrow q \equiv p \leftrightarrow \neg q$ (j) $p \leftrightarrow (q \leftrightarrow r) \equiv (p \leftrightarrow q) \leftrightarrow r$

1.3.3 Determine which of the following pairs of formulae are logically equivalent:
    (a) $p \rightarrow q$ and $\neg p \vee q$ (d) $p \rightarrow \neg q$ and $q \rightarrow \neg p$
    (b) $\neg(p \rightarrow q)$ and $p \wedge \neg q$ (e) $\neg(p \rightarrow \neg q)$ and $p \wedge q$
    (c) $\neg p \rightarrow \neg q$ and $q \rightarrow p$ (f) $((p \rightarrow q) \rightarrow q) \rightarrow q$ and $p \vee q$

(g) $(p \to r) \land (q \to r)$ and $(p \land q) \to r$ (j) $p \to (q \to r)$ and $(p \to q) \to r$

(h) $(p \to r) \lor (q \to r)$ and $(p \lor q) \to r$ (k) $p \leftrightarrow (q \leftrightarrow r)$ and $q \leftrightarrow (p \leftrightarrow r)$

(i) $((p \land q) \to r)$ and $(p \to r) \lor (q \to r)$ (l) $p \to (q \to r)$ and $(p \to q) \to (p \to r)$

1.3.4 Negate each of the following propositional formulae and transform the result to an equivalent formula in a negation normal form.

(a) $(p \lor \neg q) \land \neg p$ (d) $p \to (\neg q \to p)$

(b) $(p \to q) \lor (\neg p \to \neg q)$ (e) $(p \leftrightarrow \neg q) \to \neg r$

(c) $(p \to \neg q) \to p$ (f) $p \to (\neg q \leftrightarrow r)$

Augustus De Morgan (27.6.1806–18.3.1871) was a British mathematician, logician and a popularizer of mathematics. Influenced by George Boole, he pioneered the application of algebraic methods to the study of logic in the mid-19th century, becoming one of the founding fathers of modern mathematical logic. In particular, he was the first to formulate the logical equivalences now known as **de Morgan's laws**.

De Morgan was born in Madura, India and became blind in one eye soon after his birth. He graduated from Trinity

College, Cambridge and in 1828 became the first Professor of Mathematics at the newly established University College of London, where he taught for most of his academic life.

He was an enthusiastic and prolific writer of over 700 popular articles in mathematics for the *Penny Cyclopedia*, aiming to promote the education of mathematics in Britain. In an 1838 publication he formally introduced the term "**mathematical induction**" and developed the so-far informally used method of mathematical induction into a precise mathematical technique. He also wrote the books *Trigonometry and Double Algebra* and *The Differential and Integral Calculus*. In 1847 he published his main work on mathematical logic, *Formal Logic: The Calculus of Inference, Necessary and Probable*, which was used for a very long time and was last reprinted in 2003. De Morgan was also a passionate collector of mathematical puzzles, curiosities, and paradoxes, many of which he included in his book *A Budget of Paradoxes* published in 1872 (now digitalized and available on the internet).

In 1866 De Morgan became one of the founders and the first president of the London Mathematical Society. There is a crater on the Moon named after him.

Hugh MacColl (1831–1909) was a Scottish mathematician, logician, and novelist who made some important early contributions to modern logic.

MacColl grew up in a poor family in the Scottish Highlands and never obtained a university education because he could not afford it and refused to accept to take orders in the Church of England, a condition under which William Gladstone was prepared to support his education at Oxford. Consequently, he never obtained a regular academic position; he was a highly intelligent person however.

During 1877–1879 MacColl published a four-part article establishing the first-known variant of the propositional calculus, which he called the "calculus of equivalent statements", preceding Gottlob Frege's Begriffschrifft. Furthermore, MacColl's work on the nature of implication was later credited by C.I. Lewis as the initial inspiration of his own innovative work in modal logic. MacColl also promoted logical pluralism by exploring on a par ideas for several different logical systems such as modal logic, logic of fiction, connexive logic, many-valued logic, and probability logic, establishing himself as a pioneer in the field known as **non-classical logics** today.

1.4 Supplementary: Inductive definitions and structural induction and recursion

In section 1.1 I defined propositional formulae using a special kind of definition, which refers to the very notion it is defining. Such definitions are called *inductive*. They are very common and important, especially in logic, because they are simple, elegant, and indispensable when an infinite set of structured objects is to be defined. Moreover, properties of an object defined by inductive definitions can be proved by a uniform method, called *structural induction*, that resembles and extends the method of mathematical induction used to prove properties of natural numbers. Here I present the basics of the general theory of inductive definitions and structural induction. Part of this section, or even all of it, can be skipped, but the reader is recommended to read it through.

1.4.1 Inductive definitions

Let us begin with well-known cases: the inductive definition of words in an alphabet and then natural numbers as special words in a two-letter alphabet. Note the pattern.

1.4.1.1 The set of all finite words in an alphabet

Consider a set $\mathcal{A}$. Intuitively, a (finite) word in $\mathcal{A}$ is any string of elements of $\mathcal{A}$. We formally define the set of (finite) words in the alphabet $\mathcal{A}$ inductively as follows.

1. The empty string ϵ is a word in $\mathcal{A}$.

2. If w is a word in $\mathcal{A}$ and $a \in \mathcal{A}$, then wa is word in $\mathcal{A}$.

The idea of this definition is that words in $\mathcal{A}$ are those, and only those, objects that can be constructed following the two rules above.

1.4.1.2 The set of natural numbers

We consider the two-letter alphabet $\{0, S\}$, where $0, S$ are different symbols, and formally define natural numbers to be special words in that alphabet, as follows.

1. 0 is a natural number.

2. If n is a natural number then Sn is a natural number.

The definition above defines the infinite set $\{0, S0, SS0, SSS0, \cdots\}$.

Hereafter we denote $S\cdots_{n\text{ times}}\cdots S0$ by **n** and identify it with the (intuitive notion of) natural number n.

1.4.1.3 The set of propositional formulae

Let us denote the alphabet of symbols used in propositional logic $\mathcal{A}_{PL}$. Note that it includes a possibly infinite set PVAR of propositional variables.

We now revisit the inductive definition of propositional formulae as special words in the alphabet of symbols used in propositional logic, by paying closer attention to the structure of the definition. I emphasize the words *is a propositional formula* so we can see shortly how the definition transforms into an explicit definition and an induction principle.

Definition 18 *The property of a word in $\mathcal{A}_{PL}$ of being a propositional formula is defined inductively as follows.*

1. *Every Boolean constant (i.e., $\top$ or $\bot$) is a propositional formula.*

2. *Every propositional variable is a propositional formula.*

3. *If (the word) A is a propositional formula then (the word) $\neg A$ is a propositional formula.*

4. *If each of (the words) A and B is a propositional formula then each of (the words) $(A \wedge B)$, $(A \vee B)$, $(A \rightarrow B)$, and $(A \leftrightarrow B)$ is a propositional formula.*

The meaning of the inductive definition above can be expressed equivalently by the following explicit definition, which essentially repeats the definition above but replaces the phrase "*is a propositional formula*" with "*is in (the set)* FOR."

Definition 19 *The set of propositional formulae FOR is the* least *set of words in the alphabet of propositional logic such that the following holds.*

1. *Every Boolean constant is in FOR.*

2. *Every propositional variable is in FOR.*

3. *If A is in FOR then $\neg A$ is in FOR.*

4. *If each of A and B is in FOR then each of $(A \wedge B)$, $(A \vee B)$, $(A \rightarrow B)$, and $(A \leftrightarrow B)$ is in FOR.*

This pattern of converting the inductive definition into an explicit definition is general and can be applied to each of the other inductive definitions presented here. However, we have not yet proved that the definition of the set FOR given above is correct in the sense that the least (by inclusion) set described above even exists. Yet, if it does exist, then it is clearly unique because of being the least set with the described properties. We will prove the correctness later.

1.4.1.4 The subgroup of a given group, generated by a set of elements

I now provide a more algebraic example. The reader not familiar with the notions of groups and generated subgroups can skip this safely.

Let $\mathbf{G} = \langle G, \circ, ^{-1}, e \rangle$ be a group and X be a subset of G. The **subgroup of G generated by** X is the least subset $[X]_{\mathbf{G}}$ of G such that:

1. e is in $[X]_{\mathbf{G}}$.

2. Every element from X is in $[X]_{\mathbf{G}}$.

3. If $a \in [X]_\mathbf{G}$ then $a^{-1} \in [X]_\mathbf{G}$.

4. If $a, b \in [X]_\mathbf{G}$ then $a \circ b \in [X]_\mathbf{G}$.

Exercise: re-state the definition above as an inductive definition.

1.4.2   Induction principles and proofs by induction

With every inductive definition, a scheme for **proofs by induction** can be associated. The construction of this scheme is uniform from the inductive definition, as illustrated in the following.

1.4.2.1   Induction on the words in an alphabet

We begin with a principle of induction that allows us to prove properties of all words in a given alphabet. Given an alphabet $\mathcal{A}$, let $\mathcal{P}$ be a property of words in $\mathcal{A}$ such that:

1. The empty string ϵ has the property $\mathcal{P}$.

2. If the word w in $\mathcal{A}$ has the property $\mathcal{P}$ and $a \in \mathcal{A}$, then the word wa has the property $\mathcal{P}$.

Then, every word w in $\mathcal{A}$ has the property $\mathcal{P}$.

1.4.2.2   Induction on natural numbers

We can now formulate the well-known principle of mathematical induction on natural numbers in terms of the formal definition of natural numbers given above.

   Let $\mathcal{P}$ be a property of natural numbers such that:

1. 0 has the property $\mathcal{P}$.

2. For every natural number n, if n has the property $\mathcal{P}$ then Sn has the property $\mathcal{P}$.

Then every natural number n has the property $\mathcal{P}$.

   Here is the same principle, stated in set-theoretic terms:
   Let $\mathcal{P}$ be a set of natural numbers such that:

1. $0 \in \mathcal{P}$.

2. For every natural number n, if $n \in \mathcal{P}$ then $Sn \in \mathcal{P}$.

Then every natural number n is in $\mathcal{P}$, that is, $\mathcal{P} = \mathbb{N}$.

1.4.2.3   Structural induction on propositional formulae

Following the same pattern, we can now state a principle of induction that allows us to prove properties of propositional formulae. Note that this principle is obtained almost automatically from the inductive definition of propositional formulae by replacing throughout that definition the words *is a propositional formula* with *satisfies the property* $\mathcal{P}$. Let $\mathcal{P}$ be a property of propositional formulae such that:

1. Every Boolean constant *satisfies the property* $\mathcal{P}$.

2. Every propositional variable *satisfies the property* $\mathcal{P}$.

3. If A *satisfies the property* $\mathcal{P}$ then $\neg A$ *satisfies the property* $\mathcal{P}$.

4. If each of A and B *satisfy the property* $\mathcal{P}$ then each of $(A \wedge B)$, $(A \vee B)$, $(A \rightarrow B)$, and $(A \leftrightarrow B)$ *satisfy the property* $\mathcal{P}$.

Then every propositional formula *satisfies the property* $\mathcal{P}$.

Again, the same principle can be formulated in set-theoretic terms by treating the property $\mathcal{P}$ as the set of those propositional formulae that satisfy it, and then replacing the phrase *satisfies the property* $\mathcal{P}$ with *is in the set* $\mathcal{P}$.

The induction principle can likewise be formulated for the elements of a subgroup of a given group, generated by a given set of elements. I leave that as an exercise.

1.4.3 Basics of the general theory of inductive definitions and principles

1.4.3.1 An abstract framework for inductive definitions

We extract the common pattern in the examples above to formulate a uniform abstract framework for inductive definitions and proofs by induction. The necessary ingredients for an inductive definition are:

- A **universe** U.

    In our examples, universes were sets of words in a given alphabet and the set of all elements of a given group.

- A subset $B \subseteq U$ of **initial (basic) elements**.

    In our examples, the sets of initial elements were: $\{\epsilon\}$; $\{0\}$; $\{\top, \bot\} \cup \mathrm{PVAR}$; and the set X of generators of a subgroup.

- A set **F** of **operations (constructors)** in U.

    In our examples, these were: the operation of appending a symbol to a word; the operation of prefixing S to a natural number; the propositional logical connectives regarded as operations on words; and the group operations.

We fix the sets U, B, **F** arbitrarily thereafter. Our aim is to define formally the set of elements of U inductively defined over B by applying the operations in F, denoted $\mathcal{C}(B, \mathbf{F})$. Intuitively, this will be the set defined by the following inductive definition:

1. Every element of B is in $\mathcal{C}(B, \mathbf{F})$.

2. For every operation $f \in \mathbf{F}$ such that $f : U^n \rightarrow U$, if every $x_1, \ldots, x_n$ is in $\mathcal{C}(B, \mathbf{F})$ then $f(x_1, \ldots, x_n)$ is in $\mathcal{C}(B, \mathbf{F})$.

We give the set $\mathcal{C}(B, \mathbf{F})$ a precise mathematical meaning by defining it in two different, yet eventually equivalent, ways.

1.4.3.2   Top-down closure construction

Definition 20 *A set $C \subseteq U$ is:*

1. ***closed under the operation*** $f \in \mathbf{F}$, *such that* $f : U^n \to U$, *if* $f(x_1, \ldots, x_n) \in C$ *for every* $x_1, \ldots, x_n \in C$.
2. ***closed***, *if it is closed under every operation* $f \in \mathbf{F}$.
3. ***inductive***, *if* $B \subseteq C$ *and* C *is closed.*

Remark 21 *The elements of B can be regarded as constant (0-argument) functions in U, and the condition $B \subseteq C$ can therefore be subsumed by closedness.*

Proposition 22 *Intersection of any family of inductive sets is an inductive set.*

Proof. I leave this as an exercise.

Definition 23 C^* *is the intersection of the family of all inductive sets.*

By Proposition 22, C^* is the smallest inductive set.

1.4.3.3   Bottom-up inductive construction

Definition 24 *A **construction tree** for an element $x \in U$ is a finite tree $T(x)$, every node of which is labeled with an element of U and the successors of every node are ordered linearly, satisfying the following conditions:*

1. *Every leaf in $T(x)$ is labeled by an element of B.*
2. *If a node in $T(x)$ labeled with y has k successors labeled by elements listed in the order of successors $y_1, \ldots, y_k$, then there is a k-ary operation $f \in \mathbf{F}$ such that $y = f(y_1, \ldots, y_k)$.*
3. *The root of $T(x)$ is labeled by x.*

Definition 25 *The **height** of a finite tree is the length (number of nodes minus 1) of the longest path in the tree.*

Definition 26 *The **rank of an element** $x \in U$ is the least height $\mathbf{r}(x)$ of a construction tree for x if it exists; otherwise, the rank is ∞.*

Definition 27 *We define a hierarchy of sets $C_0 \subseteq C_1 \subseteq \cdots \subseteq C_*$ as follows.*

1. C_n *is the set of all elements of U with rank $\leq n$.*
2. $C_* := \bigcup_{n \in \mathbf{N}} C_n.$

Proposition 28 *The following holds.*

1. $C_0 = B$.
2. $C_{n+1} = C_n \cup \{f(x_1, \ldots, x_n) \mid f \in \mathbf{F} \text{ is an } n\text{-ary operation on } U \text{ and } x_1, \ldots, x_n \in C_n\}$

Proof. Easy induction on the rank n, which I leave as an exercise for the reader. ∎

1.4.3.4   Inductive definitions

Proposition 29 C_* *is an inductive set.*

Proof. Exercise. ∎

Corollary 30 $C^* \subseteq C_*$, *since C^* is the least inductive set.*

Proposition 31 $C_* \subseteq C^*$.

Proof. We prove by induction on n that $C_n \subseteq C^*$. Exercise. ∎

Definition 32 $C^*(=C_*)$ *is the **set inductively defined over B and F**, denoted $\mathcal{C}(B, \mathbf{F})$.*

1.4.3.5   Induction principle for inductively defined sets

We can easily generalize the ordinary principle of mathematical induction on natural numbers to induction in the set $\mathcal{C}(B, \mathbf{F})$.

Proposition 33 (Induction principle for $\mathcal{C}(B, \mathbf{F})$) *Let $\mathcal{P}$ be a property of elements of U, such that:*

1. *Every element of B has the property $\mathcal{P}$.*
2. *For every operation $f \in \mathbf{F}$, such that $f : U^n \to U$, if every $x_1, \ldots, x_n$ has the property $\mathcal{P}$ then $f(x_1, \ldots, x_n)$ has the property $\mathcal{P}$.*

Then every element of $\mathcal{C}(B, \mathbf{F})$ has the property $\mathcal{P}$.

The proof is straightforward. The two conditions above state precisely that the set of elements of $\mathcal{C}(B, \mathbf{F})$ that have the property $\mathcal{P}$ is inductive; it therefore contains the least inductive set, that is, $\mathcal{C}(B, \mathbf{F})$.

It is quite easy to see that the induction principle above generalizes all those formulated earlier for our examples.

1.4.4   Inductive definitions and proofs in well-founded sets

Here I briefly generalize the induction principle from inductively defined sets to the more abstract notion of *well-founded sets*.

1.4.4.1   Well-founded sets

Definition 34 *A partially ordered set (poset)* $(X, <)$ *is* **well-founded** *if it contains no infinite strictly descending sequences* $x_1 > x_2 > \ldots$.
   A well-founded linear ordering is called a **well-ordering**.

Example 35

- *Every finite poset is well-founded.*
- $\langle \mathbf{N}, < \rangle$ *is well-ordered, while* $\langle \mathbf{Z}, < \rangle$ *and* $\langle \mathbf{Q}, < \rangle$ *are not.*
- *The poset* $(\mathbf{P}(X), \subseteq)$, *where* X *is any infinite set, is not well-founded.*
- *The* **lexicographic ordering** *in* $\mathbf{N}^2$ *defined by* $\langle x_1, y_1 \rangle \leq \langle x_2, y_2 \rangle$ *iff* $x_1 < x_2$ *or* $(x_1 = x_2$ *and* $y_1 \leq y_2)$ *is a well-ordering in* $\mathbf{N}^2$.
- *The relation "A is a strict subformula of B" in the set* FOR *is a well-ordering.*

Proposition 36 *A poset* $\langle X, < \rangle$ *is well-founded iff every non-empty subset of* X *has a minimal element. Respectively, a linear ordering* $\langle X, < \rangle$ *is a well-ordering iff every non-empty subset of* X *has a least element.*

1.4.4.2   Induction principle for well-founded sets

Let $(X, <)$ be a well-founded poset. The **induction principle** for $(X, <)$ states the following:

   Let $\mathcal{P} \subseteq X$ be such that for every $x \in X$, if all elements of X less than x belong to $\mathcal{P}$ then x itself belongs to $\mathcal{P}$. Then $\mathcal{P} = X$.

Proof. Assume the contrary, that is, $X - \mathcal{P} \neq \phi$. Then $X - \mathcal{P}$ has a minimal element x. Then all elements of X less than x belong to $\mathcal{P}$, hence x must belong to $\mathcal{P}$: a contradiction. ∎

Example 37 *Let* $\mathcal{P}$ *be a property (set) of propositional formulae such that for every formula* $A \in FOR$, *if all strict subformulae of* X *have the property (belong to the set)* $\mathcal{P}$, *then* A *itself has the property (belongs to the set)* $\mathcal{P}$.
   Then every formula $A \in FOR$ *has the property (belongs to the set)* $\mathcal{P}$.

1.4.5   Recursive definitions on inductively definable sets

We now consider the following general problem: given an inductively defined set $\mathcal{C}(B, \mathbf{F})$, how should we define a function h on that set by using the inductive definition? The idea is to first define h on the set B, and then provide rules prescribing how the definition of that function propagates over all operations. Formally, in order to define by recursion a mapping $h : \mathcal{C}(B, \mathcal{F}) \rightarrow X$ where X is a fixed target set, we need:

1. A mapping $h_0 : B \rightarrow X$.
2. For every n-ary operation $f \in \mathcal{F}$ a mapping $F_f : X^{2n} \rightarrow X$.

We now define the mapping h as follows:

1. If $a \in B$ then $h(a) := h_0(a)$.
2. For every n-ary operation $f \in \mathcal{F}$:

$$h(f(a_1, \ldots, a_n)) := F_f(a_1, \ldots, a_n, h(a_1), \ldots, h(a_n)).$$

We will soon discuss the meaning and correctness of such definitions, but let us first look at some important particular cases.

1.4.5.1   Primitive recursion on natural numbers

Functions on natural numbers can be defined by so-called *primitive recursion* using the inductive definition provided earlier in this section.

1. The basic scheme of primitive recursion is:

$$h(0) = a$$
$$h(n + 1) = h(Sn) = F_S(n, h(n)).$$

   For example, the scheme

$$h(0) = 1$$
$$h(Sn) = (n + 1)h(n)$$

   defines the factorial function $h(n) = n!$.

2. The more general scheme of primitive recursion with parameters is:

$$h(\mathbf{m}, 0) = F_0(\mathbf{m}).$$
$$h(\mathbf{m}, Sn) = F_S(\mathbf{m}, n, h(\mathbf{m}, n)).$$

   For example, the scheme

$$h(m, 0) = m,$$
$$h(m, n + 1) = h(m, n) + 1$$

   defines the function addition $h(m, n) = m + n$.

1.4.5.2   Truth valuations of propositional formulae

Recall that the set of propositional formulae FOR is built on a set of propositional variables PVAR and a truth assignment is a mapping $s : \text{PVAR} \to \{\text{T}, \text{F}\}$. Now, given any truth assignment $s : \text{PVAR} \to \{\text{T}, \text{F}\}$ we can define a mapping $\alpha : \text{FOR} \to \{\text{T}, \text{F}\}$ that extends it to a **truth valuation**, a function computing the truth values of all formulae in FOR by recursion on the inductive definition of FOR as follows:

1. $\alpha(\mathbf{t}) = \text{T}, \alpha(\mathbf{f}) = \text{F}$.
2. $\alpha(p) = s(p)$ for every propositional variable p.
3. $\alpha(\neg A) = F_\neg(\alpha(A))$,
   where $F_\neg : \{\text{T}, \text{F}\} \to \{\text{T}, \text{F}\}$ is defined as follows: $F_\neg(\text{T}) = \text{F}, F_\neg(\text{F}) = \text{T}$.
4. $\alpha(A \wedge B) = F_\wedge(\alpha(A), \alpha(B))$,
   where $F_\wedge : \{\text{T}, \text{F}\}^2 \to \{\text{T}, \text{F}\}$ is defined as follows:
   $F_\wedge(\text{T}, \text{T}) = \text{T}$ and $F_\wedge(\text{T}, \text{F}) = F_\wedge(\text{F}, \text{T}) = F_\wedge(\text{F}, \text{F}) = \text{F}$.
   (That is, $F_\wedge$ computes the truth table of $\wedge$.)
5. $\alpha(A \vee B) = F_\vee(\alpha(A), \alpha(B))$,
   where $F_\vee : \{\text{T}, F\}^2 \to \{\text{T}, \text{F}\}$ is defined as follows:
   $F_\vee(\text{T}, \text{T}) = F_\vee(\text{T}, \text{F}) = F_\vee(\text{F}, \text{T}) = \text{T}$ and $F_\vee(\text{F}, \text{F}) = \text{F}$.
6. $\alpha(A \to B) = F_\to(\alpha(A), \alpha(B))$,
   where $F_\to : \{\text{T}, F\}^2 \to \{\text{T}, \text{F}\}$ is defined according to the truth table of $\to$.
7. $\alpha(A \leftrightarrow B) = F_\leftrightarrow(\alpha(A), \alpha(B))$,
   where $F_\leftrightarrow : \{\text{T}, F\}^2 \to \{\text{T}, \text{F}\}$ is defined according to the truth table of $\leftrightarrow$.

The mapping α so defined is called *the truth valuation* of the propositional formulae generated by the truth assignment s.

Using such recursive definitions, we can likewise define various other natural functions associated with propositional formulae such as length, number of occurrences of logical connectives, and set of occurring propositional variables. I leave these as exercises.

1.4.5.3   Homomorphisms on freely generated groups

Given a group $\mathbf{G}$ with a set of free generators B and any group $\mathbf{H}$, every mapping $h_0 : B \to \mathbf{H}$ can be (uniquely) extended to a homomorphism $h : \mathbf{G} \to \mathbf{H}$. The definition of h is essentially by recursion on the inductive definition of $\mathbf{G}$ as generated by B, and I leave it as an exercise.

1.4.5.4   Other inductive definitions and recursion on natural numbers

Consider the following inductive definitions:

Definition 38

1. *0 is a natural number.*
2. *If n is a natural number then $2n + 1$ is a natural number.*
3. *If n is a natural number and $n > 0$ then $2n$ is a natural number.*

Definition 39

1. *0 is a natural number.*
2. *If n is a natural number then $n + 2$ is a natural number.*
3. *If n is a natural number then $2n + 1$ is a natural number.*

I leave it as exercises for the reader to show that each of these inductive definitions defines the set of all natural numbers.

We now consider the recursive definitions:

1. $h_1(0) = 1$;
2. $h_1(2n + 1) = 3n + h_1(n)$;
3. $h_1(2n) = h_1(n) + 1$, for $n > 0$.

and

1. $h_2(0) = 0$;
2. $h_2(n + 2) = 2h_2(n) + 3$;
3. $h_2(2n + 1) = h_2(n) + 1$;

They look similar, and yet there is something wrong with the second definition. What? First, note that $h_2(1) = h_2(2 \times 0 + 1) = h_2(0) + 1 = 1$. Now, let us compute $h_2(3)$. On the one hand, $h_2(3) = h_2(1 + 2) = 2h_2(1) + 3 = 5$. On the other hand, $h_2(3) = h_2(2 \cdot 1 + 1) = h_2(1) + 1 = 2$. Thus, we have obtained two different values, which is definitely bad. The problem comes from the fact that the second definition allows for essentially different generations of the same object, leading us to define the notion of **unique generation**.

1.4.5.5 Unique generation

For the correctness of recursive definitions it should be required that every element of $C(B, \mathbf{F})$ can be constructed uniquely (up to the order of the steps). Otherwise, definitions can lead to problems as above. More formally, the elements of $C(B, \mathbf{F})$ are represented by expressions (terms) built from the elements of B by applying the operations from $\mathbf{F}$. Unique generation means that every element of $C(B, \mathbf{F})$ can be represented by a unique expression.

Example 40

1. *The standard definition of natural numbers and Definition 1 above have the unique generation property. These can be proved by induction on natural numbers.*
2. *Definition 2 given above does not satisfy the unique generation property.*
3. *The set FOR of propositional formulae satisfies the unique generation property, also known as* **unique readability property**.
4. *If the subgroup H of a group G is freely generated by a set of generators B, then it satisfies the unique generation property.*
5. *If the subgroup H of a group G is not freely generated by a set of generators B, then it does not satisfy the unique generation property.*

Theorem 41 *If $C(B, \mathbf{F})$ satisfies the unique generation property then for every mapping $h_0 : B \to X$ and mappings $\{F_f : X^{2n} \to X \mid f \in \mathcal{F}\}$ there exists a unique mapping $h : C(B, \mathcal{F}) \to X$ defined by the recursive scheme in Section 1.4.5.*

I do not give a proof here, but refer the reader to Enderton (2001).

References for further reading
For further reading in inductive definitions, structural induction, and recursion see Tarski (1965), Shoenfield (1967), Barwise and Echemendy (1999), Enderton (2001), Hedman (2004), Nederpelt and Kamareddine (2004), and Makinson (2008).

Exercises

1.4.1 Given a set Z, give an inductive definition of the set of *lists of elements of Z*. Formally, these are special words in the alphabet $Z \cup \{[\ , \ ; \ , \ ]\}$ of the type $[\cdots []; z_1]; z_2]; \cdots z_n]$, where $[]$ is the empty list and $z_1, z_2, \ldots, z_n \in Z$.
Then formulate the induction principle for the set of lists of elements of Z, both for properties of lists and for sets of lists, and use it to prove that every list of elements of Z has equal numbers of occurrences of [and of].

1.4.2 Rephrase the definition of the subgroup $[X]_\mathbf{G}$ of a group $\mathbf{G}$ generated by a set X, given in this section, as an inductive definition.
Then formulate the induction principle for the elements of $[X]_\mathbf{G}$, both for properties and for sets of elements of the group.

1.4.3 Prove by structural induction that every propositional formula has equal numbers of occurrences of (and).

1.4.4 Prove Proposition 22.

1.4.5 Prove Proposition 28.

1.4.6 Prove Proposition 31.

1.4.7 Use addition and primitive recursion to define multiplication on natural numbers.

1.4.8 Use multiplication and primitive recursion to define exponentiation on natural numbers.

1.4.9 Complete the definition of the mappings $F_\rightarrow$ and $F_\leftrightarrow$ in the recursive definition of truth valuations of propositional formulae.

1.4.10 Given a group $\mathbf{G}$ with a set of free generators B, a group $\mathbf{H}$, and a mapping $h_0 : B \rightarrow \mathbf{H}$, define the unique homomorphism $h : \mathbf{G} \rightarrow \mathbf{H}$ extending h_0 by recursion on the inductive definition of $\mathbf{G}$ as generated by B.

1.4.11 Show that each of the alternative inductive definitions of the set of natural numbers given in Section 1.4.5.4 is correct.

1.4.12 Prove by induction on natural numbers that each of the standard definitions of natural numbers and Definition 1 in Section 1.4.5.4 satisfy the unique generation property.

1.4.13 Prove by structural induction on FOR that the set FOR of propositional formulae satisfies the unique generation property.

1.4.14 Using the inductive definition of FOR, give recursive definitions of the following functions associated with propositional formulae:
   (a) the length of a formula, being the number of symbols occurring in it;
   (b) the number of occurrences of logical connectives in a formula;
   (c) the set of propositional variables occurring in a formula;
   (d) the set of logical connectives occurring in a formula; and
   (e) the nesting depth of a formula, being the largest number of logical connectives occurring in the scope of one another in the formula.

Ludwig Wittgenstein (26.04.1889–29.04.1951) was one of the most prominent philosophers and logicians of the 20th century. His work spanned logic, philosophy of mathematics, philosophy of mind, and philosophy of language. During his lifetime he published just one book in philosophy, the extremely influential (and quite enigmatic) 75-page long *Tractatus Logico-Philosophicus* in 1921, but he produced and wrote much more, mostly published only in 1953 after his death, in the book *Philosophical Investigations*.

Wittgenstein was born in Vienna into one of Europe's richest families. He completed high school in Linz, Austria and went on to study mechanical engineering in Berlin and then aeronautics in Manchester. He gradually developed a strong interest in logic and the foundations of mathematics, influenced by Russell's *Principia Mathematica* and Frege's *Grundgesetze der Arithmetik*. In 1911 he visited Frege in Jena and wanted to study with him; they did not quite match each other however. He went on to Cambridge to study logic with Russell, who quickly recognized Wittgenstein as a genius. However, Wittgenstein became depressed in Cambridge and in 1913 went to work in isolation in Skjolden, Norway, where he conceived much of *Tractatus*. When World War I broke out, Wittgenstein returned to Vienna to join the army and fight in the war for 4 years. During that period he received numerous military decorations for his courage.

During the war, Wittgenstein completed the manuscript of his *Tractatus*, where he presented his philosophical theory on how logic, language, mind, and the real world relate. In a simplified summary, he argued that words are just representations of objects and propositions are just words combined to make statements (pictures) about reality, which may be true or false, while the real world is nothing but the facts that are in it. These facts can then be reduced to *states of affair* (later leading to the concept of "possible worlds"), which in turn can be reduced to combinations of objects, thus eventually creating a precise correspondence between language and the world. Wittgenstein believed he had solved all philosophical problems in his *Tractatus*. In any case, he had essentially invented propositional formulae and truth tables. In his later period, Wittgenstein deviated from some of his views in *Tractatus*. In particular, he came to believe that the meaning of words was entirely in their use.

Wittgenstein sent his *Tractatus* to Russell from Italy in 1918, where he was detained as a prisoner of war, but only submitted it as his doctoral thesis when he returned to Cambridge in 1929. During a period of depression after the war in 1919, he gave the fortune that he had inherited from his father to his brothers and sisters. He then he had to support himself by working at various jobs, including as maths teacher in a remote Austrian village, a gardener at a monastery, and an architect for his sister's house. He remained in Cambridge to teach during 1929–1947, which was his longest period in academia. Interrupted by World War II, he decided that it was not morally justifiable to stay in the comfortable academic world in such troubled times, worked anonymously as a hospital porter in London (where he advised patients not to take the drugs they were prescribed). He resigned from Cambridge in 1947 and retired in isolation to a remote cottage on the coast of Ireland, where he died of cancer in 1951. Despite his worsening health he worked until the very end, which he is said to have accepted quite willingly as he was apparently never quite happy with his life and felt he did not really fit into this world.

To the present day, Wittgenstein remains one of the most studied and discussed, but also probably one of the least understood, modern thinkers.

2

Deductive Reasoning in Propositional Logic

This chapter presents the deductive side of propositional logic: deductive systems and formal derivations of logical consequences in them. First, I explain the concept and purpose of a deductive system and provide some historical background. I then introduce, discuss, and illustrate with examples the most popular types of deductive systems for classical logic: *Axiomatic Systems*, *Semantic Tableaux*, *Natural Deduction*, and *Resolution*. In a supplementary section at the end of the chapter I sketch generic proofs of soundness and completeness of these deductive systems. In another supplementary section I discuss briefly the computational complexity of the Boolean satisfiability problem and the concept of NP-completeness.

The deductive systems introduced in this chapter are specifically designed for classical propositional logic, but the concept is universal and applies to almost all logical systems that have been introduced and studied. In Chapter 4 I extend each of these deductive systems with additional axioms and rules for the quantifiers, so that they also work for first-order logic.

2.1 Deductive systems: an overview

2.1.1 The concept and purpose of deductive systems

The fundamental concept in logic is that of *logical consequence*. It extends the concept of logical validity and is the basis of logically correct reasoning. Verifying logical consequence in propositional logic is conceptually simple and technically easy (although possibly computationally expensive), but this is no longer the case for full *first-order logic*, which I introduce in Chapter 3, or for the variety of *non-classical logics* which I do not discuss here. In fact, verifying logical consequence (in particular, logical validity) in first-order logic is usually an *infinite task*, as it generally requires checking infinitely many possible models rather than a finite number of simple truth assignments. A different approach for proving logical validity and consequence, which is not based on the semantic definition, is therefore necessary. Such a different approach is provided by the notion of a *deductive system*, a formal, mechanical – or, at least mechanizable – procedure

Logic as a Tool: A Guide to Formal Logical Reasoning, First Edition. Valentin Goranko.
© 2016 John Wiley & Sons, Ltd. Published 2016 by John Wiley & Sons, Ltd.

for *derivation (inference, deduction)* of formulae or lists ("sequents") of formulae, by applying precise *inference rules* and possibly using some formulae called *axioms* that are postulated as derived. Deductive systems therefore *substitute* the semantic notion of logical consequence with the formal, syntactic notion of *deductive consequence*, based only on the syntactic shape of the formulae to which inference rules are applied, but not explicitly on their meaning. This means that, in theory, someone without knowledge and understanding of the meaning of logical formulae and logical consequence, or even a suitably programmed computer, should be able to perform derivations in a given deductive system.

The purpose and fundamental importance of a deductive system is not only to serve as a mechanical procedure for deriving logical consequences, but also to provide simple and intuitively acceptable principles and rules of logical inference which can be used to demonstrate or verify logically correct reasoning. The underlying idea is that if we start with premises that are known or accepted as true, and apply only rules of inference that are known to preserve truth, then we are guaranteed that every reasoning performed within such a deductive system is logically correct and that every conclusion reached by a chain of inferences within that system must also be true.

There are different types of deductive systems based on different ideas, but they all share several common features and principles. To begin with, a deductive system works within a **formal logical language** where derivations are performed on formal expressions – usually formulae or "sequents" (finite sequences of formulae) – which are special "words", that is, finite strings of symbols in the formal language, having precise **syntax**[1]. In this chapter I discuss deductive systems acting on propositional formulae. In Chapter 4 I extend these to full first-order logic.

The main component of a deductive system is the notion of **derivation (inference, deduction) from a given set of assumptions**, based on a set of precisely specified **rules of inference**. The idea is that by systematically applying these rules, we can **derive (infer, deduce)** formulae from other formulae that are already derived, or *assumed* as given, called **assumptions (premises)**. In addition, axiomatic systems also allow for an initial set of formulae, called **axioms**, to be accepted as derived without applying any rules of inference. The axioms can therefore always be used as premises in derivations.

If a formula A can be derived from a set of assumptions Γ in a deductive system $\mathcal{D}$, we denote this

$$\Gamma \vdash_{\mathcal{D}} A$$

and say that A **is a deductive consequence from** Γ **in** $\mathcal{D}$.

In particular, if $\Gamma = \emptyset$, we write $\vdash_{\mathcal{D}} A$ and say that A is a **theorem of** $\mathcal{D}$.

2.1.2   Brief historical remarks on deductive systems

The idea of a proof or derivation as demonstrative formal argumentation and of the concept of logical deductive systems go back to antiquity, and most notably to the book *Organon* of the ancient Greek philosopher Aristotle (384–322 BC), founding father of (among many scientific disciplines) formal logic. The first formal logical system, called *Syllogistic* (see Section 3.5), was developed by Aristotle, but perhaps the first real prototype of a deductive system can be found in the fundamental work *Elements* of the "Father of Geometry", the great Ancient Greek mathematician Euclid (*c.* 300 BC), who provided a systematic

---

[1] In this aspect a deductive system is much like a programming language.

development of elementary geometry based on several simple assumptions about points and lines (such as "every two different points determine exactly one line", "for every line there is a point not belonging to that line", etc.). Using these *postulates* and some informal logical reasoning, other geometric facts are derived and thus the entire body of Euclidean geometry is eventually built.

The *logical* concept of the deductive system gradually emerged much later, notably through the ideas of Gottfried Leibniz (1646–1716), one of the greatest philosophers and mathematicians of all times, of a *characteristica universalis* (universal language) and a *calculus ratiocinator* (calculus of reasoning). The first formal deductive system for modern logic, however, was only constructed in the late 19th century by the mathematician and philosopher Gottlob Frege (1848–1925) in his seminal book *Begriffsschrift* (1879), where he developed the prototype of classical first-order logic. Later, David Hilbert (1862–1943), one of the leading mathematicians of the late 19th and early 20th century, reworked Euclid's system of geometry developed in the *Elements* into a rigorous and mathematically precise treatment that resulted in his book *Foundations of Geometry* (1899). The mathematician Giuseppe Peano (1858–1932) developed a formal system of arithmetic, still known as *Peano's axiomatic system*, in his most important work *Formulario Mathematico*. The concept of formal deductive system was further developed by the philosophers Bertrand Russell (1872–1970) and Alfred North Whitehead (1861–1947) in their three-volume book *Principia Mathematica* (1910–1913), and by David Hilbert and Wilhelm Ackermann (1896–1962) in their book *Principles of Mathematical Logic* (1928). These books were most influential for the development of logic and foundations of mathematics in the first half of the 20th century. In particular, Hilbert strongly promoted the idea of building the whole body of mathematics as a formal *axiomatic system*, a deductive system mainly based on axioms and on very few and simple rules of inference. Hilbert was the leading proponent of the development of the axiomatic approach in mathematics, meant to replace the semi-formal notion of mathematical proof by the completely formalized notion of derivation in an axiomatic system. A major purpose of the formal axiomatic approach was to avoid the occurrence of any paradoxes and contradictions in mathematics by performing all mathematical proofs within such a formal system of deduction which has been proved to be *consistent*, that is, free from contradictions. Such a system, based on first-order logic, was proposed by Hilbert and Ackermann in *Principles of Mathematical Logic*, and Hilbert's ultimate goal was to prove that that system was both consistent and complete, that is, capable of deriving every mathematical truth, but not deriving any contradictions. Hilbert's idea of a deductive system was further developed by several logicians, including Jacques Herbrand, Emil Post, Alfred Tarski, and others.

In the late 1920s–early 1930s the concept of a deductive system was extended in the works of Gerhard Gentzen and Stanisław Jaśkowski by adding the possibility of introducing and withdrawing assumptions in derivations, that is, formulae that are not axioms and have not been derived. Thus, the more intuitive and practically more efficient *rule-based* type of deductive system, called *Natural Deduction* and the closely related *Sequent Calculus*, were developed, essentially founding the major field of logic known as proof theory. In the 1950s–1960s, the the *refutation-based* system of *Semantic Tableaux* emerged, implementing the idea of a systematic and exhaustive search for a falsifying model of the assumptions plus the negation of the desired conclusion, where a derivation consists of an established failure to construct such a falsifying model. The idea of refutation-based deductive systems led to the development in the 1960s–1970s of the method of *resolution*, which turned out to be very suitable for computer-aided automated deduction.

Meanwhile, in the early 1930s some groundbreaking results in logic were announced by a young logician having just completed his doctoral studies, Kurt Gödel. Those results, that made Gödel the most famous logician of the 20th century, showed that Hilbert's idea was only partly realizable, in a sense that purely logical validity and consequence in first-order logic can be axiomatized completely by *Gödel's completeness theorem* for first-order logic; it was not, however, realizable for the richer notion of mathematical consequence, even in the relatively simple mathematical systems of arithmetic of natural numbers with addition and multiplication. More precisely, in 1931 he proved his two celebrated *Gödel's incompleteness theorems*. The first of these stated that no sufficiently expressive and reasonably axiomatized (with an effectively recognizable notion of axioms) deductive system, such as Peano's system of arithmetic, can be complete. The second incompleteness theorem claimed that such a theory cannot even prove its own consistency – suitably encoded as a formula in that system – unless it is inconsistent (in which case it can derive *any formula* by using the sound logical rule *Ex Falso Quodlibet*). No absolute and finitary proof of consistency of such system is therefore possible. Hilbert's dream of formalizing the whole of mathematics in a provably consistent way turned out to be unattainable. Still, deductive systems and their proof theory have remained one of the main directions for development of modern logic.

2.1.3   Soundness, completeness and adequacy of deductive systems

A very important aspect of deductive systems is that derivations in them are *completely mechanizable procedures* that in principle do not require any intelligence or understanding of the meaning of the formulae or rules involved; in fact, such a meaning need not be specified at all. Derivations in a given deductive system can therefore be performed by a mechanical device such as a computer without any human intervention, as long as the axioms and rules of inference of the deductive system have been programmed into it.

While deductive systems are not explicitly concerned with the meaning (*semantics*) of the formulae they derive, they are designed with the purpose of deriving *only valid* logical consequences from the assumptions. A deductive system with this property is called *sound*. In particular, every theorem of a sound deductive system must be a valid formula, that is, in the case of propositional logic it must be a tautology.

Formally, a deductive system $\mathcal{D}$ is **sound** (or **correct**) for a given logical semantics (that is, well-defined notions of logical validity and consequence) if $\mathcal{D}$ can *only* derive logically valid consequences, that is:

$$A_1, \ldots, A_n \vdash_{\mathcal{D}} C \text{ implies } A_1, \ldots, A_n \models C.$$

In particular: $\vdash_{\mathcal{D}} C$ implies $\models C$.

A deductive system $\mathcal{D}$ is **complete** for a given logical semantics if $\mathcal{D}$ can derive *every* valid logical consequence (as defined in that semantics), that is:

$$A_1, \ldots, A_n \models C \text{ implies } A_1, \ldots, A_n \vdash_{\mathcal{D}} C.$$

In particular, $\models C$ implies $\vdash_{\mathcal{D}} C$, that is, a complete deductive system can derive every logically valid formula.

A deductive system $\mathcal{D}$ is **adequate** for a given semantics if it is both sound and complete for it, i.e.:

$$A_1, \ldots, A_n \models C \quad \text{if and only if} \quad A_1, \ldots, A_n \vdash_{\mathcal{D}} C$$

In particular: $\models C$ if and only if $\vdash_{\mathcal{D}} C$.

The soundness of a deductive system can be guaranteed and proved easily in principle as long as the following two conditions hold:

(i) *All axioms (if any) must be true.*

(ii) *All rules of inference must be sound*, that is, they must always produce true conclusions when applied to true assumptions. The truth therefore propagates from the axioms to all theorems.

I present here some of the most popular types of deductive systems for classical logic, namely *Axiomatic Systems, Semantic Tableaux, Natural Deduction*, and *Resolution*.[2] Each of these bears a different idea and has advantages and shortcomings compared to the others. The use of each of these is illustrated with several examples and many more exercises, and in Chapter 5 they will be put to real work for performing concrete mathematical reasoning.

References for further reading

For further general discussion and historical notes on deductive systems, see the classical masterpiece Tarski (1965) and well as Nerode and Shore (1993), Jeffrey (1994), Fitting (1996), Barwise and Echemendy (1999), Hodges (2001), and Ben-Ari (2007).

Aristotle (384–322 BC) was a great Ancient Greek philosopher, "the first genuine scientist in history" according to Encyclopaedia Britannica.

He was born in the Macedonian city of Stagira in the Chalkidiki peninsula, and joined Plato's Academy in Athens at the age of 18 where he carried out scholarly work for around 20 years. After the death of Plato he left Athens and, at the request of King Philip of Macedonia, became the private teacher of his son Alexander the Great during 356–323 BC.

It is believed that only about one-third of Aristotle's original works have survived. In these he made important contributions to just about every field of science and arts that existed in his time or was founded by him: philosophy, poetry, theatre, music, rhetoric, politics, government and ethics, physics, geology and biology. Aristotle's legacy, not only in philosophy, continues to be the object of active academic study today. For instance, the idea of systematic classification of plants and animals, developed in the 18th century by Linnaeus, originate from him.

---

[2] I do not present Sequent Calculus in this book, but it is easily reducible to Natural Deduction. See Ebbinghaus *et al.* (1996) and Boolos *et al.* (2007).

Among his many other seminal contributions, Aristotle is regarded as the founding father of logic as a scientific discipline. He was the first to propose the use of
formal logical languages in the study of reasoning and to undertake and accomplish
a systematic study of the principles of correct reasoning, unsurpassed for more than
2000 years. His treatises were collected by his followers under the name *Organon*
(from the Greek word meaning "instrument" or "tool"), consisting of six books:
Categories, *On Interpretation*, *Prior Analytics*, *Posterior Analytics*, *Topics* and *On
Sophistical Refutations*. In *Organon* Aristotle developed a theory of deduction for a
special kind of logical arguments called **syllogisms** (see Section 3.5) built on several
forms of expressions of the type "all *A*s are (not) *B*s", "some *A*s are (not) *B*s", or "no
*A*s are *B*s". A typical example of a syllogism is "All humans are mortal. All Greeks
are humans. Therefore, all Greeks are mortal." In his logical system, now called the
Syllogistic, Aristotle introduced the so-called **square of opposition**, identified and
classified all syllogistic forms into logically valid and invalid forms, proposed a system for deriving some valid syllogistic forms from others, and provided convincing
counter-examples to the invalid forms. The main rival of Aristotelian logic, which in
modern terms is a proper fragment of first-order logic, was the logic of propositions
of the Stoic school.

Aristotle also studied different **modes of truth** – possible, necessary, and
contingent – laying the foundations of **modal logic**. He also proved many
"meta-properties" of the correct logical reasoning, anticipating the rigorous,
mathematical study of modern logic.

2.2   Axiomatic systems for propositional logic

2.2.1   *Description*

Axiomatic systems are deductive systems that are mainly based on *axioms* and use very
few and simple rules of inference. Axiomatic systems are the oldest and simplest to
describe (but not to use!) of all deductive systems. The idea of basing the mathematical
reasoning on formal axiomatic systems was strongly promoted by David Hilbert in the
early 20th century. In his honor, axiomatic systems are also commonly called **Hilbert
(-style) systems**.

Here we build a hierarchy of several axiomatic systems for propositional logic by gradually adding new logical connectives and axioms for these. Before that, we need the
notion of **formula scheme** which is used in these axiomatic systems. This is the set of all
instances of propositional formulae obtained from a given formula by applying **uniform
substitution** as follows.

Let $F = F(p_1, \ldots, p_n)$ be a formula built over the propositional variables $p_1, \ldots, p_n$
and let $B_1, \ldots, B_n$ be any formulae. Then $F' = F[B_1/p_1, \ldots, B_n/p_n]$ is the formula
obtained from F by replacing simultaneously every occurrence of p_i by B_i for each
$i = 1, \ldots, n$. We say that F' is a **substitution instance** (or just an **instance**) of F. For
example,

$$((p \wedge q) \wedge \neg\neg p) \rightarrow (\neg(p \wedge q) \vee \neg p)$$

is the substitution instance of

$$(p \wedge \neg q) \rightarrow (\neg p \vee q)$$

where p is substituted by $(p \wedge q)$ and q is substituted by $\neg p$. The *scheme* of all substitution instances of the formula $F(p_1, \ldots, p_n)$ can be written simply as $F[B_1/p_1, \ldots, B_n/p_n]$, where $B_1, \ldots, B_n$ are regarded not as specific formulae but as *metavariables* (i.e., variables in our metalanguage) for formulae. The formulae of the scheme are the instances obtained by substituting formulae for these metavariables. The scheme generated from the formula $(p \wedge \neg q) \rightarrow (\neg p \vee q)$ can therefore be written $(B_1 \wedge \neg B_2) \rightarrow (\neg B_1 \vee B_2)$.

We are now ready to start introducing axiomatic systems for propositional logic. First, we assume that the only logical connectives are $\rightarrow$ and $\neg$, and all others are definable in terms of these. The axiomatic system for that language comprises the following axioms and rules.

Axiom schemes

($\rightarrow$ 1) $A \rightarrow (B \rightarrow A)$;

($\rightarrow$ 2) $(A \rightarrow (B \rightarrow C)) \rightarrow ((A \rightarrow B) \rightarrow (A \rightarrow C))$;

($\rightarrow$ 3) $(\neg B \rightarrow \neg A) \rightarrow ((\neg B \rightarrow A) \rightarrow B)$.

The only **rule of inference** is **Modus ponens:**

$$\frac{A, A \rightarrow B}{B}$$

which now reads: "If the formulae A and $A \rightarrow B$ are derived, then the formula B is also derived."

We denote this axiomatic system $\mathbf{H}(\rightarrow, \neg)$.

If we also consider the conjunction as a primitive, rather than definable connective, then it turns out that the following axiom schemes completely capture its deductive properties.

($\wedge$1) $(A \wedge B) \rightarrow A$;

($\wedge$2) $(A \wedge B) \rightarrow B$;

($\wedge$3) $(A \rightarrow B) \rightarrow ((A \rightarrow C) \rightarrow (A \rightarrow B \wedge C))$.

Added to $\mathbf{H}(\rightarrow, \neg)$, these schemes produce the system $\mathbf{H}(\rightarrow, \neg, \wedge)$.

Likewise, the following axiom schemes suffice for the disjunction:

($\vee$1) $A \rightarrow A \vee B$;

($\vee$2) $B \rightarrow A \vee B$;

($\vee$3) $(A \rightarrow C) \rightarrow ((B \rightarrow C) \rightarrow (A \vee B \rightarrow C))$.

These schemes, added to $\mathbf{H}(\rightarrow, \neg)$ produce the system $\mathbf{H}(\rightarrow, \neg, \vee)$.

Finally, the axiomatic system $\mathbf{H}(\rightarrow, \neg, \wedge, \vee)$ combining $\mathbf{H}(\rightarrow, \neg, \wedge)$ and $\mathbf{H}(\rightarrow, \neg, \vee)$ is simply denoted $\mathbf{H}$.

2.2.2   Derivations in the axiomatic system **H**

I only consider the full system **H** here, but all definitions apply to all subsystems defined above.

A **derivation in H from a set of assumptions** Γ is a finite sequence of formulae $C_1, \ldots, C_n$ such that for every $i \in \{1, \ldots, n\}$, either $C_i \in \Gamma$ or C_i is an axiom of **H**, or C_i is obtained by applying the rule Modus ponens to previously derived formulae, that is, there are $j, k < i$ such that C_k is $C_j \to C_i$. We write $\Gamma \vdash_{\mathbf{H}} C_n$ and say that C_n is **derived in H from the set of assumptions** Γ.

Note that every initial subsequence of a derivation is a derivation, so *every* formula occurring in a derivation is derived and not just the last one.

We often list the assumptions explicitly and write $A_1, \ldots, A_k \vdash_{\mathbf{H}} C$.

A formula A is said to be **derivable in H** or a **theorem of H**, denoted $\vdash_{\mathbf{H}} A$, if $\emptyset \vdash_{\mathbf{H}} A$.

The notion of derivation from a set of assumptions is meant to be the deductive analog in **H** of a logical consequence. In order to prove that "$A_1, \ldots, A_n$ logically imply B", we add the assumptions $A_1, \ldots, A_n$ to the set of axioms of **H** and try to derive B from these. Here is an example of a derivation in **H**. We list the formulae in the derivation and provide a brief justification of how they have been derived.

Example 42 $\vdash_{\mathbf{H}} (p \wedge (p \to q)) \to q$:

1. $(p \wedge (p \to q)) \to p$,   *instance of Axiom* $(\wedge 1)$;

2. $(p \wedge (p \to q)) \to (p \to q)$,   *instance of Axiom* $(\wedge 2)$;

3. $((p \wedge (p \to q)) \to (p \to q)) \to (((p \wedge (p \to q)) \to p) \to ((p \wedge (p \to q)) \to q))$,

   instance of Axiom $(\to 2)$;

4. $((p \wedge (p \to q)) \to p) \to ((p \wedge (p \to q)) \to q)$,   *by 2,3 and Modus ponens*;

5. $(p \wedge (p \to q)) \to q$,   *by 1,4 and Modus ponens*.

This example looks overly complicated for the simple tautology that it derives, but it is illustrative of how complex derivations in axiomatic systems can be. Still, we prove in Section 2.7 that **H** and each of its extensions above can derive every valid logical consequence. More precisely, the following holds.

Theorem 43 (Adequacy theorem for H) *The axiomatic system **H** is sound and complete, that is:*

$$A_1, \ldots, A_n \vdash_{\mathbf{H}} B \quad \textit{if and only if} \quad A_1, \ldots, A_n \models B.$$

In particular, $\vdash_{\mathbf{H}} B$ *if and only if B is a tautology.*

The same holds for each of the subsystems $\mathbf{H}(\to, \neg)$, $\mathbf{H}(\to, \neg, \wedge)$, and $\mathbf{H}(\to, \neg, \vee)$.

Despite being adequate, the axiomatic system **H** is not very suitable for practical derivations, as it can be judged from the example above. The derivations in it are not very well structured and often difficult to construct, because even simple cases may require involved derivations. For instance, as a challenge, try deriving the tautology $p \to p$.

Still, the derivations in **H** can be simplified significantly by using the following important result which allows us to introduce, and later eliminate, auxiliary assumptions in the derivations.

Theorem 44 (Deduction theorem) *For any formulae A, C, and set of formulae* Γ:

$$\Gamma \cup \{A\} \vdash_{\mathbf{H}} C \text{ iff } \Gamma \vdash_{\mathbf{H}} A \to C.$$

One direction of this theorem is very easy. If $\Gamma \vdash_{\mathbf{H}} A \to C$, then $\Gamma \cup \{A\} \vdash_{\mathbf{H}} A \to C$. We also have that $\Gamma \cup \{A\} \vdash_{\mathbf{H}} A$. Now, by application of the rule Modus ponens, we obtain that $\Gamma \cup \{A\} \vdash_{\mathbf{H}} C$.

The proof of the other direction (from right to left) requires a more involved argument by *induction on derivations*, associated with the inductive definition of derivations in **H** as in Section 1.4. It goes as follows.

Lemma 45 (Principle of induction on derivations in H) *Let* $\mathcal{P}$ *be a property of propositional formulae and* Γ *be a set of formulae, such that:*

1. *Every axiom of* **H** *has the property* $\mathcal{P}$.

2. *Every formula in* Γ *has the property* $\mathcal{P}$.

3. *Whenever the formulae* Q, R *are such that both* Q *and* $Q \to R$ *have the property* $\mathcal{P}$, *then* R *also has the property* $\mathcal{P}$.
   Then every formula C *such that* $\Gamma \vdash_{\mathbf{H}} C$ *has the property* $\mathcal{P}$.

The proof is left as an exercise. It is an application of the general observation about inductive definitions and proofs by induction, made in Section 1.4.

The direction from right to left of the Deduction Theorem can now be proved as follows. Let the property $\mathcal{P}$ state for a formula C that "For every set of formulae Γ and a formula A, if $\Gamma \cup \{A\} \vdash_{\mathbf{H}} C$ then $\Gamma \vdash_{\mathbf{H}} A \to C$."

We leave the proof as an exercise of application of the principle of induction on derivations in **H**. Note that you will need to use the axioms for the implication.

Using the Deduction Theorem, derivations such as $\vdash_{\mathbf{H}} p \to p$ become straightforward.

Example 46 *The derivation in Example 46 can also be substantially simplified.*

1. $p \wedge (p \to q) \vdash_{\mathbf{H}} p$,   *by Axiom* $(\wedge 1)$ *and the Deduction Theorem;*
2. $p \wedge (p \to q) \vdash_{\mathbf{H}} p \to q$,   *by Axiom* $(\wedge 2)$ *and the Deduction Theorem;*
3. $p \wedge (p \to q) \vdash_{\mathbf{H}} q$,   *by 1,2, and Modus ponens;*
4. $\vdash_{\mathbf{H}} (p \wedge (p \to q)) \to q$,   *by 3 and the Deduction Theorem.*

Here is another example of a derivation in **H** using the Deduction Theorem.

Example 47 $p, \neg p \vdash_{\mathbf{H}} q$:

1. $\neg p \vdash_{\mathbf{H}} \neg q \to \neg p$,   *by Axiom* $(\to 1)$ *and the Deduction Theorem;*
2. $\neg p \vdash_{\mathbf{H}} (\neg q \to p) \to q$,   *by 1, Axiom* $(\to 3)$, *and Modus ponens;*
3. $p \vdash_{\mathbf{H}} \neg q \to p$,   *by Axiom* $(\to 1)$ *and the Deduction Theorem;*
4. $p, \neg p \vdash_{\mathbf{H}} q$,   *by 2,3, and Modus ponens.*

Note that adding more premises does not affect the validity of derivations, but can possibly add more derivable formulae. This justifies step 4 above.

References for further reading

For further discussion and examples on derivations in axiomatic systems for propositional logic, see Tarski (1965), Shoenfield (1967), Hamilton (1988), Fitting (1996), and Mendelson (1997).

Exercises

2.2.1 Show that the result of uniform substitution of a propositional formula for a propositional variable in a propositional formula is again a propositional formula. (Hint: use induction on the formula in which the substitution is performed.)

2.2.2 Derive the following in **H** *without* using the Deduction Theorem. (NB: some of these are simple, others are quite tricky. The purpose of this exercise is to make you appreciate the Deduction Theorem.)

(a) $p \vdash_{\mathbf{H}} p \vee q; q \vdash_{\mathbf{H}} p \vee q$
(b) $p, q \vdash_{\mathbf{H}} p \wedge q$
(c) $\vdash_{\mathbf{H}} (p \wedge q) \to (q \wedge p)$
(d) $\vdash_{\mathbf{H}} (p \vee q) \to (q \vee p)$
(e) $\vdash_{\mathbf{H}} p \to p$

(f) $\vdash_{\mathbf{H}} (p \to q) \to ((q \to r) \to (p \to r))$
(g) $\vdash_{\mathbf{H}} (\neg p \to \neg q) \to (q \to p)$
(h) $\vdash_{\mathbf{H}} (q \to p) \to (\neg p \to \neg q)$

2.2.3 Prove the principle of induction on derivations in **H** by using induction on lengths of derivations.

2.2.4 Complete the proof of the Deduction Theorem by using the principle of induction on derivations in **H**.

2.2.5 Prove that if $\Gamma, A \vdash_{\mathbf{H}} B$ and $\Gamma, B \vdash_{\mathbf{H}} C$ then $\Gamma, A \vdash_{\mathbf{H}} C$. (Hint: use the Deduction Theorem.)

2.2.6 If $A \vdash_{\mathbf{H}} B$ then $B \to C \vdash_{\mathbf{H}} A \to C$.

2.2.7 Prove, by induction on derivations in **H**, that every theorem of **H** is a tautology.

2.2.8 More generally, using induction on derivations in **H**, prove that **H** is *sound*, that is, for every set of formulae Γ and a formula A, if $\Gamma \models A$ then $\Gamma \vdash_{\mathbf{H}} A$.

2.2.9 Derive the following in the axiomatic system **H** using the Deduction Theorem, where P, Q, R are any formulae. (Hint: for some of these exercises you may use the previous exercises.)

(a) $P \vdash_{\mathbf{H}} P \vee Q; Q \vdash_{\mathbf{H}} P \vee Q$
(b) $P, Q \vdash_{\mathbf{H}} P \wedge Q$
(c) $\vdash_{\mathbf{H}} (P \wedge Q) \to (Q \wedge P)$
(d) $\vdash_{\mathbf{H}} (P \vee Q) \to (Q \vee P)$
(e) $P \to (Q \to R) \vdash_{\mathbf{H}} Q \to (P \to R)$
(f) $(P \wedge Q) \to R \vdash_{\mathbf{H}} P \to (Q \to R)$
(g) $P \to (Q \to R) \vdash_{\mathbf{H}} (P \wedge Q) \to R$
(h) If $\neg P \vdash_{\mathbf{H}} \neg Q$ then $Q \vdash_{\mathbf{H}} P$

(i) $(\neg P \to \neg Q) \vdash_{\mathbf{H}} (Q \to P)$
(j) If $\neg P, Q \vdash_{\mathbf{H}} \neg Q$ then $Q \vdash_{\mathbf{H}} P$
(k) If $Q \vdash_{\mathbf{H}} P$ then $\neg P \vdash_{\mathbf{H}} \neg Q$
(l) $(Q \to P) \vdash_{\mathbf{H}} (\neg P \to \neg Q)$
(m) $\neg\neg P \vdash_{\mathbf{H}} P$
(n) $P \vdash_{\mathbf{H}} \neg\neg P$
(o) $\vdash_{\mathbf{H}} P \vee \neg P$
(p) $\vdash_{\mathbf{H}} \neg(P \wedge \neg P)$

(q) $\neg P \vee Q \vdash_{\mathbf{H}} P \to Q$

(r) $P \to Q \vdash_{\mathbf{H}} \neg P \vee Q$

(s) $\neg(P \vee Q) \vdash_{\mathbf{H}} \neg P \wedge \neg Q$

(t) $\neg P \wedge \neg Q \vdash_{\mathbf{H}} \neg(P \vee Q)$

(u) $\neg(P \wedge Q) \vdash_{\mathbf{H}} \neg P \vee \neg Q$

(v) $\neg P \vee \neg Q \vdash_{\mathbf{H}} \neg(P \wedge Q)$

(w) $\vdash_{\mathbf{H}} (Q \to P) \to ((Q \to \neg P) \to \neg Q)$

(x) $\vdash_{\mathbf{H}} (\neg Q \to P) \to ((\neg Q \to \neg P) \to Q)$

(y) If $P \vdash_{\mathbf{H}} Q$ and $P \vdash_{\mathbf{H}} \neg Q$ then $\vdash_{\mathbf{H}} \neg P$.

(z) If $P, Q \vdash_{\mathbf{H}} R$ and $P, \neg Q \vdash_{\mathbf{H}} R$ then $P \vdash_{\mathbf{H}} R$.

2.2.10 Using the Deduction Theorem, show that if any of $\wedge$, $\vee$, or $\to$ is considered definable in terms of the others, the corresponding axioms in **H** can be derived from the others.

Euclid of Alexandria (*c.* 325–265 BC) was the most prominent and influential mathematician of ancient Greece. He wrote a monumental 13-book work, known as the *Elements*, in which he laid the systematic foundations of both geometry and arithmetic. *Elements* encompassed the system of postulates and theorems for what is now called Euclidean geometry and is probably the most influential book written in the history of mathematics. It was still used as *the* classic textbook for the study of geometry until the end of the 19th century, when David Hilbert revised and modernized it.

Furthermore, the book is written in a distinctly axiomatic style, so Euclid's system of geometry can be regarded as the first axiomatic system of a mathematical theory. The first book of *Elements* contains five postulates for points and lines in the plane. The fifth postulate is the famous **Parallel Postulate**, stating that for every line and for every point that does not lie on that line there exists a unique line through the point that is parallel to the given line. Since it was not as simple to state or as obvious as the other postulates, Euclid himself and many mathematicians after him tried for two millennia to derive it from the other four postulates. However, all these attempts were futile. It was only in the early 19th century that the mathematicians Gauss, Lobachevski, and Bolyai showed independently that negations of this postulate lead to the development of consistent, alternative **Non-Euclidean geometries**, such as elliptic and hyperbolic geometries. The axiomatic approach of Euclid therefore turned out to be much more than just a matter of style and illustrated the great importance of the axiomatic method in mathematics, taken up later by Dedekind, Peano, Frege, Russell, Whitehead, Hilbert, Bernays, and their followers.

Besides geometry, Euclid also made fundamental contributions to number theory in *Elements*, including the study of prime numbers, the celebrated proof of existence of infinitely many prime numbers, Euclid's lemma on factorization, and, of course, the Euclidean procedure for computing the greatest common divisor of two integers. That procedure was one of the earliest and most famous instances of what is today called an **algorithm**.

See on the following page a copy of the first few definitions and postulates from an 1838 edition of Euclid's *Elements*, published by Robert Simson in Philadelphia, USA.

THE

ELEMENTS OF EUCLID.

BOOK I.

DEFINITIONS.

I.
A POINT is that which hath no parts, or which hath no magnitude.
II.
A line is length without breadth.
III.
The extremities of a line are points.
IV.
A straight line is that which lies evenly between its extreme points.
V.
A superficies is that which hath only length and breadth.
VI.
The extremities of a superficies are lines.
VII.
A plane superficies is that in which any two points being taken,* the straight line between them lies wholly in that superficies.
VIII.
" A plane angle is the inclination of two lines to one another* in a plane, which meet together, but are not in the same direction."
IX.
A plane rectilineal angle is the inclination of two straight lines to one another, which meet together, but are not in the same straight line.

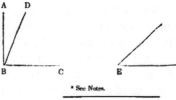

* See Notes.

POSTULATES.

I.
LET it be granted that a straight line may be drawn from any one point to any other point.
II.
That a terminated straight line may be produced to any length in a straight line.
III.
And that a circle may be described from any centre, at any distance from that centre.

AXIOMS.

I.
THINGS which are equal to the same are equal to one another.
II.
If equals be added to equals, the wholes are equals.

2.3   Semantic Tableaux

Here I develop the idea of testing the validity of propositional formulae and inferences by systematically searching for a falsifying truth assignment into a formal deductive system called **Semantic Tableaux** (hereafter denoted **ST**). The first versions of Semantic Tableaux were designed in the 1950s independently by Evert Beth and Jaakko Hintikka, and further developed in the 1960s and later by Raymond Smullyan, Melvin Fitting, and others. Since then, systems of Semantic Tableaux have also been developed for a great variety of *non-classical* logical systems. Read more on Beth and Hintikka in the biographic boxes at the end of this section, and on Smullyan and Fitting at the end of Section 4.2.

In order to prove the validity of the consequence $A_1, \ldots, A_n \models B$, the method of Semantic Tableaux involves demonstrating that there is no truth assignment which falsifies that consequence, which is equivalent to showing that the formulae $A_1, \ldots, A_n$ and $\neg B$ *cannot be satisfied simultaneously.*

2.3.1   Description of the deductive system **ST** of Semantic Tableaux

The deductive system of **ST** is based on **formula decomposition rules** which reduce the truth or falsity of a formula to the truth or falsity of its main subformulae. These rules can be extracted from the truth tables of the propositional connectives as follows.

1. $\neg$:
   (a) For $\neg A$ to be true, A must be false.
   (b) For $\neg A$ to be false, A must be true.

2. $\wedge$:
   (a) For $A \wedge B$ to be true, A must be true *and* B must be true.
   (b) For $A \wedge B$ to be false, A must be false *or* B must be false.

3. $\vee$:
   (a) For $A \vee B$ to be true, A must be true *or* B must be true.
   (b) For $A \vee B$ to be false, A must be false *and* B must be false.

4. $\rightarrow$:
   (a) For $A \rightarrow B$ to be true, A must be false *or* B must be true.
   (b) For $A \rightarrow B$ to be false, A must be true *and* B must be false.

5. $\leftrightarrow$:
   ($\leftrightarrow$ T) For $A \leftrightarrow B$ to be true, A and B must be true *or* A and B must be false.
   ($\leftrightarrow$ F) For $A \leftrightarrow B$ to be false, A must be true and B must be false *or* A must be false and B must be true.

These rules are formalized in Table 2.1, presented as rules for the **signed version** of the system of Semantic Tableaux. The signed version works with formulae that are labeled with the truth values they are required to obtain.

Using these rules, one can search *systematically* for an assignment falsifying a given formula or a set of formulae. In the process of this systematic application of the rules we build a "search tree" called **tableau**. If we are to derive the consequence $A_1, \ldots, A_n \models B$, that is, to prove its validity, we start by placing the signed formulae

$$A_1 : \mathrm{T}, \ldots, A_n : \mathrm{T}, B : \mathrm{F}$$

at the **root** of the tableau and then extend it downward by repeatedly applying the decomposition rules to signed formulae appearing on the tree as follows.

1. A branch on the tree is selected.

2. A signed formula occurring on that branch, to which a decomposition rule has not yet been applied, is selected.

Table 2.1   Rules for Semantic Tableaux in propositional logic: signed version

Non-branching rules (α-rules)	Branching rules (β-rules)
($\wedge$T) $\begin{array}{c} A \wedge B : \text{T} \\ \downarrow \\ A : \text{T}, B : \text{T} \end{array}$	($\wedge$F) $\begin{array}{c} A \wedge B : \text{F} \\ \swarrow \qquad \searrow \\ A : \text{F} \qquad B : \text{F} \end{array}$
($\vee$F) $\begin{array}{c} A \vee B : \text{F} \\ \downarrow \\ A : \text{F}, B : \text{F} \end{array}$	($\vee$T) $\begin{array}{c} A \vee B : \text{T} \\ \swarrow \qquad \searrow \\ A : \text{T} \qquad B : \text{T} \end{array}$
($\rightarrow$F) $\begin{array}{c} A \rightarrow B : \text{F} \\ \downarrow \\ A : \text{T}, B : \text{F} \end{array}$	($\rightarrow$T) $\begin{array}{c} A \rightarrow B : \text{T} \\ \swarrow \qquad \searrow \\ A : \text{F} \qquad B : \text{T} \end{array}$
($\neg$T) $\begin{array}{c} \neg A : \text{T} \\ \downarrow \\ A : \text{F} \end{array}$	($\leftrightarrow$T) $\begin{array}{c} A \leftrightarrow B : \text{T} \\ \swarrow \qquad \searrow \\ A : \text{T}, B : \text{T} \qquad A : \text{F}, B : \text{F} \end{array}$
($\neg$F) $\begin{array}{c} \neg A : \text{F} \\ \downarrow \\ A : \text{T} \end{array}$	($\leftrightarrow$F) $\begin{array}{c} A \leftrightarrow B : \text{F} \\ \swarrow \qquad \searrow \\ A : \text{T}, B : \text{F} \qquad A : \text{F}, B : \text{T} \end{array}$

3. The decomposition rule that corresponds to that signed formula is applied to it. As a result, one (for the α-rules) or two (for the β-rules) successor nodes are added, labeled with the signed formulae which that rule introduces.

We repeat this procedure along every branch of the tableau until:

- either a **contradictory pair of signed formulae** of the type $A : \text{T}, A : \text{F}$ appears on the branch, in which case we declare that branch **closed** (meaning that no assignment can ever be found which satisfies all signed formulae on that branch), mark it $\times$ and do not extend it any further;

- or no *new* applications of decomposition rules are possible on that branch (i.e., every signed formula appearing on the branch has already been decomposed further on that branch); we then say that the branch is **saturated** and, if a saturated branch is not yet closed, we declare it **open** and mark it $\bigcirc$.

Note that an open branch defines a truth assignment for the variables occurring on that branch as follows: a variable p is true under that assignment if $p : \text{T}$ appears on the branch, and is false otherwise (in which case, $p : \text{F}$ may or may not appear on the branch). It is easy to check that such an assignment will satisfy *all* signed formulae on the branch. That can be done step-by-step by going up the branch. All signed formulae at the root will eventually be satisfied, which will falsify the logical consequence $A_1, \ldots, A_n \models B$ and therefore prove it invalid.

On the other hand, if all branches of the tableau close, then the entire tableau is declared **closed**. That means that the systematic search for an assignment satisfying the formulae at the root has failed. We then say that B **is derived from** $A_1, \ldots, A_n$ **in ST**, which we denote by $A_1, \ldots, A_n \vdash_{\mathbf{ST}} B$. In particular, if $\emptyset \vdash_{\mathbf{ST}} B$ then we say that B is a **theorem of ST**.

Theorem 48 (Adequacy Theorem for ST) *The system of Semantic Tableaux is sound and complete, that is:*

$$A_1, \ldots, A_n \vdash_{\mathbf{ST}} B \quad \text{if and only if} \quad A_1, \ldots, A_n \models B.$$

In particular, $\vdash_{\mathbf{ST}} B$ *if and only if* B *is a tautology.*

I only sketch the proof idea here. The soundness of **ST** means that if the input formula is satisfiable, then the tableau remains open. This can be proved by fixing a satisfying truth assignment at the beginning and tracing the branch that corresponds to it. For the completeness, it is sufficient to note that if the tableau remains open then a satisfying truth assignment can be extracted from any saturated open branch. For a more detailed, generic proof of soundness and completeness see Section 2.7.

2.3.2 Some derivations in ST

Let us demonstrate the method of Semantic Tableaux on some of the examples included informally in Section 1.1.

1. Show that $\neg(p \to \neg q) \to (p \vee \neg r)$ is a tautology:

$$\neg(p \to \neg q) \to (p \vee \neg r) : \mathbf{F}$$
$$\downarrow$$
$$\neg(p \to \neg q) : \mathbf{T}, (p \vee \neg r) : \mathbf{F}$$
$$\downarrow$$
$$p \to \neg q : \mathbf{F}$$
$$\downarrow$$
$$p : \mathbf{T}, \neg q : \mathbf{F}$$
$$\downarrow$$
$$q : \mathbf{T}$$
$$\downarrow$$
$$p : \mathbf{F}, \neg r : \mathbf{F}$$
$$\times$$

Thus, $\vdash_{\mathbf{ST}} \neg(p \to \neg q) \to (p \vee \neg r)$ and hence $\models \neg(p \to \neg q) \to (p \vee \neg r)$.

Note that no branching rules were applicable in this example; the tableau consists of only one branch which closes because the complementary pair $p : \mathbf{T}, p : \mathbf{F}$ appears on it. The superscript [1] indicates where the decomposition rule is applied to the formula $(p \lor \neg r) : \mathrm{F}$. This is not needed for the derivation, but has only been included for completeness and to make the tableau easier to read, and will not be used further.

2. Check if $\neg p \models \neg (p \lor \neg q)$.

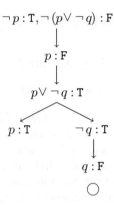

The left-hand branch closes but the right-hand branch does not, and no more decomposition rules are applicable on it which means it is open. The whole tableau is therefore open, which means that $\neg p \not\models \neg (p \lor \neg q)$. Indeed, the right-hand branch provides a falsifying valuation for that consequence: $p : \mathrm{F}, q : \mathrm{F}$. (Check it!)

3. Check whether $(\neg p \rightarrow q), \neg r \models p \lor \neg q$.

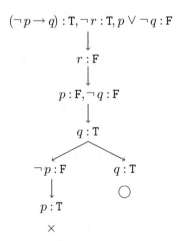

The tableau is again open, and a falsifying valuation is provided by the open branch $p : \mathrm{F}, q : \mathrm{T}, r : \mathrm{F}$. Note that we choose to first decompose the subformulae $\neg r : \mathrm{T}$ and $(p \lor \neg q) : \mathrm{F}$ because they are non-branching (α-formulae) and then $\neg p \rightarrow q : \mathrm{T}$ which is branching (β-formula). This does not make a difference to the outcome of the tableau but saves some work, so it is generally the preferable strategy.

4. Check whether $((p \wedge \neg q) \rightarrow \neg r) \leftrightarrow ((p \wedge r) \rightarrow q)$ is a tautology.

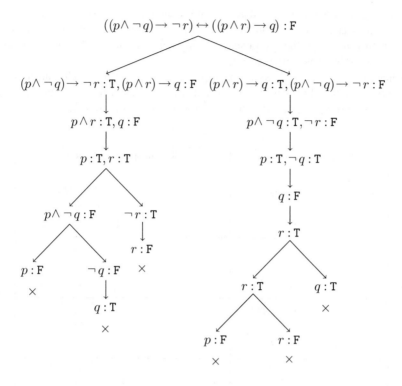

The tableau closes, and so the formula is a tautology.

I end this section with a few remarks.

1. At any stage during the construction of a Semantic Tableau, more than one decomposition rule may be applicable. Depending on the order of their application the tableau tree may develop differently, but the final result – whether the tableau closes or not – will always be the same. The order of rule application is therefore unimportant. However, it may affect the size of the tableau tree. As a general guideline, it is better to first apply the non-branching rules before applying the branching rules (if there is a choice), as this would defer the branching and reduce the size of the tableau.

2. Note that no node in a tableau can have more than two children nodes. If a node in your tableau has three or more children, then there is something wrong!

3. Although we can use equivalences such as $p \wedge (q \vee p) \equiv p$ and $\neg(p \wedge q) \equiv \neg p \wedge \neg q$ when reasoning informally, these equivalences do not form part of the rules of the deductive system **ST** and should *not* be used in the construction of the tableaux graphs. As emphasized in Section 2.1, derivations within a deductive system may only use the rules of that system in order to be completely mechanizable.

2.3.3   Unsigned version of the system of Semantic Tableaux

Unlike the signed version presented above, in the unsigned version of Semantic Tableaux all formulae appearing on the branches are assumed to be true; there is therefore no need to assign truth values. For that purpose, every signed formula $A : \text{T}$ is replaced simply by A, whereas $A : \text{F}$ is transformed to $\neg A$.

The rules for the unsigned version can be produced by a simple modification of the rules for the signed version, by using the basic equivalences used for importing the negation inside the other propositional connectives. They are listed in Table 2.1.

Non-branching rules (α-rules)		Branching rules (β-rules)	
($\wedge$)	$A \wedge B$ $\downarrow$ A, B	($\neg\wedge$)	$\neg(A \wedge B)$ $\neg A \qquad \neg B$
($\neg\vee$)	$\neg(A \vee B)$ $\downarrow$ $\neg A, \neg B$	($\vee$)	$A \vee B$ $A \qquad B$
($\neg\rightarrow$)	$\neg(A \rightarrow B)$ $\downarrow$ $A, \neg B$	($\rightarrow$)	$A \rightarrow B$ $\neg A \qquad B$
($\neg\neg$)	$\neg\neg A$ $\downarrow$ A	($\leftrightarrow$)	$A \leftrightarrow B$ $A, B \qquad \neg A, \neg B$
		($\neg\leftrightarrow$)	$\neg(A \leftrightarrow B)$ $A, \neg B \qquad \neg A, B$

Figure 2.1   Rules for Semantic Tableaux in propositional logic: unsigned version

A branch of the unsigned tableau is **closed** if a "complementary pair" of formulae $A, \neg A$ appears on it; otherwise, it remains **open**. The rest is the same as for the signed version. The two versions are essentially equivalent and it is only a matter of preference or convenience when deciding which to use.

References for further reading

D'Agostino *et al.* (1999) is a comprehensive handbook on tableaux methods, not only in classical logic. Fore more details on theory and examples of derivations in Semantic Tableaux see Nerode and Shore (1993), Jeffrey (1994, who used the expression "analytic trees"), Smullyan (1995, one of the first versions of Semantic Tableaux), Fitting (1996), Smith (2003, who called them "trees"), Ben-Ari (2012), and van Benthem *et al.* (2014).

Exercises

For each of the following exercises use Semantic Tableaux (the version you prefer).

2.3.1 Check which of the following formulae are tautologies.
  (a) $((p \to q) \to q) \to q$
  (b) $((q \to p) \to q) \to q$
  (c) $((p \to q) \land (p \to \neg q)) \to \neg p$
  (d) $((p \lor q) \to \neg r) \to \neg(\neg q \land r)$
  (e) $((p \to q) \land (p \to r)) \to (p \to (q \land r))$
  (f) $((p \to r) \land (q \to r)) \to ((p \lor q) \to r)$
  (g) $((p \to r) \lor (q \to r)) \to ((p \lor q) \to r)$
  (h) $((p \to r) \land (q \to r)) \to ((p \land q) \to r)$
  (i) $p \to ((q \to r) \to ((p \to q) \to r))$
  (j) $(p \to (q \to r)) \to ((p \to q) \to (p \to r))$
  (k) $((\neg p \land q) \to \neg r) \to (\neg q \to \neg(p \land \neg r))$
  (l) $(p \to (q \lor \neg r)) \to (((q \to \neg p) \land r) \to \neg p)$

2.3.2 Check which of the following logical consequences are valid.
  (a) $\neg p \to q, \neg p \to \neg q \models p$
  (b) $(\neg p \land q) \to \neg r, r \models p \lor \neg q$
  (c) $(p \land q) \to r \models (p \to r) \lor (q \to r)$
  (d) $(p \land q) \to r \models (p \to r) \land (q \to r)$
  (e) $(p \lor q) \to r \models (p \to r) \land (q \to r)$
  (f) $p \to (q \lor r) \models (p \to q) \lor (p \to r)$
  (g) $p \to (q \land r) \models (p \to q) \land (p \to r)$
  (h) $p \to q, r \to s \models (p \lor r) \to (q \lor s)$
  (i) $(p \to q) \land (r \to s) \models (p \land r) \to (q \land s)$
  (j) $(p \to q) \lor (r \to s) \models (p \lor r) \to (q \lor s)$
  (k) $(p \to q) \lor (r \to s) \models (p \to s) \lor (r \to q)$
  (l) $(p \to q) \land (r \to s) \models (p \to s) \land (r \to q)$
  (m) $\neg((p \to r) \lor (q \to r)) \models \neg((p \lor q) \to r)$
  (n) $\neg((p \lor q) \to r) \models \neg((p \to r) \lor (q \to r))$

2.3.3 Let P, Q, R, and S be propositions. Check whether:
  (a) if $P \to Q, Q \to R$, and $P \land S$ are true, then $R \land S$ is true;
  (b) if R and $(P \land \neg R)$ are false and $\neg P \to \neg Q$ is true, then Q is false;
  (c) if $Q \to (R \land S)$ is true and $Q \land S$ is false, then $R \land Q$ is false.
  (d) if $P \to Q$ and $Q \to (R \lor S)$ are true and $P \to R$ is false, then S is true;
  (e) if $Q \to (R \land S)$ is true and $Q \land S$ is false, then Q is false.

2.3.4 Check the soundness of each of following rules of propositional inference.

  (a)
  $$\frac{(p \lor q) \to r, p \lor r}{r}$$

  (c)
  $$\frac{p \to q, p \lor \neg r}{q \to r}$$

  (b)
  $$\frac{p \to q, \neg q \lor r, \neg r}{\neg p}$$

  (d)
  $$\frac{\neg p \to \neg q, \neg(p \land \neg r), \neg r}{\neg q}$$

(e)   (f)

$$\frac{p \to q, r \vee (s \wedge \neg q), q}{\neg p} \qquad\qquad \frac{p \to q, q \to (r \vee s), \neg.(p \to r)}{s}$$

2.3.5 Check the logical correctness of each of the following propositional arguments.

(a)

If the triangle ABC has a right angle, then it is not equilateral.
The triangle ABC does not have a right angle.

Therefore, the triangle ABC is equilateral.

(b) In the following argument n is a certain integer.

If n is divisible by 2 and n is divisible by 3, then n is divisible by 6.
If n is divisible by 6, then n is divisible by 2.
n is not divisible by 3.

Therefore n is not divisible by 6.

(c)

If Thomas likes logic, then he is clever.
If Thomas is clever, then he is rich.

Therefore, if Thomas likes logic, then he is rich.

(d)

Socrates is wise or not happy.
If Socrates is not wise then Socrates is not a philosopher.

Therefore, if Socrates is a philosopher then Socrates is not happy.

(e)

Alexis smiles if she sings and is happy.
Alexis cries if she is not happy.

Therefore, if Alexis sings, then she smiles or cries.

(f)

Victor will go to a university if he studies hard and he is clever.
Victor will not get a good job if he does not study hard.

Therefore, if Victor is clever, then he will go to a
university or will not get a good job.

(g) Replace the second premise in the previous argument with

"Victor will not get a good job only if he does not study hard"

and determine if the resulting argument is logically correct.

(h)

Johnnie will not learn logic if he skips lectures.
Johnnie will pass the exam if he learns logic and does not skip lectures.

Therefore, if Johnnie skips lectures or does not pass the exam,
then he will not learn logic.

(i) Replace the second premise in the previous argument with

"Johnnie will pass the exam only if he learns logic and does not skip lectures"

and determine if the resulting argument is logically correct.

(j)

If Bonnie does not come to the party, then Clyde will not come, either, or Alice will come.
If Clyde comes to the party, then Alice will not come.

Therefore, if Clyde comes to the party, then Bonnie will come, too.

(k) Replace the second premise in the previous argument with

"Alice will come to the party if Clyde does not come."

and determine if the resulting argument is logically correct.

Evert Willem Beth (7.7.1908–12.4.1964) was a Dutch philosopher and logician who made some seminal contributions to mathematical logic.

Beth studied mathematics and physics at Utrecht University, as well as philosophy and psychology, and obtained a PhD in philosophy in 1935. He became Professor of Logic and the Foundations of Mathematics at the University of Amsterdam in 1946 – the first such professor in the Netherlands – and worked there until his death with two brief interruptions (the first was in 1952, when he was a Research Associate at the University of California in Berkeley with Alfred Tarski).

Beth is best known for the so-called **Beth definability theorem** (stating that implicit and explicit definability of functions and predicates in first-order logic are equivalent), for his construction of what is now called **Beth models** for the intuitionistic logic, and for the development of the method of **semantic tableaux**, originally devised independently by him (1955), Hintikka (1955), and Schütte (1956). The tableau method is a systematic search for refutations, usually presented in an upside-down tree form, similar to Gentzen's method for natural deduction (a systematic search for proofs presented in a tree-like form).

During his last years, Beth tried to relate his research in logic to various applications including automated theorem proving, mathematical education, and heuristics, as well as translation methods in natural languages.

An E. W. Beth Dissertation Prize named in his honour is awarded annually by the Association for Logic, Language and Information (FoLLI) to outstanding interdisciplinary PhD theses in these fields.

 **Jaakko Hintikka** (12.01.1929–12.08.2015) was a Finnish philosopher and logician, well known for several important contributions to a very wide range of topics including mathematical and philosophical logic, philosophy of logic and mathematics, philosophy of language philosophy of science, epistemology, and history of philosophy.

Hintikka was a philosophy student of G. H. von Wright at the University of Helsinki and obtained his PhD in 1953 on distributive normal forms. Following three years as a post-doctoral researcher at Harvard, his 60-year-long academic career included professorial positions at the University of Helsinki, the Academy of Finland, Florida State University, and finally Boston University from 1990.

Hintikka introduced or discovered independently and developed several now fundamental logical concepts and methods, including: the **method of semantic tableaux** (which he referred to as "tree-method") independently of Beth in 1955; **possible-worlds semantics** for modal logic in 1957; **epistemic logic** (logic of knowledge and belief) in his ground-breaking 1962 book *Knowledge and Belief*; **game-theoretical semantics**; **infinitary logics**; and **independence-friendly logic**. He was active in research until his final days, and participated in the Congress of Logic, Methodology and Philosophy of Science in Helsinki just a few days before his death at the age of 86.

2.4 Natural Deduction

The system of Natural Deduction aims to provide a well-structured way of formal reasoning that represents a systematic, intuitively natural, and, of course, logically correct argumentation. It was originally developed in the late 1920s – early 1930s by Gerhard Gentzen and a version of it independently by Stanisław Jaśkowski[3].

Building on Hilbert's idea of axiomatic system as a paradigm of a deductive system, Natural Deduction was the first *rule-based* deductive system, adding the possibility for *making and withdrawing additional assumptions* in derivations, that is, formulae that are neither axioms nor have been derived. It involves two main kinds of inference rules – *introduction rules* and *elimination rules* – for each logical connective, an idea that turned out to be very important in proof theory. Gentzen's motivation for developing Natural Deduction was to provide a purely syntactic, combinatorial proof of the consistency of Peano's axiom system of arithmetic, and he was partly successful in that pursuit. In order to analyze derivations better and to prove a special property of derivations that could be used for the proof of consistency, Gentzen also introduced *Sequent Calculus*, a deduction system based not just on formulae but on *sequents* – pairs of lists of formulae, being assumptions on the left and conclusions on the right – with a set of rules for manipulating and deriving such sequents.

---

[3] Read more on Gentzen's and Jaśkowski's work in the biographic boxes at the end of this section.

2.4.1   Description

The deductive system of **Natural Deduction**, hereafter denoted **ND**, is based on a set of inference rules which naturally reflect the logical meaning of the propositional connectives.

ND has no axioms, but several inference rules (listed further) including a pair of rules for each logical connective: an **introduction rule** which produces a conclusion containing that connective as the main one; and an **elimination rule** in which the connective occurs as the main connective of a premise. To reduce the number of rules, we will assume $\leftrightarrow$ to be defined in terms of $\rightarrow$ and $\wedge$. Note that, since $\neg A \equiv A \rightarrow \bot$, the rules for $\neg$ can be regarded as particular cases of the corresponding rules for $\rightarrow$. There are also two additional rules, $(\bot)$ and (RA), which will be discussed later in this section.

The derivation in **ND** consists of successive application of the inference rules, using as premises the initial assumptions or already derived formulae as well as *additional assumptions* which can be added at any step of the derivation. Some of the rules allow for **cancellation** (or **discharge**) **of assumptions**, which is indicated by putting them in square brackets. The idea of the additional assumptions is that they only play an auxiliary role in the derivation; when no longer needed they are canceled, but *only* with an application of an appropriate rule which allows such a cancellation. Note that the cancellation of an assumption, when the rule allows it, is an option, not an obligation, so an assumption can be re-used several times before being canceled. However, all assumptions which have not been canceled during the derivation *must be declared* in the list of assumptions from which the conclusion is proved to be a logical consequence. If we want to prove that a formula C is a logical consequence from a set of assumptions Γ, then any assumption which is not in Γ must therefore be canceled during the derivation.

Here is a brief explanation of the rules of **ND** presented in Table 2.2.

$(\wedge I)$ To prove the truth of a conjunction $A \wedge B$, we have to prove the truth of each of A and B.

$(\wedge E)$ The truth of a conjunction $A \wedge B$ implies the truth of each of the conjuncts.

$(\vee I)$ The truth of a disjunction $A \vee B$ follows from the truth of either disjunct.

$(\vee E)$ If the premise is a disjunction $A \vee B$ we reason *per cases*, that is, we consider separately each of the two possible cases for that disjunction to be true: *Case 1: A* is true; and *Case 2: B* is true. If we succeed in proving that in each of these cases the conclusion C follows, then we have a proof that C follows from $A \vee B$.

$(\rightarrow I)$ To prove the truth of an implication $A \rightarrow B$, we assume that (besides all premises) the antecedent A is true and try to prove that the consequent B is true.

$(\rightarrow E)$ This is the detachment rule (Modus ponens) which we have already discussed in Section 2.2.

$(\neg I)$ To prove a negation $\neg A$ we can apply *proof by contradiction*, that is, assume A (which is equivalent to the negation of $\neg A$) and try to reach a contradiction.

$(\neg E)$ The falsum follows from any contradiction.

$(\bot)$ "**Ex falso sequitur quodlibet**": from a false assumption anything can be derived.

(RA) This rule is called '**Reductio ad absurdum**'. It formalizes the method of proof by contradiction or *proof from the contrary*: if the assumption that A is false (i.e., $\neg A$ is true) leads to a contradiction, then A must be true, formalized as $\neg \neg A \rightarrow A$.

Table 2.2   Rules for Natural Deduction in propositional logic.
The vertical dots indicate derivations

Introduction rules	Elimination rules

$(\wedge I) \dfrac{A, B}{A \wedge B}$

$(\wedge E) \dfrac{A \wedge B}{A}, \dfrac{A \wedge B}{B}$

$(\vee I) \dfrac{A}{A \vee B}, \dfrac{B}{A \vee B}$

$$[A][B]$$
$$\vdots \ \ \vdots$$
$(\vee E) \dfrac{A \vee B \quad C \quad C}{C}$

$$[A]$$
$$\vdots$$
$(\to I) \dfrac{B}{A \to B}$

$(\to E) \dfrac{A, A \to B}{B}$

$$[A]$$
$$\vdots$$
$(\neg I) \dfrac{\bot}{\neg A}$

$(\neg E) \dfrac{A, \neg A}{\bot}$

$$[\neg A]$$
$$\vdots$$
$(\bot) \dfrac{\bot}{A}$

$(RA) \dfrac{\bot}{A}$

The rule RA formalizes a typical pattern of *non-constructive reasoning*. The formula $\neg\neg A \to A$ is a tautology in the classical logic that we study here, but not in an important non-classical logical system called **intuitionistic logic**, developed in the early 20th century and meant, *inter alia*, to capture and formalize constructive reasoning where RA is not an admissible inference rule. In fact, a sound and complete system of Natural Deduction for the intuitionistic logic can be obtained from **ND** by simply removing the rule RA. The intuitionistic logic also rejects the validity of other classical tautologies, expressing non-constructive claims and derived by using RA, such as the law of excluded middle $A \vee \neg A$ and the implication $\neg(A \wedge B) \to \neg A \vee \neg B$, hence also rejecting the respective Morgan's law $\neg(A \wedge B) \equiv \neg A \vee \neg B$.

Theorem 49 (Adequacy Theorem for ND) *The system of Natural Deduction is sound and complete, that is:*

$$A_1, \ldots, A_n \vdash_{\textbf{ND}} B \quad \text{if and only if} \quad A_1, \ldots, A_n \models B.$$

In particular, $\vdash_{\textbf{ND}} B$ *if and only if B is a tautology.*

The soundness of **ND** can be proved by induction on the derivations, and is a fairly routine exercise. The proof of completeness is more difficult, but it can easily be derived from the completeness of **H** by showing that every logical consequence that is derivable in **H** is also derivable in **ND**. This will be sketched in Section 2.7.

2.4.2   Examples of derivations in Natural Deduction

1. Commutativity of the conjunction $A \wedge B \vdash_{\textbf{ND}} B \wedge A$:

$$(\wedge I)\dfrac{(\wedge E)\dfrac{A \wedge B}{B} \quad (\wedge E)\dfrac{A \wedge B}{A}}{B \wedge A}$$

Note that an assumption $A \wedge B$ was used twice.

2. $\vdash_{\textbf{ND}} A \rightarrow \neg\neg A$:

$$(\rightarrow I)\dfrac{(\neg I)\dfrac{(\neg E)\dfrac{[A]^2, [\neg A]^1}{\bot}}{\neg\neg A}1}{A \rightarrow \neg\neg A}2$$

In this derivation two additional assumptions have been made: A and $\neg A$. The label 1 indicates the application of the rule which allows the cancellation of the assumption $\neg A$, while label 2 indicates the cancellation of A.

3. $\vdash_{\textbf{ND}} \neg\neg A \rightarrow A$:

$$(\rightarrow I)\dfrac{(RA)\dfrac{(\neg E)\dfrac{[\neg\neg A]^2, [\neg A]^1}{\bot}}{A}1}{\neg\neg A \rightarrow A}2$$

Note the application of (RA) in this derivation. It can be proved that the derivation cannot be made without this or an equivalent rule.

4. $\vdash_{\textbf{ND}} (A \rightarrow (B \rightarrow C)) \rightarrow ((A \wedge B) \rightarrow C)$:

$$(\rightarrow I)\dfrac{(\rightarrow I)\dfrac{(\rightarrow E)\dfrac{(\wedge E)\dfrac{[A \wedge B]^1}{B} \quad (\rightarrow E)\dfrac{(\wedge E)\dfrac{[A \wedge B]^1}{A}, [A \rightarrow (B \rightarrow C)]^2}{B \rightarrow C}}{C}}{(A \wedge B) \rightarrow C}1}{(A \rightarrow (B \rightarrow C)) \rightarrow ((A \wedge B) \rightarrow C)}2$$

From this point onward I will usually omit the rule labels of the steps in the derivations, but the reader should be able to figure out which rule is being applied at each step.

5. $A \rightarrow B \vdash_{\textbf{ND}} \neg B \rightarrow \neg A$:

$$\dfrac{\dfrac{\dfrac{[\neg B]^1 A \rightarrow B}{B}, [\neg B]^2}{\dfrac{\bot}{\neg A}1}}{\neg B \rightarrow \neg A}2$$

6. $\neg B \to \neg A \vdash_{\mathbf{ND}} A \to B$:

$$\cfrac{\cfrac{\cfrac{[\neg B]^1, \neg B \to \neg A}{[\neg A]}, \quad [A]^2}{\cfrac{\bot}{B}1}}{(\to I)\cfrac{}{A \to B}2}$$

Again, this derivation requires the use of (RA).

7. $A \vee B \vdash_{\mathbf{ND}} \neg A \to B$:

$$\cfrac{A \vee B \quad \cfrac{\cfrac{[\neg A]^1, [A]^3}{\bot}}{\cfrac{B}{\neg A \to B}1} \quad \cfrac{[\neg A]^2, [B]^3}{\neg A \to B}2}{\neg A \to B}3$$

Note the application of the rule $(\vee E)$ (reasoning per cases) in the last step of this derivation.

References for further reading
For more details on the theory and examples of derivations in Natural Deduction see van Dalen (1983), Jeffrey (1994) (who referred to "synthetic trees"), Smullyan (1995), Fitting (1996), Huth and Ryan (2004), Nederpelt and Kamareddine (2004), Prawitz (2006, the original development of the modern version of Natural Deduction), Chiswell and Hodges (2007), and van Benthem *et al.* (2014), as well as Kalish and Montague (1980) and Bornat (2005), who present Natural Deduction derivations in a boxed form rather than in tree-like shape.

Exercises

2.4.1 Prove that all inference rules of **ND** are logically sound.

2.4.2 Construct logically sound rules for the introduction and elimination of the biconditional $\leftrightarrow$.

2.4.3 Show that the following tautologies are theorems of **ND**.
(Hint: for some of these you will have to use the rule (RA).)
(a) $(p \wedge (p \to q)) \to q$
(b) $p \to \neg\neg p$
(c) $\neg\neg p \to p$
(d) $p \vee \neg p$
(e) $((\neg p \to q) \wedge (\neg p \to \neg q)) \to p$
(f) $((p \to q) \wedge (p \to \neg q)) \to \neg p$
(g) $((p \to q) \wedge (q \to r)) \to (p \to r)$
(h) $((p \to q) \wedge (p \to r)) \to (p \to (q \wedge r))$
(i) $((p \to r) \wedge (q \to r)) \to ((p \vee q) \to r)$
(j) $p \to ((q \to r) \to ((p \to q) \to (p \to r)))$
(k) $(p \to (q \vee \neg r)) \to (((q \to \neg p) \wedge r) \to \neg p)$

2.4.4 Show that for any formulae A, B:
    (a) If $A \vdash_{\text{ND}} B$ then $\neg B \vdash_{\text{ND}} \neg A$.
    (b) If $\neg B \vdash_{\text{ND}} \neg A$ then $A \vdash_{\text{ND}} B$.
    (c) If $\neg B \vdash_{\text{ND}} \neg A$ then $A \vdash_{\text{ND}} B$.

2.4.5 Derive each of the following for any formulae A, B, C.
    (Hint: for some of these exercises use derivations of previous exercises.)
    (a) $A \rightarrow C, B \rightarrow C \vdash_{\text{ND}} (A \vee B) \rightarrow C$
    (b) $(A \rightarrow B) \vee C \vdash_{\text{ND}} A \rightarrow (B \vee C)$
    (c) $(A \vee B) \rightarrow C \vdash_{\text{ND}} (A \rightarrow C) \wedge (B \rightarrow C)$
    (d) $A \rightarrow (B \vee C) \vdash_{\text{ND}} (A \rightarrow B) \vee C$
    (e) $A \rightarrow \neg A \vdash_{\text{ND}} \neg A$
    (f) $\neg A \rightarrow A \vdash_{\text{ND}} A$
    (g) $\neg A \rightarrow B, \neg A \rightarrow \neg B \vdash_{\text{ND}} A$
    (h) $\neg A \rightarrow B \vdash_{\text{ND}} A \vee B$
    (i) $(\neg A \wedge B) \rightarrow \neg C, C \vdash_{\text{ND}} A \vee \neg B$
    (j) $\neg(A \wedge B) \vdash_{\text{ND}} \neg A \vee \neg B$
    (k) $\neg(A \vee B) \vdash_{\text{ND}} \neg A \wedge \neg B$
    (l) $\neg A \vee \neg B \vdash_{\text{ND}} \neg(A \wedge B)$
    (m) $\neg A \wedge \neg B \vdash_{\text{ND}} \neg(A \vee B)$
    (n) $\neg(A \rightarrow B) \vdash_{\text{ND}} A \wedge \neg B$
    (o) $A \wedge \neg B \vdash_{\text{ND}} \neg(A \rightarrow B)$

2.4.6 Derive the following logical consequences in Natural Deduction.
    (a) $p \rightarrow \neg q, \neg r \vee p \models (\neg p \rightarrow r) \rightarrow \neg q$
    (b) $\neg p \rightarrow \neg q, \neg(p \wedge \neg r), \neg r \models \neg q$
    (c) $\neg p \vee \neg r, r \rightarrow \neg q, \neg q \rightarrow p \models \neg r$
    (d) $(\neg p \wedge q) \rightarrow \neg r, r \models p \vee \neg q$

2.4.7 Formalize the following propositional arguments and prove their logical correctness by using Natural Deduction.
    (a)

> Alice is at home or at work.
> Alice is not at home.
> ―――――――――――――――
> Therefore, Alice is at work.

    (b)

> Nina will go to a party or will not go to the office.
> Nina will not go to a party or will not go to the office.
> ―――――――――――――――――――――――――――――――
> Therefore, Nina will not go to the office.

    (c)

> Victor is lazy or Victor is clever.
> If Victor is not successful then Victor is lazy.
> Victor is not lazy.
> ―――――――――――――――――――――――
> Therefore, Victor is clever and successful.

(d)

Socrates is happy or not stupid.
If Socrates is happy then Socrates is not a philosopher.

Therefore, if Socrates is a philosopher then Socrates is not stupid.

(e)

If the court is wet then it rains or the sprinkler is on.
The court is not closed if the sprinkler is on.

Therefore, if the court is closed and wet then it rains.

(f)

Olivia is not sleeping if she is not smiling.
Olivia is not eating or she is smiling.

If Olivia is sleeping whenever she is not eating, then Olivia is smiling.

(g)

Alexis cries if she is sad.
Alexis sulks or does not cry.
Alexis does not sulk.

Therefore, Alexis is not sad.

(h)

If Bill smokes regularly, then Bill is a heavy smoker.
It is not true that if Bill smokes regularly then Bill will quit smoking.
If Bill is a heavy smoker then Bill will quit smoking or get lung cancer.

Therefore, Bill will get lung cancer.

(i)

Bonnie was an accomplice if Alice is innocent and
Alice witnessed the robbery.
Clyde was not an accomplice if Alice is not innocent.

Therefore, if Alice witnessed the robbery,
then Bonnie was an accomplice or Clyde was not an accomplice.

(j)

If Kristina is not a good student, then she is not clever or she is lazy.
Kristina is not clever or she is not lazy.

Therefore, if Kristina is clever, then she is a good student.

(k)

It will not rain if it is sunny.
It will be misty if it rains and is not sunny.

Therefore, if it is sunny or is not misty then it will not rain.

 **Gerhard Karl Erich Gentzen** (24.11.1909–4.8.1945) was a German mathematician and logician who made pioneering contributions to the foundations of mathematics, especially in **proof theory** of which he is one of the founders.

Gentzen was a student at the University of Göttingen of Paul Bernays, a Swiss logician and collaborator of David Hilbert. Gentzen completed his doctoral thesis in 1933 under Bernays' supervision, but when Bernays was fired in April 1933 for being "non-Aryan" Hermann Weyl formally acted as Gentzen's supervisor until its defence. Nevertheless, Gentzen kept in contact with Bernays until the beginning of the World War II. During 1935–1939 he was an assistant of Hilbert in Göttingen.

In the early 1930s Gentzen developed his logical deductive systems **natural deduction** and **sequent calculus** for both classical and intuitionistic logic. In his thesis *Untersuchungen über das logische Schliessen* (*Investigations into Logical Inference*), he wrote that his aim was formal analysis of mathematical proofs that occur in practice. The main result he proved there, known as *Gentzen's Hauptsatz*, was the **cut-elimination** property of his sequent calculus, that is, elimination of the rule "cut" (essentially, the sequent version of Modus Ponens), which allows for easier proofs of consistency. Gentzen's ultimate goal was to provide a purely syntactic, combinatorial proof of the consistency of Peano's axiom system of arithmetic, and he succeeded in that to some extent (by invoking transfinite induction). However, later that goal was proved unattainable by Gödel's Second Incompleteness Theorem, stating that no sufficiently expressive deductive system, such as Peano's arithmetic, can prove its own consistency. Still, proof theory flourished as one of the main branches of mathematical logic after Gentzen's ground-breaking work, and it has achieved much in analyzing the notion of formal derivation and in proving relative consistency results.

Gentzen joined the Nazi paramilitary wing Sturmabteilung in November 1933 and the NSDAP in 1937. In April 1939 he swore the oath of loyalty to Hitler as part of his academic appointment. Under a contract from the SS, Gentzen evidently worked for the V-2 project.

From 1943 Gentzen lectured at the University of Prague. Towards the end of the war, on 7 May 1945, he was arrested with all other Germans in Prague, and starved to death in early August at the age of 35.

 **Stanisław Jaśkowski** (22.04.1906–16.11.1965) was a Polish logician who made important contributions to proof theory and formal semantics. He was a student of Jan Łukasiewicz and a member of the famous Lwów-Warsaw School of Logic.

Jaśkowski was one of the founders of the system of natural deduction, which he developed independently from (and possibly before) Gerhard Gentzen in the late 1920s, developed in his doctoral thesis defended in 1932 and published in the 1934 paper *On the Rules of Suppositions in Formal Logic*. Gentzen's approach eventually became more popular, in particular because Gentzen proved the cut-elimination property for it, but Jaśkowski's method was considered by some logicians to be closer to the way that proofs are made in practice.

Jaśkowski was one of the first logicians to propose a formal calculus of *inconsistency-tolerant logic* (now also known as **paraconsistent logic**) in his 1948 paper *A propositional Calculus for Inconsistent Deductive Systems*. He was also one of the pioneers in the study of both intuitionistic logic, for which he developed adequate semantics based on logical matrices, and **free logic**, which allows empty domains.

Jaśkowski was a President (Rector) of the Nicolaus Copernicus University in Toruń.

2.5 Normal forms and Propositional Resolution

The deductive systems presented so far are, more or less, suitable for human use but not for computer-based derivations (with the exception of Semantic Tableaux). Here I present yet another deductive system which is particularly suitable for automation as it is based on no axioms and uses only one simple rule of inference, called **(Propositional) Resolution**. Originally introduced by Martin Davis and Hilary Putnam[4] around 1960, it was later improved and refined into the Davis–Putnam–Logemann–Loveland (DPLL) algorithm. This backtracking-based search algorithm for determining the satisfiability of propositional formulae in conjunctive normal form (see below) is still the basis of most efficient complete Boolean satisfiability problem (SAT-) solving algorithms (see more on these in Section refLB12:sect6:AddBoolSat). As for the previously studied deductive systems, the method of Propositional Resolution is sound and complete for propositional logic.

The rule of Propositional Resolution is very simple:

$$\frac{A \vee C, \quad B \vee \neg C}{A \vee B},$$

where A, B, C are any propositional formulae. The formula $A \vee B$ is called a **resolvent** of $A \vee C$ and $B \vee \neg C$, and we write $A \vee B = Res(A \vee C, B \vee \neg C)$.

This rule is sound in the usual sense: $A \vee C, B \vee \neg C \models A \vee B$. The proof of this is left as an easy exercise for the reader.

2.5.1 Conjunctive and disjunctive normal forms of propositional formulae

The resolution rule is particularly efficient when applied to formulae that are pre-processed in so-called *clausal form*, which can be obtained immediately from *conjunctive normal form*. Normal forms are also useful for other technical purposes, and I devote this subsection to them. First, I provide some basic definitions.

Definition 50

1. A *literal* is a propositional variable or its negation.

2. An *elementary disjunction* (respectively, *elementary conjunction*) is any literal or a disjunction (respectively, conjunction) of two or more literals.

3. A *disjunctive normal form (DNF)* is any elementary conjunction or a disjunction of two or more elementary conjunctions.

4. A *conjunctive normal form (CNF)* is any elementary disjunction or a conjunction of two or more elementary disjunctions.

Examples:

- $p, \neg q, p \vee \neg q, p \vee \neg p \vee q \vee \neg r$ are elementary disjunctions;

- $p, \neg q, \neg p \wedge q, \neg p \wedge q \wedge \neg r \wedge \neg p$ are elementary conjunctions;

---

[4] Read more on Davis and Putnam in the biographic boxes at the end of this section.

- p, $\neg q$, $p \wedge \neg q$, $p \vee \neg q$, $(p \wedge \neg p) \vee \neg q$, $(r \wedge q \wedge \neg p) \vee (\neg q \wedge p) \vee (\neg r \wedge p)$ are disjunctive normal forms;

- p, $\neg q$, $p \wedge \neg q$, $p \vee \neg q$, $p \wedge (\neg p \vee \neg q)$, $(r \vee q \vee \neg r) \wedge \neg q \wedge (\neg p \vee r)$ are conjunctive normal forms.

The normal forms are convenient for various symbolic manipulations and logical computations. They are also used for construction and optimization of logical circuits in computer systems design, which is why the following fact is very important.

Theorem 51 *Every propositional formula is equivalent to a disjunctive normal form and to a conjunctive normal form.*

We give two methods for construction of such equivalent normal forms.

The first method is based on the following algorithm which transforms any formula into a DNF, respectively CNF, using some of the logical equivalences listed above.

1. Eliminate all occurrences of $\leftrightarrow$ and $\rightarrow$ using the logical equivalences

$$A \rightarrow B \equiv \neg A \vee B,$$

$$A \leftrightarrow B \equiv (A \rightarrow B) \wedge (B \rightarrow A).$$

2. Import all negations in front of the propositional variables, using the logical equivalences listed above.

3. For a DNF: distribute all conjunctions over disjunctions using $p \wedge (q \vee r) \equiv (p \wedge q) \vee (p \wedge r)$.

4. Respectively, for a CNF: distribute all disjunctions over conjunctions using $p \vee (q \wedge r) \equiv (p \vee q) \wedge (p \vee r)$.

Throughout this process the formulae can be simplified by using commutativity, associativity, and idempotency of $\vee$ and $\wedge$ as well as:

$p \vee \neg p \equiv \top$; $p \wedge \neg p \equiv \bot$;

$p \wedge \top \equiv p$; $p \wedge \bot \equiv \bot$;

$p \vee \top \equiv \top$; $p \vee \bot \equiv p$.

Example 52 $(p \wedge \neg r) \rightarrow (p \leftrightarrow \neg q)$

$\equiv \neg(p \wedge \neg r) \vee ((p \rightarrow \neg q) \wedge (\neg q \rightarrow p))$

$\equiv (\neg p \vee \neg\neg r) \vee ((\neg p \vee \neg q) \wedge (\neg\neg q \vee p))$

$\equiv \neg p \vee r \vee ((\neg p \vee \neg q) \wedge (q \vee p))$

For a DNF we further distribute $\wedge$ over $\vee$ and simplify:

$\equiv \neg p \vee r \vee (((\neg p \vee \neg q) \wedge q) \vee ((\neg p \vee \neg q) \wedge p))$

$\equiv \neg p \vee r \vee ((\neg p \wedge q) \vee (\neg q \wedge q)) \vee ((\neg p \wedge p) \vee (\neg q \wedge p))$

$\equiv \neg p \vee r \vee ((\neg p \wedge q) \vee \bot) \vee (\bot \vee (\neg q \wedge p))$

$\equiv \neg p \vee r \vee (\neg p \wedge q) \vee (\neg q \wedge p).$

For a CNF we distribute $\lor$ over $\land$ and simplify:

$$\equiv (\neg p \lor r \lor \neg p \lor \neg q) \land (\neg p \lor r \lor q \lor p)$$
$$\equiv (\neg p \lor r \lor \neg q) \land (\top \lor r \lor q)$$
$$\equiv (\neg p \lor r \lor \neg q) \land \top$$
$$\equiv \neg p \lor r \lor \neg q.$$

As can be seen, in this case the CNF also turns out to be a DNF, even simpler than the one we obtained above. The problem of *minimization* of normal forms, which is of practical importance, will not be discussed here.

The second method constructs the normal forms directly from the truth table of the given formula. I outline it for a DNF.

Given the truth table of the formula A we consider all rows (i.e., all assignments of truth values to the occurring variables) where the truth value of A is true. If there are no such rows, the formula is a contradiction and a DNF for it is, for example, $p \land \neg p$. Otherwise, with every such row we associate an elementary conjunction in which all variables assigned value $\top$ occur positively, while those assigned value F occur negated. For instance, the assignment $\mathrm{F}, \mathrm{T}, \mathrm{F}$ to the variables p, q, r is associated with the elementary conjunction $\neg p \land q \land \neg r$. Note that such an elementary conjunction is true only for the assignment with which it is associated.

As an exercise, show that the disjunction of all elementary conjunctions associated with the rows in the truth table of a formula A is logically equivalent to A.

Example 53 *The formula $p \leftrightarrow \neg q$ has a truth table*

p	q	$p \leftrightarrow \neg q$
T	T	F
T	F	T
F	T	T
F	F	F

The corresponding DNF is $(p \land \neg q) \lor (\neg p \land q)$. Check this!

As an exercise, outline a similar method for construction of a CNF from the truth table of a formula and prove your claim.

2.5.2 Clausal Resolution

Definition 54

1. *A **clause** is essentially an elementary disjunction $l_1 \lor \ldots \lor l_n$ but written as a set of literals $\{l_1, \ldots, l_n\}$.*
2. *The **empty clause** $\{\}$ is a clause containing no literals; a **unit clause** is a clause containing only one literal.*
3. *A **clausal form** is a (possibly empty) set of clauses, written as a list: $C_1 \ldots C_k$. It represents the* conjunction *of these clauses.*

Every CNF can therefore be rewritten in a clausal form, and every propositional formula is therefore equivalent to a formula in a clausal form.

As an example, the clausal form of the CNF formula $(p \vee \neg q \vee \neg r) \wedge \neg p \wedge (\neg q \vee r)$ is $\{p, \neg q, \neg r\}\{\neg p\}\{\neg q, r\}$.

The resolution rule can be rewritten for clauses as follows:

$$\frac{\{A_1, \ldots, C, \ldots, A_m\}\{B_1, \ldots, \neg C, \ldots, B_n\}}{\{A_1, \ldots, A_m, B_1, \ldots, B_n\}}.$$

The rule above is called **Clausal Resolution**. The clause $\{A_1, \ldots, A_m, B_1, \ldots, B_n\}$ is a **resolvent** of the clauses $\{A_1, \ldots, C, \ldots, A_m\}$ and $\{B_1, \ldots, \neg C, \ldots, B_n\}$.

The soundness of the Clausal Resolution rule is again left as an easy exercise.

Example 55 *Here are some examples of applying Clausal Resolution:*

$$\frac{\{p, q, \neg r\}\{\neg q, \neg r\}}{\{p, \neg r\}}, \quad \frac{\{\neg p, q, \neg r\}\{r\}}{\{\neg p, q\}}, \quad \frac{\{\neg p\}\{p\}}{\{\}}.$$

Note that two clauses can have more than one resolvent, for example:

$$\frac{\{p, \neg q\}\{\neg p, q\}}{\{p, \neg p\}}, \quad \frac{\{p, \neg q\}\{\neg p, q\}}{\{\neg q, q\}}.$$

However, it is *wrong* to apply the resolution rule for both pairs of complementary literals concurrently and obtain

$$\frac{\{p, \neg q\}\{\neg p, q\}}{\{\}}.$$

Why? Because the conclusion here is the empty clause, which is false, does not follow logically from the premises.

2.5.3 Resolution-based derivations

The method of resolution works similarly to Semantic Tableaux. In order to determine whether a logical consequence $A_1, \ldots, A_n \models B$ holds using the method of resolution, we negate the conclusion B and transform each of the formulae $A_1, \ldots, A_n, \neg B$ into clausal form. We then test whether the resulting set of clauses is unsatisfiable by looking for a resolution-based proof of the empty clause from that set of clauses. Formally:

Definition 56 *A **resolution-based derivation** of a formula B from a list of formulae $A_1, \ldots, A_n$ is a derivation of the empty clause $\{\}$ from the set of clauses obtained from $A_1, \ldots, A_n, \neg B$ by successive applications of the rule of Propositional (Clausal) Resolution. We will denote the system of Propositional Resolution by **RES** and the claim that there is a resolution-based derivation of B from $A_1, \ldots, A_n$ by $A_1, \ldots, A_n \vdash_{\mathbf{RES}} B$.*

Example 57 *Check whether $p \rightarrow q, q \rightarrow r \vdash_{\mathbf{RES}} p \rightarrow r$.*

First, transform $p \rightarrow q, q \rightarrow r, \neg(p \rightarrow r)$ to clausal form:

$$C_1 = \{\neg p, q\}, C_2 = \{\neg q, r\}, C_3 = \{p\}, C_4 = \{\neg r\}.$$

Now, applying resolution successively:

$$C_5 = \{q\} = Res(C_1, C_3);$$
$$C_6 = \{r\} = Res(C_2, C_5);$$
$$C_6 = \{\} = Res(C_4, C_6).$$

The derivation of the empty clause is a proof that $p \to q, q \to r \vdash_{\text{RES}} p \to r.$

Example 58 *Check whether* $\neg p \to q, \neg r \vdash_{\text{RES}} p \vee (\neg q \wedge \neg r).$

First, transform $(\neg p \to q), \neg r, \neg(p \vee (\neg q \wedge \neg r))$ *to clausal form:*

$$C_1 = \{p, q\}, C_2 = \{\neg r\}, C_3 = \{\neg p\}, C_4 = \{q, r\}.$$

Now, applying resolution successively:

$$C_5 = Res(C_1, C_3) = \{q\};$$
$$C_6 = Res(C_2, C_4) = \{q\}.$$

At this stage no new applications of the Clausal Resolution rule are possible; the empty clause is therefore not derivable. We therefore have $\neg p \to q, \neg r \nvdash_{\text{RES}} p \vee (\neg q \wedge \neg r).$

Deletion of repeating literals

Recall that clauses are sets of literals but are recorded as lists, meaning that if more than one copy of a given literal is listed in a clause, the extra copies should be deleted. From this point onwards we assume that such deletion of repeating literals is done automatically in every new resolvent. For example, instead of

$$\frac{\{p, \neg q, \neg r\}\{q, \neg r\}}{\{p, \neg r, \neg r\}}$$

we derive directly

$$\frac{\{p, \neg q, \neg r\}\{q, \neg r\}}{\{p, \neg r\}}.$$

Theorem 59 (Adequacy theorem for RES) *The system of Clausal Resolution is sound and complete, that is:*

$$A_1, \ldots, A_n \vdash_{\text{RES}} B \quad \text{if and only if} \quad A_1, \ldots, A_n \models B.$$

In particular, $\vdash_{\text{RES}} B$ *if and only if* B *is a tautology.*

The soundness of **RES** means that if the empty clause is derivable then the input set of formulae is unsatisfiable. It follows easily from the soundness of the Clausal Resolution rule: every truth assignment satisfying the input set of formulae must also satisfy the initial set of clauses produced from them, but then, by soundness of the rule, it must eventually satisfy the empty clause.

A more detailed generic proof of completeness is sketched in Section 2.7.

2.5.4 Optimizing the method of resolution

The method of resolution is amenable to various optimizations by means of special additional rules and derivation strategies that go beyond the scope of this book. I only mention some simple rules that remove redundant clauses.

- **Tautology deletion**: if a clause contains a complementary pair of literals, then it is a tautology and is therefore of no use for deriving a contradiction, that is, the empty clause, so it can be removed.

- **Subsumption deletion**: if a clause C contains (as a set of literals) a clause C', then it is *subsumed* by C', and hence can be safely removed from the set of clauses because every derivation of the empty clause using C can be reduced to a possibly shorter derivation using C' instead.

- **Removal of clauses with mono-polar literals**: if a clause C contains a literal l such that its complementary literal $\bar{l}$ does not occur in any clause of the current set, then C can be removed from the set because every derivation of the empty clause using C can be simplified to a possibly shorter derivation not using C. The reason is simple: the literal l will be inherited in every resolvent clause produced by using C as a parent clause because the complementary literal $\bar{l}$ can never appear in a clause produced from the current clause set.

References for further reading

Fore more details on the theory and examples of derivations in Propositional Resolution, see Nerode and Shore (1993), Ebbinghaus et al. (1996), Fitting (1996), Chang and Lee (1997), Hedman (2004), and Ben-Ari (2012).

Exercises

2.5.1 Prove that the rule of Propositional Resolution is logically sound.

2.5.2 Show that the second method for constructing DNF, presented here and illustrated in Example 53, is correct in the sense that the disjunction of all elementary conjunctions associated with the rows in the truth table of a formula A is logically equivalent to A.

2.5.3 Develop a method for construction of a CNF from the truth table of a formula, analogous to that for DNF. Prove your claim.

2.5.4 Construct a DNF and a CNF, equivalent to each of the following formulae, using both methods.
   (a) $\neg(p \leftrightarrow q)$
   (b) $((p \rightarrow q) \wedge \neg q) \rightarrow p$
   (c) $(p \leftrightarrow \neg q) \leftrightarrow r$
   (d) $p \rightarrow (\neg q \leftrightarrow r)$
   (e) $(\neg p \wedge (\neg q \leftrightarrow p)) \rightarrow ((q \wedge \neg p) \vee p)$

2.5.5 Using the method of Clausal Resolution check which of the following formulae are tautologies.

(a) $((p \to q) \to q) \to q$

(b) $((q \to p) \to q) \to q$

(c) $((p \to q) \land (p \to \neg q)) \to \neg p$

(d) $((p \lor q) \to \neg r) \to \neg(\neg q \land r)$

(e) $((p \to q) \land (p \to r)) \to (p \to (q \land r))$

(f) $((p \to r) \land (q \to r)) \to ((p \land q) \to r)$

(g) $((p \to r) \lor (q \to r)) \to ((p \lor q) \to r)$

(h) $((p \to r) \land (q \to r)) \to ((p \lor q) \to r)$

(i) $p \to ((q \to r) \to ((p \to q) \to r))$

(j) $(p \to (q \to r)) \to ((p \to q) \to (p \to r))$

2.5.6 Using the method of Clausal Resolution, check the validity of the following logical consequences:

(a) $\neg p \to q, \neg p \to \neg q \models p$

(b) $p \to r, q \to r \models (p \lor q) \to r$

(c) $(p \lor q) \to r \models (p \to r) \land (q \to r)$

(d) $(p \land q) \to r \models (p \to r) \lor (q \to r)$

(e) $p \to q, p \lor \neg r, \neg q \models \neg r$

(f) $(\neg p \land q) \to \neg r, r \models p \lor \neg q$

(g) $p \to q, r \lor (s \land \neg q) \models \neg r \to \neg p$

(h) $p \to q, q \to (r \lor s), \neg(p \to r) \models s$

(i) $p \to (q \land r) \models (p \to q) \land (p \to r)$

(j) $(p \land q) \to r \models (p \to r) \land (q \to r)$

2.5.7 Using Clausal Resolution, check the logical correctness of the following propositional arguments. If not logically correct, find a falsifying truth assignment.

(a)

> If Socrates is a philosopher then Socrates is not happy.
> If Socrates is wise and happy then Socrates is not a philosopher.
> ___
> Therefore, if Socrates is a philosopher then Socrates is wise.

(b)

> Nina wears a red dress or will not wear a silk scarf at the dinner.
> If Nina wears a red dress at the dinner, then she will wear high heels.
> Nina does not wear a red dress at the dinner.
> ___
> Therefore, Nina will wear high heels and will not wear a
> silk scarf at the dinner.

(c)

> If Olivia is not sleeping, then she is crying or she is not hungry.
> If Olivia is crying, then she is hungry.
> ___
> Therefore, if Olivia is not crying, then she is sleeping.

(d)

The property prices increase if the interest rates go down.
The interest rates go down or the economy is not doing well.
The property prices do not increase.
_____
Therefore the economy is not doing well.

(e) Replace the first premise of the propositional argument in the previous question with

"The property prices increase only if the interest rates go down"

and check the correctness of the revised argument.

(f)

Alice will not come to the party if Bonnie comes to the party.
Clyde will come to the party only if Alice comes and
Bonnie does not come.
_____
Therefore Alice will not come to the party if Bonnie comes or
Clyde does not come.

(g) Replace the second premise of the propositional argument in the previous question with

"Clyde will come to the party if Alice comes and Bonnie does not come"

and check the correctness of the revised argument.

(h)

Hans will be promoted if he is clever and does his job well.
Hans will be fired if he does not do his job well.
_____
Therefore, if Hans is clever, then he will be promoted or will be fired.

(i) Replace the second premise of the propositional argument in the previous question with

"Hans will be fired only if he does not do his job well"

and check the correctness of the revised argument.

2.5.8 Prove that each of the simplification rules – *deletion of repeating literals, tautology deletion, subsumption deletion,* and *removal of clauses with mono-polar literals* – is based on a logically sound rule.

Martin Davis (b. 1928) is an American mathematician known for several important ideas and results in mathematics, logic, and computability theory.

Davis was an undergraduate student of Emil Post in New York and later completed his doctoral study under the supervision of Alonso Church at Princeton University in 1950. He is currently a Professor Emeritus at New York University.

One of Davis' most important work, together with Hilari Putnam, led to the solution of **Hilbert's tenth problem**, namely the proof of the algorithmic unsolvability of Diophantine equations. That proof was eventually completed by the young Russian mathematician Yuri Matiyasevich in 1970, essentially using results by Davis, Putnam, and Julia Robinson.

In around 1960, Davis and Putnam invented the **Davis–Putnam algorithm** for checking the validity of a first-order logic formula (which only terminates on valid formulae), using a **resolution-based decision procedure** for propositional logic. Later, their algorithm was improved to the **Davis–Putnam–Logemann–Loveland (DPLL) algorithm** which is still the basis for the currently most efficient complete SAT-solvers.

Davis is also known for inventing his models of **Post-Turing machines**.

Hilary Whitehall Putnam (31.07.1926 – 13.03.2016) is an American philosopher, mathematician, and computer scientist, a leading figure in analytic philosophy since the 1960s, especially in philosophy of mind, of language, of logic, of mathematics, and of science.

Putnam studied mathematics and philosophy at the University of Pennsylvania and received his PhD in philosophy at UCLA in 1951 for a dissertation on *The Meaning of the Concept of Probability in Application to Finite Sequences*, working with Reichenbach. He then taught philosophy at Northwestern University, Princeton University, MIT and at Harvard University from 1965 until the end of his active academic career in 2000.

Putnam is well known for his many influential works on theories of mind and meaning and, in particular, for his arguments in defence of scientific realism and objectivity of truth, knowledge, and mathematical reality.

Putnam also made important contributions to mathematics and logic. In around 1960, he and Martin Davis developed the **Davis–Putnam algorithm** for solving the Boolean satisfiability problem. Putnam also contributed to the eventual solution of Hilbert's tenth problem.

2.6 Supplementary: The Boolean satisfiability problem and NP-completeness

The Boolean satisfiability problem (SAT) is the problem of deciding whether a given input propositional (Boolean) formula is satisfiable, that is, whether there is a truth assignment to the propositional (Boolean) variables occurring in the formula that makes it true. It is the most popular **NP-complete decision problem**, the meaning of which I explain in the following.

First, NP stands for "non-deterministic polynomial" time. An algorithmic decision problem (that requires an answer Yes/No for every input) is in the **complexity class NP** if, intuitively, for every input where the answer is Yes there is an evidence (proof or witness) that can be verified "quickly and efficiently", which is assumed to mean in **polynomial time**, that is, in a number of steps that is bounded by a fixed polynomial in the length of the input. Equivalently, and again intuitively, a problem is in NP if there is an algorithm that can run on a hypothetical **non-deterministic** computing device (e.g., non-deterministic Turing machine) to always solve the problem and, when the answer is Yes, to produce that answer in polynomial time. Simply put, an NP problem may be computationally difficult (slow) to *find* a solution, but it is easy (quick) to *check* whether a proposed, or guessed, solution is correct.

Of course, every decision problem that can be solved (for any answer) by an algorithm that works in **deterministic polynomial time** (i.e., one that runs on a usual digital computer and always takes a number of steps that is bounded by a fixed polynomial in the length of the input) is in NP. Besides, there may be problems that are in NP but for which there is no algorithm that can solve them in polynomial time. The SAT problem seems to be a case in point. Indeed, it is in NP because whenever a Boolean formula is satisfiable, any given satisfying truth assignment can be verified very quickly in a number of steps which is linear in the length of the input formula. Alternatively, the SAT problem can be solved efficiently with a non-deterministic computing device that guesses a satisfying assignment, if there is one, and then verifies in linear time that it indeed works. Equivalently, we can think of solving SAT by using a hypothetical *unboundedly parallel* computing device, simply by determining *simultaneously all truth assignments* whether any of them renders the formula true.

There is a number of practically very efficient algorithms for solving the SAT problem, known as *SAT solvers*, which usually solve most of the input SAT problems involving thousands of variables and millions of clauses within seconds. On the other hand, there is currently no known algorithm run on a digital computer that can *always* solve the SAT problem efficiently, that is, in polynomial time. Indeed, all methods for solving the SAT problem that are known so far – including truth tables, Semantic Tableaux, Propositional Resolution, and many more modern, extremely clever and practically efficient algorithms – require *in the worst case* a number of steps which is exponential in the number of variables occurring in the input formula, and therefore exponential in the length of the formula. This is because there are 2^n possible truth assignments for a set of n propositional variables.

There are many hundreds of other very important algorithmic problems that are in NP, including problems of scheduling, coloring of maps with a given number of colors, or finding the prime factorization of a given integer.

Some of the problems in NP, such as SAT, are also known to be **NP-complete**. An algorithmic decision problem is NP-complete if solving any problem that is in NP can be reduced efficiently (i.e., in a number of steps bounded by some fixed polynomial in the length of the input) to solving the problem P. Any algorithm solving an NP-complete problem can therefore be transformed in an at most "polynomially slower" (i.e., not *much* slower) algorithm solving any given problem in NP. If an *efficient algorithm* (always taking a number of steps polynomially bounded in the size of the input) solving SAT is found, it can therefore be suitably modified to efficiently solve any other problem in NP.

Can there be an efficient algorithm solving SAT, or any other NP-complete problem? This has been an open question since 1971 when it was first stated by Stephen Cook in his seminal paper *The Complexity of Theorem Proving Procedures*. It is literally a million-dollar question, being one of the seven *Millennium Prize Problems* stated by the Clay Mathematics Institute in 2000, each bearing a 1,000,000 US$ prize tag for the first solution which is officially accepted as correct by the scientific community. The problem is known as "P=NP?" and is currently considered to be the most challenging and important unsolved problem in computer science. Most computer scientists believe that such a miracle algorithm, solving NP-complete problems in polynomial time, does not exist at all, but there are some who have justified reasons to be more optimistic. I can only conjecture here that we will know the answer before the turn of this century.

References for further reading
For more on the Boolean satisfiability problem and NP-completeness from a logical perspective, see Nerode and Shore (1993), Fitting (1996), Hedman (2004), and Ben-Ari (2012).

Stephen Cook (b. 1939) is an American–Canadian computer scientist and mathematician, one of the founders of the theory of computational complexity and the study of proof complexity.

Cook completed his PhD at Harvard in 1966 as a student of Hao Wang. In his seminal 1971 paper *The Complexity of Theorem Proving Procedures* he formalized the notions of polynomial-time reduction and NP-completeness and proved that the Boolean satisfiability problem (SAT) is NP-complete. He also formulated the most famous problem in computer science, **P v. NP**, and conjectured that there are no polynomially fast algorithms solving NP-complete problems, that is, that P$\neq$ NP.

In 1982, Cook received the prestigious ACM Turing award for his fundamental contributions to complexity theory.

Leonid Levin (b. 1948) is a Soviet–American computer scientist, known for his work in the theory of computing, algorithmic complexity, and intractability.

Levin studied at Moscow University in 1970 where he completed doctoral studies under the supervision of Andrey Kolmogorov in 1972. In 1977 he emigrated to the US, where he completed another PhD at the Massachusetts Institute of Technology (MIT) in 1979 under the supervision of Albert R. Meyer.

During his doctoral studies he discovered the existence of NP-complete problems independently of Stephen Cook, and published this result in 1973. It is now often called the **Cook–Levin Theorem**. Levin was awarded the prestigious Knuth Prize in 2012 for his discovery of NP-completeness and the development of the theory of average-case complexity.

2.7 Supplementary: Completeness of the propositional deductive systems

Here I summarize the important concepts and claims related to proving soundness and completeness of a deductive system for propositional logic and outline a generic proof of completeness which can be applied *mutatis mutandis* to each of the deductive systems introduced here. The proofs of most of the claims are left as exercises but they can be found in many excellent textbooks on logic, some of which are listed as references at the end of this chapter. For some of these claims the proofs are generic, that is, essentially the same for each deductive system, while for others the proofs are essentially different as they make use of the specific deductive machinery – axioms (if any) and inference rules – of the system.

Hereafter, **D** denotes any Axiomatic System (**H**), Semantic Tableau (**ST**), Natural Deduction (**ND**), or Resolution (**RES**). Derivability in **D** is denoted $\vdash_{\mathbf{D}}$.

First, we need some more terminology.

By a **(propositional) theory** I mean any set of propositional formulae. Recall that a set of formulae is satisfiable if there is a truth assignment that makes all formulae in the set true; otherwise it is unsatisfiable. Here are some important properties relating satisfiability and logical consequence that we will need.

Proposition 60 (Satisfiability and logical consequence) *For any formula* A:

1. $\Gamma \cup \{A\}$ *is satisfiable iff* $\Gamma \nvDash \neg A$.

2. $\Gamma \models A$ *iff* $\Gamma \cup \{\neg A\}$ *is unsatisfiable*.

3. *If* $\Gamma \cup \{A\}$ *is unsatisfiable and* $\Gamma \cup \{\neg A\}$ *is unsatisfiable then* Γ *is also unsatisfiable*.

I leave the proofs of these as easy exercises.

Definition 61 (Deductive consistency) *A theory* Γ *is:*

1. ***consistent in* D** *(or simply **D**-consistent) if there is no formula* A *such that* $\Gamma \vdash_{\mathbf{D}} A$ *and* $\Gamma \vdash_{\mathbf{D}} \neg A$; *otherwise,* Γ *is **D**-inconsistent; or*

2. *a **maximal D-consistent theory** if it is **D**-consistent and cannot be extended to a larger* **D**-*consistent theory, that is, adding to* Γ *any formula that is not already in* Γ *results in a* **D**-*inconsistent theory.*

Proposition 62 (D-inconsistency) *The following are equivalent for any theory* Γ:

1. Γ *is **D**-inconsistent.*

2. $\Gamma \vdash_{\mathbf{D}} \bot$.

3. $\Gamma \vdash_{\mathbf{D}} A$ *for every formula* A.

4. *There are formulae* $A_1, \ldots, A_n \in \Gamma$ *such that* $\vdash_{\mathbf{D}} \neg(A_1 \wedge \cdots \wedge A_n)$.

The proofs differ for each deductive system as they use the specific notion of derivation in each, but every one of them is a useful and not very difficult exercise.

Let us now observe that deductive consequence in each of our systems **D** and the notion of **D**-consistency have the properties of logical consequence and satisfiability (i.e., semantic consistency), respectively, stated earlier in Proposition 60:

Proposition 63 (Consistency and deductive consequence) *For any formula* A:

1. $\Gamma \cup \{A\}$ *is **D**-consistent iff* $\Gamma \nvdash_{\mathbf{D}} \neg A$.

2. $\Gamma \vdash_{\mathbf{D}} A$ *iff* $\Gamma \cup \{\neg A\}$ *is **D**-inconsistent.*

3. *If* $\Gamma \cup \{A\}$ *is* **D**-*inconsistent and* $\Gamma \cup \{\neg A\}$ *is* **D**-*inconsistent then* Γ *is* **D**-*inconsistent.*

The proofs differ again for each deductive system as they use the specific notion of derivation in each of them. I leave them as useful exercises.

Next, let us revisit the notion of soundness by redefining it in two different ways.

Definition 64 (Soundness 1) *A deductive system* **D** *is* **sound₁** *if for every theory* Γ *and a formula A,*

$$\Gamma \vdash_{\mathbf{D}} A \ \text{implies} \ \Gamma \models A.$$

Definition 65 (Soundness 2) *A deductive system* **D** *is* **sound₂** *if for every theory* Γ,

if Γ *is satisfiable then* Γ *is* **D***-consistent.*

We therefore now have two different definitions of soundness: one in terms of logical and deductive consequence, and the other in terms of satisfiability and deductive consistency. These definitions are in fact equivalent and the proof of that, left as an exercise, is quite generic for all deductive systems using Propositions 60 and 63.

As I have already stated in the respective sections, each of our deductive systems is sound (in either sense). I leave the proofs of these as exercises. Let us now turn to completeness.

Definition 66 (Completeness 1) *A deductive system* **D** *is* **complete₁** *if for every theory* Γ *and a formula A,*

$$\Gamma \models A \ \text{implies} \ \Gamma \vdash_{\mathbf{D}} A.$$

Definition 67 (Completeness 2) *A deductive system* **D** *is* **complete₂** *if for every theory* Γ,

if Γ *is* **D***-consistent then* Γ *is satisfiable.*

Again, we have two different definitions of completeness, one in terms of logical and deductive consequence and the other in terms of satisfiability and deductive consistency. As for soundness, these definitions are equivalent and the proof is again generic for all deductive systems, using Propositions 60 and 63. These proofs are left as an exercise.

Proposition 68 (Maximal consistent theory 1) *Every maximal* **D***-consistent theory* Γ *is closed under deductive consequence in* **D**, *that is, for any formula A, if* $\Gamma \vdash_{\mathbf{D}} A$ *then* $A \in \Gamma$.

Proof: Exercise.

Definition 69 (D-completeness) *A theory* Γ *is* **D***-complete if it is* **D***-consistent and for every formula A,* $\Gamma \vdash_{\mathbf{D}} A$ *or* $\Gamma \vdash_{\mathbf{D}} \neg A$.

Proposition 70 (Maximal consistent theory 2) *A theory* Γ *is a maximal* **D***-consistent theory iff it is closed under deductive consequence in* **D** *and is* **D***-complete.*

The proof is generic, similar for each **D**, and left as an easy exercise.

The next theorem shows that membership of a given maximal consistent theory has the same properties as a truth assignment (just replace membership in the theory with truth in each of the clauses below.)

Theorem 71 (Maximal consistent theory 3) *For every maximal **D**-consistent theory* Γ *and formulae* A, B, *the following hold:*

1. $\neg A \in \Gamma$ *iff* $A \notin \Gamma$.
2. $A \wedge B \in \Gamma$ *iff* $A \in \Gamma$ *and* $B \in \Gamma$.
3. $A \vee B \in \Gamma$ *iff* $A \in \Gamma$ *or* $B \in \Gamma$.
4. $A \rightarrow B \in \Gamma$ *iff* $A \in \Gamma$ *implies* $B \in \Gamma$ *(i.e.,* $A \notin \Gamma$ *or* $B \in \Gamma$*).*

The proof is specific to each deductive system as it uses its specific deductive machinery. Each proof is left as an exercise.

Given a theory Γ, consider the following truth assignment:

$$S_\Gamma(p) := \begin{cases} \text{T}, & \text{if } p \in \Gamma; \\ \text{F}, & \text{otherwise.} \end{cases} \text{ for every propositional variable } p.$$

The truth assignment S_Γ extends to a truth valuation of every formula by applying a recursive definition according to the truth tables (see Section 1.4.5.2). The truth valuation is denoted S_Γ.

Lemma 72 (Truth Lemma) *If* Γ *is a maximal **D**-consistent theory, then for every formula* A, $S_\Gamma(A) = T$ *iff* $A \in \Gamma$.

Proof. Exercise. Use Theorem 71. ∎

Corollary 73 *Every maximal **D**-consistent theory is satisfiable.*

Lemma 74 (Lindenbaum's Lemma) *Every **D**-consistent theory* Γ *can be extended to a maximal **D**-consistent theory.*

Proof. Let $A_0, A_1,\ldots$ be a list of all propositional formulae. (NB: they are countably many, so we can list them in a sequence.) I will define a chain by inclusion of theories $\Gamma_0 \subseteq \Gamma_1 \subseteq \ldots$ defined by recursion on n as follows:

- $\Gamma_0 := \Gamma$;
- $\Gamma_{n+1} := \begin{cases} \Gamma_n \cup \{A_n\}, & \text{if } \Gamma_n \cup \{A_n\} \text{ is } \mathbf{D}\text{-consistent;} \\ \Gamma_n \cup \{\neg A_n\}, & \text{otherwise.} \end{cases}$

Note that every Γ_n is an **D**-consistent theory. Prove this by induction on n, using the properties of the deductive consequence from Proposition 63.

Now, we define

$$\Gamma^* := \bigcup_{n \in \mathbf{N}} \Gamma_n.$$

Clearly, $\Gamma \subseteq \Gamma^*$. Γ^* is a maximal **D**-consistent theory. Indeed:

- Γ^* is **D**-consistent. Otherwise, $\Gamma^* \vdash_\mathbf{D} A$ and $\Gamma^* \vdash_\mathbf{D} \neg A$. Since a derivation in each **D** uses only finitely many assumptions (we say that the deductive consequence in **D** is **compact**), it follows that $\Gamma_n \vdash_\mathbf{D} A$ and $\Gamma_n \vdash_\mathbf{D} \neg A$ for some large enough index n, which contradicts the consistency of Γ_n (fill in the details here).

- Γ^* is **maximal D-consistent**. Indeed, take any formula A. Let $A = A_m$. Then $A_m \in \Gamma_{m+1}$ or $\neg A_m \in \Gamma_{m+1}$, so $A \in \Gamma^*$ or $\neg A \in \Gamma^*$; hence, Γ^* cannot be extended to a larger **D**-consistent theory. ∎

Theorem 75 (Completeness of D) *The axiomatic system* **D** *is complete.*

Proof. Let Γ be a **D**-consistent theory. Then, by Lindenbaum's Lemma 74, Γ can be extended to a maximal **D**-consistent theory Γ^*. That theory is satisfiable by the Truth Lemma 72. ∎

The outlined proof above applies to each of the deductive systems that we have studied. However, in order to prove their completeness we do not have to follow the same scheme of steps for each of them. It is sufficient to prove the soundness and completeness for one of them, say **D**, and then reduce the proofs for any of the other to this result, as follows.

Proposition 76 (Relative soundness and completeness) *For any deductive systems for propositional logic* **D** *and* **D**′*, the following hold.*

1. *To prove the soundness of* **D**′ *given the soundness of* **D**, *it is sufficient to show that for every theory* Γ *and a formula* A,

$$\Gamma \vdash_{\mathbf{D}'} A \ \text{ implies } \ \Gamma \vdash_\mathbf{D} A.$$

2. *To prove the completeness of* **D**′ *given the completeness of* **D**, *it is sufficient to show that for every theory* Γ *and a formula* A,

$$\Gamma \vdash_\mathbf{D} A \ \text{ implies } \ \Gamma \vdash_{\mathbf{D}'} A.$$

I leave the proof as an easy exercise.

Using the above proposition, the soundness and completeness of any of our deductive systems can be reduced to the soundness and completeness of any other (see exercises).

References for further reading
For more details and complete proofs of soundness and completeness of propositional deductive systems, see van Dalen (1983) for a completeness proof for **ND**; Hamilton (1988) for completeness of **H**; Nerode and Shore (1993) and Fitting (1996) and for completeness of **ST** and **RES**; Smullyan (1995) for completeness of **ST** and **ND**; Ebbinghaus *et al.* (1996), Hedman (2004), and Ben-Ari (2012) for completeness of **RES**; and Tarski (1965) for a general discussion and methodology.

Exercises

No solutions are provided for these exercises, but most of these proofs can be found in the references listed above.

2.7.1 Prove Proposition 60.

2.7.2 Prove Proposition 62 for **H**.

2.7.3 Prove Proposition 62 for **ST**.

2.7.4 Prove Proposition 62 for **ND**.

2.7.5 Prove Proposition 62 for **RES**.

2.7.6 Prove Proposition 63 for **H**.

2.7.7 Prove Proposition 63 for **ST**.

2.7.8 Prove Proposition 63 for **ND**.

2.7.9 Prove Proposition 63 for **RES**.

2.7.10 Using Propositions 60 and 63, prove that the two definitions of soundness are generically equivalent for each deductive system **D**.

2.7.11 Using Propositions 60 and 63, prove that the two definitions of completeness are generically equivalent for each deductive system **D**.

2.7.12 Prove Proposition 68 generically for each deductive system **D**.

2.7.13 Prove Proposition 70 generically for each deductive system **D**.

2.7.14 Prove Theorem 71 for **H**.

2.7.15 Prove Theorem 71 for **ST**.

2.7.16 Prove Theorem 71 for **ND**.

2.7.17 Prove Theorem 71 for **RES**.

2.7.18 Show that if the construction in Lindenbaum's lemma is modified as follows:

$$\Gamma_{n+1} := \begin{cases} \Gamma_n \cup \{A_n\}, & \text{if } \Gamma_n \cup \{A_n\} \text{ is } \mathbf{D}\text{-consistent;} \\ \Gamma_n, & \text{otherwise.} \end{cases}$$

then the resulting theory Γ^* is still a maximal **D**-consistent theory.

2.7.19 Prove Proposition 76.

2.7.20 Assuming the soundness and completeness of **H**, prove the soundness and completeness of each of **ST**, **ND**, and **RES**, by using Proposition 76.

2.7.21 Assuming the soundness and completeness of **ST**, prove the soundness and completeness of each of **H**, **ND**, and **RES** by using Proposition 76.

2.7.22 Assuming the soundness and completeness of **ND**, prove the soundness and completeness of each of **ST**, **H**, and **RES** by using Proposition 76.

2.7.23 Assuming the soundness and completeness of **RES**, prove the soundness and completeness of each of **H**, **ST**, and **ND** by using Proposition 76.

Jan Leopold Łukasiewicz (21.12.1878–13.02.1956) was a Polish logician and philosopher who introduced mathematical logic in Poland and made notable contributions to analytical philosophy, mathematical logic, and history of logic.

Łukasiewicz studied first law and then mathematics and philosophy at the University of Lwów where he achieved a PhD in 1902 under the supervision of Kazimierz Twardowski for a dissertation *On induction as the inverse of deduction*. He taught at the University of Lwów before WW I then joined the University of Warsaw in 1915, where he held the position of rector in 1922–23 and 1931–32. He also served as a minister of education in 1919. Together with another prominent logician, Stanislaw Lesniewski, Łukasiewicz founded the world-famous Warsaw School of Logic. Alfred Tarski, a student of Lesniewski but also strongly influenced by Łukasiewicz, also contributed to the reputation of the school. Łukasiewicz fled from Poland during WW II. In 1946 he was appointed Professor of Mathematical Logic at the Royal Irish Academy in Dublin, where he worked until his retirement in 1953.

Łukasiewicz did important work in modernizing formal logic. He developed propositional logic and its implicational and equivalential fragments, for all of which he obtained some elegant short axiomatizations. Notably, he introduced **many-valued logics** (partly as an alternative to the Aristotelian 2-valued logic) in 1917. Łukasiewicz also introduced the **Polish notation** which allowed logical formulae to be written unambiguously without the use of brackets. For instance, the formula $(p \rightarrow (\neg p \rightarrow q))$ is written in Polish notation as $CpCNpq$. He also developed a theory of axiomatic rejection, wrote a book on *Logical Foundations of Probability Theory*, and conducted very important work in the history of logic by studying and popularizing both Aristotle's syllogistic and Stoic's propositional logic.

Emil Leon Post (11.02.1897–21.04.1954) was a Polish-born American mathematician and logician, regarded as one of the founders of both **computability and recursion theory**, along with Alan Turing and Alonso Church, and of **proof theory**.

Post was born in Augustów, then in the Russian Empire (now in Poland), into a Polish–Jewish family that emigrated to America when he was 7 years old. When he was a child he lost an arm in an accident in New York, which prevented him from studying astronomy (his favourite subject at that time). He graduated from City College of New York in 1917 and completed his PhD in mathematics at Columbia University in 1920. He then completed a post-doctorate at Princeton, where he anticipated the incompleteness phenomena that Gödel discovered several years later, as well as the undecidability results of Church and Turing. He wrote about his incompleteness ideas to Church but did not publish them, planning to analyze them better. It was only when Kurt Gödel published his famous proof in 1931 that Post shared his ideas with him. It is believed that the excitement from these discoveries probably triggered Post's manic-depressive attacks, a condition from which he suffered all his life.

After returning from Princeton, Post became a very successful and popular high school mathematics teacher at the City College of New York.

Post published some mathematical works as an undergraduate student, but his first publication in logic was a shortened version of his doctoral dissertation. This contained a precise formulation and systematic study of the propositional fragment of Russell and Whitehead's *Principia Mathematica*, and provided the first published proof of its completeness and decidability by showing that any formula in the system is provably equivalent to one in a full disjunctive normal form. Notably, Post introduced truth tables for the propositional connectives even though, as he noted, they were implicit in the *Principia*. He then generalized his two-valued truth-table method to one which had an arbitrary finite number of truth values. He also promoted in his thesis the pioneering idea of the development of systems for logical inference based on a finite process of manipulation of symbols. Such systems produce, in today's terminology, **recursively enumerable sets** of words in a finite alphabet, just like formal systems of deduction, so Post can also be credited with laying the foundations of both proof theory and recursion theory in his thesis.

In 1936 he developed and published, independently of Turing, an abstract mathematical model of computation, now known as **Post machine** or **Post–Turing machine**, essentially equivalent to the Turing machine.

Post is also well known for his work on polyadic groups, recursively enumerable sets, and degrees of unsolvability, as well as for his results on the algorithmic unsolvability of certain combinatorial problems, notably the **Post correspondence problem** he introduced in 1946. It is equivalent to the **Halting problem** for Turing machines, but simpler to use in proofs of undecidability. In 1947 Post showed that the word problem for semigroups was recursively unsolvable, therefore solving a problem posed by Thue in 1914.

Post continued to suffer from manic-depressive illness throughout his life, and died in 1954 at the age of 57 from a heart attack induced by electro-shock treatment at a mental hospital in New York.

3

Understanding First-order Logic

Propositional logic can only formalize some patterns of reasoning, but it cannot grasp the logical structure or the truth behavior of very simple sentences such as:

- "$x + 2$ is greater than 5;"
- "there exists y such that $y^2 = 2$;"
- "for every real number x, if x is greater than 0 then there exists a real number y such that y is less than 0 and y^2 equals x;" or, for a non-mathematical example;
- "every man loves a woman."

Indeed, note that an expression such as "$x + 2$ is greater than 5" is *not* a proposition, for it can be true or false depending on the choice of x. Neither is "there exists y such that $y^2 = 2$" a proposition until the range of possible values of y is specified: if y is an integer, then the statement is false; but if y can be any real number, then it is true. As for the third sentence above, it *is* a proposition but its truth depends heavily on its internal logical structure and mathematical meaning of all phrases involved, and those are not tractable on a propositional level.

All these sentences take us out of the simple world of propositional logic into the realm of **first-order logic** (the term "first-order" will be explained soon), also known as **classical predicate logic** or just **classical logic**, the basic concepts of which we introduce and discuss here.

First-order logic (just like every formal logical system) has two major aspects:

- precise **syntax**, involving a formal language called a **first-order language**, that enables us to express statements in a uniform way by means of **logical formulae**;

- formal **logical semantics**, specifying the meaning of all components of the language by means of their **interpretation** into suitable models called **first-order structures** and formal **truth definitions**, extending the truth tables for the propositional connectives[1].

Here we discuss the basic components of the syntax and semantics of first-order logic and their relevance to mathematical reasoning.

---

[1] The reader who has some experience with programming languages should find the concepts of formal syntax and semantics familiar.

Logic as a Tool: A Guide to Formal Logical Reasoning, First Edition. Valentin Goranko.
© 2016 John Wiley & Sons, Ltd. Published 2016 by John Wiley & Sons, Ltd.

3.1 First-order structures and languages: terms and formulae of first-order logic

3.1.1 First-order structures

First-order logic is a formal logical language for formalizing statements and reasoning about universes represented as so-called **first-order structures**. Here we first discuss the basic components of a first-order structure and then give a more formal definition.

Domains

When we reason and make statements about objects, we have in mind a certain **domain of discourse**. In mathematics, that domain includes mathematical objects such as numbers (integers, rationals, reals, etc.), vectors, geometric figures (points, lines, triangles, circles, etc.), sets and graphs. A domain in a non-mathematical discourse may consist of material objects, human beings, ideas, or anything else. In either case, it is important that we have specified our domain of discourse, so that we know what we are talking about. Moreover, as one of the sentences mentioned above suggests, a statement can be true or false depending on the domain in which it is considered.

Predicates

When we reason about the objects from our domain of discourse, we usually make statements about properties they have or do not have. For instance, talking about integers, we may discuss properties such as being "positive", "divisible by 3", "not greater than 1999", etc. Triangles can be "obtuse", "equilateral", etc. A human being can be "female", "male", "young", "blue-eyed", etc.

The logical term for a property is **predicate**. Predicates need not only concern one object, as for the examples mentioned above which are called **unary predicates**. We also deal with **binary predicates**, relating two objects such as: "_ is less than _" or "_ is divisible by _" (for numbers), or "_ is a son of _" or "_ loves _" (for humans), etc. **Ternary predicates** relate three objects, such as "_ is between _and _" (for points on a line) or "_ and _ are the parents of _" (for humans), etc. In general, we can talk about **n-ary predicates**, relating n objects at a time.

As long as we specify the meaning (semantics) of a predicate and the objects which it relates, that predicate becomes true or false, that is, represents a proposition. For example, the propositions "12 is divisible by 6" and "13 is divisible by 7" are both instances of the binary predicate "x is divisible by y." Furthermore, we can connect predicates by using propositional connectives in compound statements, just like we did earlier with propositions.

Functions

Typically in mathematics (and not only there) we use **functions** to represent operations which, applied to one or several objects, determine an object. Depending on the number of arguments, we talk about **unary functions, binary functions**, etc. Standard examples are all arithmetic operations and, more generally, all algebraic functions that we have studied at school. Examples such as "the mother of _" or "the father of _" are in the domain of humans.

Note that in our logical framework all functions will be considered **total**, that is, defined for all possible values of the argument (or all possible tuple values of the arguments) in the domain. This is quite often not the case, for example subtraction in the domain of

natural numbers, division in the domain of reals, or the function "the daughter of _"
in the domain of humans. This problem has an easy (albeit artificial) fix, good enough
for our formal purposes: we designate an element u (for undefined) from the domain and
make the function total by assigning u to be its value for all (tuples of) arguments for
which it is not defined. For instance, we can extend division by 0 in the domain of reals
by putting $x/0 = 17$ for any real x. This may sound reckless, but if proper care is taken
when reasoning about division, it will not lead to confusion or contradiction.

Constants

Some objects in the domain can be distinguished in a way that would allow us to make
direct references to them, by giving them **names**. Such distinguished objects are called
constants[2]. Examples in the domain of real numbers are $0, 1, 3/7, \sqrt{2}, \pi, e$, etc. Note that
the notion of a constant pertains to the language rather than to the domain. For instance,
in common calculus we do not have a special name for the least positive solution of
the equation $\cos x = x$, but if for some purposes we need to refer directly to it then we
can extend our mathematical language by giving it a name, that is, make it a constant in
our domain.

　　We are now ready to define the general notion of a first-order structure.

Definition 77 *A **first-order structure** (hereafter, just **structure**) consists of a non-empty
domain and a family of distinguished functions, predicates, and constants in that domain.*

Example 78 *Here are some examples of structures that will be used further.*

- *$\mathcal{N}$, with domain being the set of natural numbers $\mathbf{N}$, the unary function s (successor, i.e.,
$s(x) = x + 1$), the binary functions $+$ (addition) and $\times$ (multiplication), the predicates
$=$, $<$, and $>$, and the constant 0.*

- *$\mathcal{Z}$: as for $\mathcal{N}$, but with the domain being the set of integers $\mathbf{Z}$ and the additional function $-$ (subtraction).*

- *With the same functions and predicates we take the domain to be the set of rational
numbers $\mathbf{Q}$ or the reals $\mathbf{R}$. The resulting structures are denoted $\mathcal{Q}$ and $\mathcal{R}$, respectively.
(For those familiar with some abstract algebra: algebraic structures such as groups,
rings, and fields are all examples of first-order structures. Algebraic structures usually
only involve functions and constants but no predicates, except $=$ and possibly $<$.)*

- *$\mathcal{H}$: the domain is the set of all humans, with functions m (for "the mother of") and f (for
"the father of"); unary predicates M (for "man") and W (for "woman"); binary predicates P (for "parent of"), C (for "child of"), L (for "loves"); and constants (names),
for example John, Mary, Adam, Eve.*

- *$\mathcal{G}$: the domain is the set of all points and lines in the plane, with unary predicates P for
"point" L for "line", and the binary predicate I for "incidence" between a point and
a line.*

- *$\mathcal{S}$: the domain is the collection of all sets[3], with binary predicates $=$ for "equality" and
$\in$ for "membership".*

---

[2] Note that the use of the word "constant" in first-order logic is not exactly the same as in the common mathematical
language.
[3] Strictly speaking, this is not a structure because the collection of all sets is not a proper set, but we can obtain a
structure if we relativize it to the family of all subsets of a given set, a "universe." See more in Section 5.2.1.

3.1.2   First-order languages

In order to refer to objects, functions and predicates in our domain, we give them *names*. In mathematics they are typically symbols or abbreviations such as $5, \pi, +, \sqrt{}, \sin, =$ or $<$; for humans they are proper names. Together with logical connectives, variables and some auxiliary symbols, these symbolic names determine formal logical languages called "first-order languages". In particular, with every structure $\mathcal{S}$ we can associate a **first-order language** $\mathcal{L}_{\mathcal{S}}$ for that structure, containing:

1. **Functional, predicate, and constant symbols**, used as names for the functions, predicates, and constants we consider in the structure. All these are referred to as **non-logical symbols**.

   I emphasize that the non-logical symbols are *just names* for functions, predicates, and constants; in principle, we should use different symbols to denote objects and their names. Later on I introduce a generic notation that will take care of that, but meanwhile I allow a little sloppiness and use the same symbols for functions, predicates, and constants in our example structures and for their names in the first-order languages designed for them.

2. **Individual variables.** Quite often, particularly in mathematics, we deal with unknown or unspecified objects (individuals) from the domain of discourse. In order to be able to reason and make statements about such objects, we use *individual variables* to denote them. For instance, talking about numbers, we use phrases such as "Take a positive integer n", "Every real number x greater than $\sqrt{2}$ ...", etc. In natural language instead of variables we usually use pronouns or other syntactic constructions but that often leads to awkwardness and ambiguity (e.g. "If a man owes money to another man then that man hates the other man"). The use of variables is therefore indispensable, not only in mathematics, and it is very important to learn how to use them properly.

   We also use variables as *placeholders* for the arguments of the various predicates and functions we deal with. We can therefore talk about the (unary) functions $f(x)$ and "the mother of z", or the (binary) predicates $P(x, y)$ and "x loves y", etc.

   We will assume that any first-order language contains an infinite set of individual variables, denoted VAR. The letters u, v, w, x, y, z, possibly indexed, are used to denote individual variables.

3. **Auxiliary symbols**, such as "(", ",", ")" etc.

4. Last but most important, **logical symbols**, including:
   (a) **Propositional connectives**, which we already know: $\neg, \wedge, \vee, \rightarrow, \leftrightarrow$; and
   (b) **Quantifiers.** Often we use special phrases to *quantify* objects of our discourse, such as:

      "(for) every (objects) x ...",
      "there is (an object) x such that ...",
      "for most (objects) x ...",
      "there are at least 5 (objects) x such that ...",
      "there are more (objects) x than y such that ...",

      etc. (Note the use of variables here.)

The first two quantifiers are particularly important and many others can be expressed by means of them, so they are given special names and notation:

The quantifier "for every" is called the **universal quantifier**, denoted ∀.

The quantifier "there exists" is called **existential quantifier**, denoted ∃.

These are the only quantifiers used in first-order languages, so they are called **first-order quantifiers**. The term "first-order" refers to the fact that variables and quantifications in these languages are *only permitted to range over individuals* in the universe of discourse, called "first-order objects"[4].

As well as the phrases above, the universal quantifier is usually represented by *"all"*, *"for all"*, and *"every"*, while the existential quantifier can appear as *"there is"*, *"a"*, *some"*, and *"for some"*, particularly in a non-mathematical discourse.

Sometimes, recognizing the correct quantification in natural language can be quite tricky, or even confusing. Take for example: "a dog ate my homework", meaning *"some dog ate my homework"* (existential quantification) v. "a dog is an animal", meaning *"every* dog is an animal" (universal quantification). Alternatively, the well-known expression[5] "All that glitters is not gold" is actually meant to mean "Not all that glitters is gold", something logically quite different, as we will see in Section 3.4. So, watch out!

3.1.3   Terms and formulae

Using the symbols in a given first-order language and following certain common syntactic rules, we can compose formal expressions which allow us to symbolically represent statements, to reason about them, and to prove them in a precise, well-structured, and logically correct way. There are two basic syntactic categories in a first-order language: **(first-order) terms** and **(first-order) formulae**.

3.1.3.1   Terms

Terms are formal expressions (think of algebraic expressions) built from constant symbols and individual variables, using functional symbols. Terms are used to denote specified or unspecified *individuals*, that is, elements of the domain.

Here is the formal inductive definition of **terms** in a first-order language $\mathcal{L}$.

Definition 79 (Terms) *Let $\mathcal{L}$ be any first-order language.*

1. *Every constant symbol in $\mathcal{L}$ is a term in $\mathcal{L}$.*

2. *Every individual variable in $\mathcal{L}$ is a term in $\mathcal{L}$.*

---

[4] First-order logic can be extended to *second-order logic*, where there are second-order variables and quantifiers ranging over sets, relations, and functions; then further to *third-order logic* with variables and quantifiers over more complex objects definable in second-order logic, etc. In this book we will not go beyond first-order logic.
[5] From Shakespeare's play *The Merchant of Venice*.

3. *If $t_1, \ldots, t_n$ are terms in $\mathcal{L}$ and f is an n-ary functional symbol in $\mathcal{L}$, then $f(t_1, \ldots, t_n)$ is a term in $\mathcal{L}$.*

We denote the set of terms of $\mathcal{L}$ by $\mathrm{TM}(\mathcal{L})$.

The set of variables occurring in a term t is denoted by $\mathrm{VAR}(t)$.

Terms that do not contain variables are called **constant terms** or **ground terms**.

Example 80 *Some examples of terms in the first-order languages for some of the structures we have seen include the following.*

1. *In the language $\mathcal{L}_{\mathcal{N}}$:*
   First, 0, $s(0)$, $s(s(0))$, etc. are constant terms. The term $s(\ldots s(0) \ldots)$, where s occurs n times, is denoted **n**. *That term is called the **numeral for** n[6].*
   We are less formal from this point onward, and allow ourselves to use the common, infix notation, as well as omitting outermost parentheses in terms whenever that does not affect the correct reading. With that in mind, other examples of terms include:
   - $+(\mathbf{2}, \mathbf{2})$, *which in a more familiar notation is written as* $\mathbf{2} + \mathbf{2}$;
   - $\times(\mathbf{3}, y)$, *written in the usual notation as* $\mathbf{3} \times y$;
   - $(x^2 + x) \times \mathbf{5}$, *where x^2 is an abbreviation of $x \times x$;*
   - $x_1 + s((y_2 + \mathbf{3}) \times s(z))$, *etc.*

2. *In the 'human' language $\mathcal{L}_{\mathcal{H}}$:*
   - Mary;
   - x;
   - m(John) *(meant to denote "the mother of John");*
   - f(m(y)) *(meant to denote "the father of the mother of y").*

The inductive definition of terms generates the respective principle of induction following the general scheme presented in Section 1.4.

Proposition 81 (Induction on terms) *Let $\mathcal{L}$ be any first-order language and $\mathfrak{P}$ a property of terms in $\mathcal{L}$, such that:*

1. *every constant symbol in $\mathcal{L}$ has the property $\mathfrak{P}$;*

2. *every individual variable in $\mathcal{L}$ has the property $\mathfrak{P}$;*

3. *if $t_1, \ldots, t_n$ are terms in $\mathcal{L}$ that have the property $\mathfrak{P}$ and f is an n-ary functional symbol in $\mathcal{L}$, then the term $f(t_1, \ldots, t_n)$ has the property $\mathfrak{P}$.*

Then every term in $\mathcal{L}$ has the property $\mathfrak{P}$.

Let us now illustrate definitions by recursion (recall Section 1.4) on the inductive definition of terms by formally defining for every term t the set of variables *VAR(t)* occurring in that term.

---

[6] The *term* **n**, which is a syntactic object, that is, just a string of symbols, must be distinguished from the *number* n, which is a mathematical entity.

Definition 82 (The set of variables in a term) *Let $\mathcal{L}$ be any first-order language. We define a mapping VAR : $\mathrm{TM}(\mathcal{L}) \to \mathcal{P}(VAR)$ recursively on the structure of terms in $\mathcal{L}$ as follows.*

1. *If t is a constant symbol in $\mathcal{L}$, then $VAR(t) = \emptyset$.*
2. *If t is the individual variable x in $\mathcal{L}$, in $\mathcal{L}$, then $VAR(t) = \{x\}$.*
3. *If $t = f(t_1, \ldots, t_n)$, where $t_1, \ldots, t_n$ are terms in $\mathcal{L}$ and f is an n-ary functional symbol in $\mathcal{L}$, then $VAR(t) = VAR(t_1) \cup \ldots \cup VAR(t_n)$.*

With every term we can associate a construction tree and a parsing tree in the same way as for propositional formulae. For instance, here is the parsing tree of the term $t = +(\times(x, +(\mathbf{1}, y)), \times(s(s(x)), y))$ (recall that $\mathbf{1} = s(\mathbf{0})$):

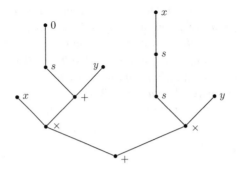

Respectively, for every term t we define the set $\mathrm{sub}(t)$ of **subterms** being all terms used in the construction of that term, that is, all terms with construction trees rooted as subtrees at nodes of the construction tree of t. The definition is by recursion on the inductive definition of terms. For instance, for the term t in the example above, $\mathrm{sub}(t) = \{t, \times(x, +(\mathbf{1}, y)), x, +(\mathbf{1}, y), \mathbf{1}, \mathbf{0}, y, \times(s(s(x)), y), s(s(x)), s(x)\}$.

3.1.3.2 Atomic formulae

By applying predicate symbols to terms we can build **atomic formulae**, the simplest first-order formulae. They have no internal logical structure and correspond to atomic propositions in propositional logic. They are formally defined as follows.

Definition 83 (Atomic formulae) *Let $\mathcal{L}$ be any first-order language. An **atomic formula** in $\mathcal{L}$ is any string $p(t_1, \ldots, t_n)$ where $t_1, \ldots, t_n$ are terms in $\mathcal{L}$ and p is an n-ary predicate symbol in $\mathcal{L}$.*
We denote the set of atomic formulae of $\mathcal{L}$ by $\mathrm{AFOR}(\mathcal{L})$.

Example 84 *Some examples of atomic formulae include the following:*

1. *In $\mathcal{L}_\mathcal{N}$:*
   - *$< (\mathbf{1}, \mathbf{2})$, or in traditional notation: $\mathbf{1} < \mathbf{2}$;*
   - *$x = \mathbf{2}$;*
   - *$\mathbf{5} < (x + \mathbf{4})$;*

- $2 + s(x_1) = s(s(x_2))$;
- $(x^2 + x) \times 5 = 0$;
- $x \times (y + z) = x \times y + x \times z$, *etc.*

2. *In* $\mathcal{L}_\mathcal{H}$:
   - $x = \mathsf{m}(\mathsf{Mary})$ *(meant to say "x is the mother of Mary").*
   - $\mathsf{L}(\mathsf{f}(y), y)$ *(meant to say "the father of y loves y").*

3.1.3.3   Formulae

Finally, using atomic formulae and logical connectives we can build **(compound) formulae** like in propositional logic, but now we can also use quantifiers.

The inductive definition of formulae of a first-order language $\mathcal{L}$ naturally extends the definition of propositional formulae as follows.

Definition 85 (Formulae) *Let $\mathcal{L}$ be any first-order language.*

1. *Every atomic formula in $\mathcal{L}$ is a formula in $\mathcal{L}$.*

2. *If A is a formula in $\mathcal{L}$ then $\neg A$ is a formula in $\mathcal{L}$.*

3. *If A and B are formulae in $\mathcal{L}$ then each of $(A \vee B)$, $(A \wedge B)$, $(A \to B)$, and $(A \leftrightarrow B)$ is a formula in $\mathcal{L}$.*

4. *If A is a formula in $\mathcal{L}$ and x is an individual variable, then each of $\forall x A$ and $\exists x A$ is a formula in $\mathcal{L}$.*
   We denote the set of formulae of $\mathcal{L}$ by $FOR(\mathcal{L})$.
   The set of variables occurring in a formula A is denoted $VAR(A)$.
   *Formulae that do not contain variables are called **ground formulae**.*

Example 86 *Some examples of formulae include the following.*

1. *In* $\mathcal{L}_\mathcal{Z}$:
   - $(5 < x \wedge x^2 + x - 2 = 0)$;
   - $\exists x(5 < x \wedge x^2 + x - 2 = 0)$;
   - $\forall x(5 < x \wedge x^2 + x - 2 = 0)$;
   - $(\exists y(x = y^2) \to (\neg x < 0))$;
   - $\forall x(\exists y(x = y^2) \to (\neg x < 0))$.

2. *In* $\mathcal{L}_\mathcal{H}$:
   - $\mathsf{John} = \mathsf{f(Mary)} \to \exists x \mathsf{L}(x, \mathsf{Mary})$;
   - $\exists x \forall z(\neg \mathsf{L}(z, y) \to \mathsf{L}(x, z))$;
   - $\forall x(\exists y(x = \mathsf{m}(y)) \to \exists z(\mathsf{L}(z, x)))$.

Think about the intended meaning of each of these formulae.

As in propositional logic, we will allow ourselves to omit parentheses wherever possible without resulting in ambiguity. For that purpose we will, again, impose a priority order among the logical connectives:

- the unary connectives – negation and quantifiers – have the strongest binding power, that is, the highest priority;

- then come the conjunction and disjunction;

- then the implication; and

- the biconditional has the lowest priority.

Example:

$$\forall x((\exists y(x = y^2)) \rightarrow ((\neg(x < \mathbf{0})) \vee (x = \mathbf{0})))$$

can be simplified to

$$\forall x(\exists y(x = y^2) \rightarrow \neg x < \mathbf{0} \vee x = \mathbf{0}).$$

On the other hand, sometimes we can use redundant parentheses in order to improve the readability.

With every first-order formula we can associate a construction tree and a parsing tree in the same way as for propositional formulae. The only differences are that the leaves of the construction/parsing tree of a formula are labeled with atomic formulae rather than propositional variables, and that internal nodes can also be labeled with pairs ⟨quantifier, individual variable⟩, such as $\forall x$ or $\exists x$, and these nodes have single successor nodes.

For every first-order formula A we define its **main connective** as the connective labeling the root of the construction/parsing tree of the formula, and the set $sub(A)$ of **subformulae** of A being all formulae with construction trees rooted as subtrees at nodes of the construction tree of A.

The inductive definition of first-order formulae generates the respective principle of induction, following the general scheme presented in Section 1.4.

Proposition 87 (Induction on formulae) *Let $\mathcal{L}$ be any first-order language and $\mathfrak{P}$ a property of formulae in $\mathcal{L}$, such that:*

1. *every atomic formula in $\mathcal{L}$ has the property $\mathfrak{P}$;*

2. *if A is a formula in $\mathcal{L}$ that has the property $\mathfrak{P}$, then $\neg A$ has the property $\mathfrak{P}$;*

3. *if A and B are formulae in $\mathcal{L}$ that have the property $\mathfrak{P}$, then each of $(A \vee B)$, $(A \wedge B)$, $(A \rightarrow B)$, and $(A \leftrightarrow B)$ has the property $\mathfrak{P}$;*

4. *if A is a formula in $\mathcal{L}$ that has the property $\mathfrak{P}$ and x is an individual variable, then each of $\forall x A$ and $\exists x A$ has the property $\mathfrak{P}$.*

Then every formula in $FOR(\mathcal{L})$ has the property $\mathfrak{P}$.

3.1.3.4   Unique readability of terms and formulae

Natural languages do not provide a reliable medium for precise reasoning because they are ambiguous: the same phrase or sentence may have several different – yet grammatically correct – readings, and therefore several different meanings. Eliminating ambiguity is one of the main reasons for using formal logical languages instead. In particular, the first-order languages introduced here are unambiguous in their formal syntax. In particular, it can be proved (although the proof is long and somewhat tedious) that the terms and formulae of any first-order language $\mathcal{L}$ have the **unique readability property** that every term or formula has an essentially unique construction tree, respectively parsing tree, up to the order of listing of the successor nodes. More precisely, for any first-order language $\mathcal{L}$ the following hold.

1. Every occurrence of a functional symbol in a term t from $TM(\mathcal{L})$ is the beginning of a unique subterm of t.

2. Every occurrence of a predicate symbol $\neg$, $\exists$, or $\forall$ in a formula A from $FOR(\mathcal{L})$ is the beginning of a unique subformula of A.

3. Every occurrence of any binary connective $\circ \in \{\wedge, \vee, \rightarrow, \leftrightarrow\}$ in a formula A from $FOR(\mathcal{L})$ is in a context $(B_1 \circ B_2)$ for a unique pair of subformulae B_1 and B_2 of A.

3.1.3.5 First-order instances of propositional formulae

Definition 88 *Given a propositional formula A, any uniform substitution of first-order formulae for the propositional variables in A produces a first-order formula, called a **first-order instance of** A.*

For example, substituting $(\mathbf{5} < x)$ for p and $\exists y(x = y^2)$ for q in the propositional formula

$$(p \wedge \neg q) \rightarrow (q \vee p)$$

produces its first-order instance

$$((\mathbf{5} < x) \wedge \neg \exists y(x = y^2)) \rightarrow (\exists y(x = y^2) \vee (\mathbf{5} < x)).$$

3.1.3.6 Many-sorted first-order structures and languages

Often the domain of discourse involves different **sorts** of objects, for example: integers and reals; scalars and vectors; or points, lines, triangles, and circles.

The notions of first-order structures and languages can be extended naturally to **many-sorted structures** and languages, with suitable inter-sort and cross-sort functions and predicates. An example is a two-sorted language for a simple geometric reasoning in the plane involving two sorts, one for points and another for lines. Each sort has its own domain and these domains are disjoint (no point is a line). The only non-logical symbol is a binary relation of *incidence* between a point and a line.

Using several sorts for individuals comes with a technical overhead, as it requires different sorts of individual variables and various syntactic restrictions. Instead, in this book we will use unary predicates to identify the different sorts within a universal domain.

References for further reading

For many more examples (mostly mathematical) and further discussion of first-order structures and languages, see Tarski (1965), Kalish and Montague (1980), van Dalen (1983), Hamilton (1988), Ebbinghaus *et al.* (1996), Barwise and Echemendy (1999), Enderton (2001), Hodges (2001), Hedman (2004), Nederpelt and Kamareddine (2004), Bornat (2005), Chiswell and Hodges (2007), Ben-Ari (2012), and van Benthem *et al.* (2014).

Exercises

3.1.1 🔵 Define suitable first-order structures and introduce appropriate first-order languages for the following types of mathematical structures: graphs, groups, rings, matrices, and vector spaces.

3.1.2 🔲 Define suitable first-order structures and introduce appropriate first-order languages for the following data types: Booleans, strings, lists, and trees.

3.1.3 Which of the following are syntactically correct terms in the language $\mathcal{L}_N$? Re-write them in the usual infix notation.

(a) $\times(x, y)$

(b) $\times(2, x, y)$

(c) $+xy$

(d) $+(5, \times(x, +(3, y))$

(e) $+(\times(x, +(3, y)), \times(+(3, y), x))$

(f) $\times(s(\times(2, x), +(3, y)))$

(g) $s(\times(\times(2, x), +(3, s)))$

(h) $s(\times(\times(2, x), +(3, y)))$

(i) $\times(s(\times(2, x)), s(+(s(3), y)))$

(j) $s(\times(s(s(x)), s(s(s(2)))), s(2))$

3.1.4 Construct the parsing tree of each of the syntactically correct terms from the previous exercise and determine the set of its subterms.

3.1.5* Prove that every occurrence of a functional symbol in a term t is the beginning of a unique subterm of t. (Hint: use induction on the definition of terms and the fact that a subterm of a subterm is a subterm.)

3.1.6 Which of the following are syntactically correct formulae in the language $\mathcal{L}_N$? Assume that terms are written in the usual infix notation and parentheses can be omitted whenever possible or added for improved readability.

(a) $\neg \forall x (x = 0)$

(b) $\forall \neg x (x = 0)$

(c) $\forall x (\neg x = 0)$

(d) $\forall x (x\neg = 0)$

(e) $\forall x (x = \neg 0)$

(f) $\neg \forall x (\neg x = 0)$

(g) $\neg \exists x (\neg x = 0.5)$

(h) $\forall x (x \lor \neg x)$

(i) $\forall x \forall y ((x + y) \rightarrow (y + x))$

(j) $\forall x ((x < 0) \rightarrow \forall x (x > 0))$

(k) $\forall x ((x + 0) \rightarrow \forall x (x > 0))$

(l) $\forall x (\forall x (x > 0) \rightarrow x < 0)$

(m) $\forall y (x < x \rightarrow \forall x (x < x))$

(n) $\forall y ((y < y) \neg \forall x (x < x))$

3.1.7 Construct the parsing tree of each of the syntactically correct formulae from the previous exercise and determine its main connective and the set of its subformulae.

3.1.8* Prove that every occurrence of a predicate symbol, $\neg$, $\exists$, or $\forall$, in a formula A from $FOR(\mathcal{L})$ is the beginning of a unique subformula of A. (Hint: use induction on the definition of formulae and the fact that a subformula of a subformula is a subformula.)

3.1.9* Prove that every occurrence of a binary connective $\circ \in \{\land, \lor, \rightarrow, \leftrightarrow\}$ in a formula A is in a context $(B_1 \circ B_2)$ for a unique pair of subformulae B_1 and B_2 of A. (Hint: use induction on the definition of the formula A.)

3.1.10   Determine the first-order instances of the propositional formula

$$A = (\neg p \rightarrow (q \vee \neg(p \wedge q)))$$

where:
(a) $\neg(f(x) = y)$ is substituted for p and $\exists x(x > y)$ is substituted for q; and
(b) $\neg(x = y)$ is substituted both for p and q.

3.1.11   Is the formula $(\neg(x = y) \rightarrow (\exists x(x > y) \vee \neg((y = x) \wedge \exists x(x > y))))$ a first-order instance of the propositional formula A from the previous exercise?

Friedrich Ludwig Gottlob Frege (8.11.1848–26.7.1925) was a German mathematician, logician, and philosopher who conducted groundbreaking studies on philosophy of mathematics, logic, and language. He is one of the founders of modern first-order logic, where his contribution is regarded by many as the most important development in logic since Aristotle.

Frege's first major scientific work, which was also the most influential in the future development of mathematical logic, was his book *Begriffsschrift, eine der Arithmetischen Nachgebildete Formelsprache des reinen Denkens* (*Concept Notation, a Formal Language for Pure Thought, Modeled on that of Arithmetic*) published in 1879. In that book he systematically developed a system of logic with rather unusual notation which is very close in spirit to modern-day first-order logic (also known as "predicate calculus"). He explicitly introduced quantifiers, quantification, and logical formulae, and formalized the notion of a "proof" in a way that is still used today. Frege had realized that natural everyday language is often imprecise and ambiguous, and therefore unreliable for performing rigorous mathematical reasoning. He wanted to develop a precise, unambiguous language for expressing mathematical statements and a formal logical system for mathematical reasoning. Frege had a *logicistic view*, that the whole of mathematics should be reducible to logic. He stated in the preface to *Begriffsschrift* that he wanted to prove all basic results in arithmetic "by means of pure logic." He pursued this idea further in his next major work, *The Foundations of Arithmetic*, published in 1884, and made huge achievements in formalizing much arithmetic in purely logical terms.

The reaction to Frege's work by the mathematicians and philosophers of his time was disappointingly weak and shallow, as very few of them were able to understand his groundbreaking ideas. Particularly devastating for Frege was the criticism of Cantor, even though Cantor's views on the foundations of mathematics were very close to Frege's philosophical views. Nevertheless, Frege went on developing and elaborating his philosophy of logic and mathematics and was about to publish the 2nd volume of *The Foundations of Arithmetic* when, in 1902, he received a letter from the young philosopher and mathematician Bertrand Russell. Russell pointed out a contradiction in Frege's system of axioms, revealed by Russell's paradoxical set of all sets that are not elements of themselves. This was a heavy blow to Frege, who attempted (as it turned out later, unsuccessfully) to fix the problem in an appendix to his book by amending one of his axioms, but he was so strongly shaken by that discovery that he stopped working on that project and never published the planned 3rd volume. Frege never succeeded in reducing the entire mathematics to logic. Much later, in the early 1920s, he came to the conclusion that this goal was impossible and decided, instead, that the whole of mathematics should be based on geometry, but never managed to develop his ideas until the end of his life.

Frege strongly influenced the views and work of many prominent followers, including Peano, Wittgenstein, Husserl, Carnap, and Russell. After his death his revolutionary works in the philosophy of mathematics, logic, and language gradually received due recognition and have since been regarded as being of major scientific importance.

3.2   Semantics of first-order logic

We now consider an arbitrarily fixed first-order language $\mathcal{L}$. The formulae of $\mathcal{L}$ are meant to express statements about structures "matching" that language $\mathcal{L}$. The meaning of the formulae is relative to the given structure of discourse and is "computed" compositionally following the structure of the formula, the values of the occurring variables, and the meaning of the logical and non-logical symbols occurring in it. The precise meaning of logical formulae is determined by their formal **semantics**, described in this section.

3.2.1   The semantics of first-order logic: an informal outline

In order to determine the meaning of the statement expressed by a given formula of our first-order language $\mathcal{L}$ we first need to fix its **interpretation**. It is a first-order structure – the **structure of discourse** – that corresponds to the language $\mathcal{L}$, in a sense that all distinguished functions, predicates, and constants in the structure should have names in $\mathcal{L}$, that is, respective functional, predicate, and constant symbols. Conversely, all non-logical symbols of the language $\mathcal{L}$ should correspond to explicitly defined functions, predicates, and constants in the structure.

Next, we need to *assign values* to the individual variables of the language, or at least to those occurring in the formula (but only to those that are not quantified over). These values are elements of the structure of discourse, so every variable refers to the element which is its value.

As a running example, consider the following formula from the language $\mathcal{L}_{\mathcal{N}}$ of arithmetic: $\exists x (\neg (x < y + z) \land (z \times x < (x \times (y + 1)))) \rightarrow (x = z + 2))$. (What it says is not really important for our purposes.)

We consider the intended interpretation, namely the structure $\mathcal{N}$ with the standard meanings of $<, +, \times, \mathbf{1}, \mathbf{2}$. Take an assignment of values in $\mathcal{N}$ to the variables as follows: $x = 5, y = 3, z = 2$ (we are not interested in the values of the other variables).

Once the structure of discourse is fixed and the individual variables are assigned values in that structure, the value of every term in $\mathcal{L}$ can be computed step-by-step following the structure of the term, in the same way as evaluating arithmetic expressions using the addition and multiplication tables. Eventually, every term in $\mathcal{L}$ is evaluated as a unique element of the structure, and that element is its value.

In our example, the terms occurring in the formula are evaluated as follows: $y + z = 5$; $z \times x = 10$; $x \times (y + 1) = 20$; and $z + 2 = 4$.

We are now ready to "compute" the meaning of every atomic formula: all we need to do is apply the predicate in the formula to the values of the terms appearing as its arguments. That meaning is a truth value – true or false – just as for propositional logic. Once these truth values are computed, we can treat atomic formulae in the same way as propositional variables.

In our example, the atomic subformulae of the formula acquire truth values as follows: $x < (y + z)$ becomes $5 < 5$, the truth value of which is false; $z \times x < (x \times y + 1)$ becomes $10 < 20$, which has truth value true; and $x = (z + 2)$ becomes $5 = 4$, which has truth value false.

Once we have computed the truth values of the atomic formulae, we can compute the truth value of any first-order formula following the structure of that formula, in the same way as computing truth values of propositional formulae; the only difference is that we also now have to deal with the quantifiers.

Let us compute the truth values of the subformulae in our example. Since $x < (y + z)$ is false for the given assignment of values to the variables, we find that $\neg(x < y + z)$ is true. Likewise, since $z \times x < (x \times (y + 1))$ is true for that assignment, we find that $\neg(x < y + z) \wedge (z \times x < (x \times (y + 1)))$ is true. Hence, since $x = z + 2$ is false, $\neg(x < y + z) \wedge (z \times x < (x \times (y + 1))) \rightarrow (x = z + 2)$ is false.

We now come to the most essential new step: the quantifiers. This is where computing the meaning, that is, the truth value, of a formula becomes generally difficult. Conceptually, there is no problem as we understand quite well what the quantifiers intuitively mean:

- for the existential quantifier: $\exists x A$ is true if *there is* a possible value of the variable x in the given structure of discourse, that is, there *is* an element of that structure assigned as a value to x that renders the formula A true; and

- for the universal quantifier: $\forall x A$ is true if *every* possible value of the variable x in the given structure of discourse, that is, *any* element of that structure assigned as a value to x renders the formula A true.

In the long run, we have informally defined the notion of a formula A true in a structure $\mathcal{S}$ under a variable assignment v, denoted

$$\mathcal{S}, v \models A.$$

Well, that sounds simple and, in a sense, it *is* simple. The difficulty is only technical: the structure may have many (possibly infinitely many) elements, and checking that all of them (respectively, some of them) make the formula A true can be a long (possibly *infinitely* long) and hard task. In practice, we therefore *guess* a suitable value of x in the case of existential quantification, or come up with a *uniform* argument that works for all possible values of x in the case of universal quantification.

In order to determine the truth value of the entire formula in our example, we need to find out whether there is any natural number that, assigned as a value to x while y and z keep the same values as before, will render the subformula $\neg(x < y + z) \wedge (z \times x < (x \times (y + 1))) \rightarrow (x = z + 2)$ true. In this case it is not difficult to guess such a value; it is sufficient to find one that makes the antecedent of the implication false. For that, we can take any value that makes $\neg(x < y + z)$ false, that is, $x < y + z$ true, for instance $x = 0$. The formula $\exists x(\neg(x < y + z) \wedge (z \times x < (x \times (y + 1)))) \rightarrow (x = z + 2))$ is therefore true in $\mathcal{N}$.

Let us now change the existential quantifier into a universal quantifier. Is the formula $\forall x(\neg(x < y + z) \wedge (z \times x < (x \times (y + 1)))) \rightarrow (x = z + 2))$ still true in $\mathcal{N}$? To determine that, in principle we have to consider all natural numbers as possible values of x and determine the truth value of $\neg(x < y + z) \wedge (z \times x < (x \times (y + 1))) \rightarrow (x = z + 2)$ for each of them. However, this is not necessary in our case since we already know a value of x that makes that subformula false; we therefore find that $\forall x(\neg(x < y + z) \wedge (z \times x < (x \times (y + 1)))) \rightarrow (x = z + 2))$ is false in $\mathcal{N}$.

What if the existentially quantified formula was false or the universally quantified formula was true, however? Would we have to perform the infinitely many checks then? Usually a uniform argument can be found that proves the case. For instance, we can prove that $\forall x((x + 1)^2 = x^2 + 2x + 1)$ by straightforward calculations, or that $\exists x(x^2 + x + 1 = 0)$ is false by using some high-school algebra.

In general, however, this question is much deeper than it looks, so I will leave it there. I will only mention that it was proved in 1973 (by the collective effort of several logicians and mathematicians, notably Martin Davis, Yuri Matiyasevich, Hilary Putnam, and Julia Robinson) that there is no general algorithm that can answer such questions, even when the formula is a very simple one of the type $\exists x(P(x) = 0)$ where $P(x)$ is a polynomial with integer coefficients[7]. The truth of some formulae in the language $\mathcal{L}_\mathcal{N}$ or simple extensions of it has been settled with a formal proof only after hundreds of years of futile attempts of many great minds. One famous example is *Fermat's Last Theorem*, which can be stated by a simple one-line formula using exponentiation in $\mathcal{N}$:

$$\forall x \forall y \forall z \forall n((x \neq 0 \wedge y \neq 0 \wedge z \neq 0 \wedge n > 2) \rightarrow x^n + y^n \neq z^n).$$

The truth of other such formulae is still unsettled after centuries of attempts, for example the famous *Goldbach conjecture* which states that "*every even integer greater than 2 equals the sum of two prime natural numbers*"[8]. Formalizing this statement in $\mathcal{L}_\mathcal{N}$ is given later as an exercise. Although we (believe to) understand very well what this statement says, "computing" its formal "meaning", that is, its truth value, appears to be a formidable task.

This is the end of the quick and informal exposition. We now embark on the more detailed and formal version.

---

[7] This was a negative answer to Hilbert's Tenth Problem, one of the 23 famous open challenges to the mathematics of the 20th century posed by David Hilbert at the World Congress of Mathematics in 1900.

[8] A positive integer greater than 1 is prime if it has no other positive integer divisors but 1 and itself.

3.2.2 Interpretations of first-order languages

An interpretation of a first-order language $\mathcal{L}$ is a *matching* first-order structure $\mathcal{S}$, that is, a structure with a family of distinguished functions, predicates, and constants that correspond to (and match the respective numbers of arguments of) the non-logical symbols in $\mathcal{L}$.

Some first-order languages, like all those that we have considered so far, are designed for specific structures which are their intended or **standard interpretations**. Other first-order languages are designed for *classes of structures*. For instance, the first-order language containing one binary relational symbol R (plus equality) can be regarded as the language of directed graphs, where the intended interpretation of R in any directed graph is the edge relation in that graph. Likewise, the first-order language of (algebraic) groups contains the following non-logical symbols: one binary functional symbol $\circ$, with intended interpretation being the group operation; one unary functional symbol $'$, with intended interpretation being the inverse operation; and one constant symbol e, with intended interpretation the identity element.

Note, however, that every first-order language may have many *unintended* interpretations. For instance, the language for directed graphs can be interpreted in the domain of integers, with R interpreted as "divisible by", or in the domain of humans, with R interpreted as "is a friend of." Indeed, most of the unintended interpretations are practically meaningless. For instance, the language $\mathcal{L}_{\mathcal{H}}$ can be interpreted in the domain of integers where, for example, the functional symbol m is interpreted as "$\mathsf{m}(n) = 2n$", f is interpreted as "$\mathsf{f}(n) = n^5 - 1$", the unary predicate M is interpreted as "is prime", and the unary predicate W is interpreted as "is greater than 2012", the binary predicate symbols P, C, and L are interpreted respectively as "is greater than", "is divisible by", and "has the same remainder modulo 11", and the constant symbols John, Mary, Adam, Eve are interpreted respectively as the numbers $-17, 99, 0$, and 10. Of course, such unintended interpretations are not interesting, but they must be taken in consideration when judging whether a given first-order formula is *logically valid*, that is, true in *every* possible interpretation. That will be discussed later in Section 3.4 however, and we now come back to the meaning of first-order terms and formulae under a given interpretation.

Once a given first-order language $\mathcal{L}$ is interpreted, that is, a matching first-order structure $\mathcal{S}$ is fixed, the **value in** $\mathcal{S}$ of every term t from $TM(\mathcal{L})$ can be "computed" as soon as all individual variables occurring in t are assigned *values* in $\mathcal{S}$, that is, elements of $\mathcal{S}$. The meaning of every formula A in $FOR(\mathcal{L})$ can also then be "computed", just as for propositional logic, from the values of the terms and the interpretation of the predicate symbols occurring in A and the standard meaning of the logical connectives. The rules for computing this meaning determine the **semantics of first-order logic**. I will spell out these rules without going into more technical detail than is really necessary.

3.2.3   Variable assignment and evaluation of terms

As I discussed above, in order to compute the truth value of a formula in a given structure S we first have to *evaluate* the terms occurring in it, that is, determine the elements of S denoted by these terms. For that, we must first assign values in S to the individual variables by means of a **variable assignment** in S, which is a mapping $v : \text{VAR} \to |S|$ from the set of individual variables VAR to the domain of S.

The evaluation of a term is now done as for the evaluation of an algebraic expression that we know from primary school: starting with the values of the variables and constant symbols we systematically apply the functions in S which interpret the respective functional symbols occurring in the term. That is, when evaluating a term $f(u_1, \ldots, u_m)$, we first compute *recursively* the values of the arguments $u_1, \ldots, u_m$ and then apply the interpretation of f in S to these values.

Formally, due to the unique readability of terms, every variable assignment $v : \text{VAR} \to |S|$ in a structure S can be uniquely extended to a mapping $v^S : TM(\mathcal{L}) \to |S|$, called **term evaluation**, such that for every n-tuple of terms $t_1, \ldots, t_n$ and an n-ary functional symbol f:

$$v^S(f(t_1, \ldots, t_n)) = f^S(v^S(t_1), \ldots, v^S(t_n))$$

where f^S is the interpretation of f in S.

In this way, once a variable assignment v in the structure S is fixed, every term t in $TM(\mathcal{L})$ is evaluated as a unique element $v^S(t)$ of S (or just $v(t)$ when S is fixed), called **the value of the term t for the variable assignment v**.

The following proposition essentially says that the value of any term only depends on the assignment of values to the variables occurring in that term. We leave the proof as an easy exercise.

Proposition 89 *For any given first-order language $\mathcal{L}$, term $t \in \text{TM}(\mathcal{L})$, and an $\mathcal{L}$-structure S, if v_1, v_2 are variable assignments in S that assign the same values to all variables in $VAR(t)$, then $v_1^S(t) = v_2^S(t)$.*

Example 90

- Let v be a variable assignment in the structure $\mathcal{N}$ such that $v(x) = 3$ and $v(y) = 5$. Here is a step-by-step computation of the value of the term $s(s(x) \times y)$:
  $v^{\mathcal{N}}(s(s(x) \times y)) = s^{\mathcal{N}}(v^{\mathcal{N}}(s(x) \times y)) = s^{\mathcal{N}}(v^{\mathcal{N}}(s(x)) \times^{\mathcal{N}} v^{\mathcal{N}}(y)) = s^{\mathcal{N}}(s^{\mathcal{N}}(v^{\mathcal{N}}(x)) \times^{\mathcal{N}} v^{\mathcal{N}}(y)) = s^{\mathcal{N}}(s^{\mathcal{N}}(3) \times^{\mathcal{N}} 5) = s^{\mathcal{N}}((3+1) \times^{\mathcal{N}} 5) = ((3+1) \times 5) + 1 = 21.$
- Likewise, $v^{\mathcal{N}}(1 + (x \times s(s(2)))) = 13$.
- If $v(x) = $ "Mary" then $v^{\mathcal{H}}(\mathbf{f}(\mathbf{m}(x))) = $ "the father of the mother of Mary".

3.2.4   Truth of first-order formulae

Eventually, we want to formally define the notion of a **formula A being true in a structure S for a variable assignment** v, denoted

$$S, v \models A.$$

The definition will be *recursive*, following the inductively defined structure of the formula A.

3.2.4.1 Atomic formulae

We begin with the simplest case, where A is an atomic formula. The truth value of the atomic formula $p(t_1, \ldots, t_n)$ is determined by the interpretation of the predicate symbol p in $\mathcal{S}$, applied to the tuple of arguments $v^{\mathcal{S}}(t_1), \ldots, v^{\mathcal{S}}(t_n)$:

$$\mathcal{S}, v \models p(t_1, \ldots, t_n) \text{ iff } p^{\mathcal{S}} \text{ holds true for } v^{\mathcal{S}}(t_1), \ldots, v^{\mathcal{S}}(t_n).$$

If $\mathcal{S}, v \models p(t_1, \ldots, t_n)$ does not hold, we write $\mathcal{S}, v \not\models p(t_1, \ldots, t_n)$.

Example 91 *If L is a binary predicate symbol interpreted in $\mathcal{N}$ as "less than", and the variables x and y are assigned values as above, then:*

- $\mathcal{N}, v \models L(1 + (x \times s(s(2))), s(s(x) \times y))$ *iff*
  $L^{\mathcal{N}}((1 + (x \times s(s(2))))^{\mathcal{N}}, (s(s(x) \times y))^{\mathcal{N}})$
  iff $13 < 21$, *which is* true.
- *likewise*, $\mathcal{N}, v \models 8 \times (x + s(s(y))) = (s(x) + y) \times (x + s(y))$ *iff*
  $(8 \times (x + s(s(y))))^{\mathcal{N}} = ((s(x) + y) \times (x + s(y)))^{\mathcal{N}}$
  iff $80 = 81$, *which is* false.

3.2.4.2 Propositional connectives

The truth values propagate over the propositional connectives according to their truth tables, as in propositional logic:

- $\mathcal{S}, v \models \neg A$ iff $\mathcal{S}, v \not\models A$;
- $\mathcal{S}, v \models (A \wedge B)$ iff $\mathcal{S}, v \models A$ and $\mathcal{S}, v \models B$;
- $\mathcal{S}, v \models (A \vee B)$ iff $\mathcal{S}, v \models A$ or $\mathcal{S}, v \models B$;
- $\mathcal{S}, v \models (A \rightarrow B)$ iff $\mathcal{S}, v \not\models A$ or $\mathcal{S}, v \models B$;
- $\mathcal{S}, v \models (A \leftrightarrow B)$ iff $(\mathcal{S}, v \not\models A$ if and only if $\mathcal{S}, v \models B)$.

3.2.4.3 Quantifiers

Finally, the truth of formulae $\forall x A(x)$ and $\exists x A(x)$ is computed according to the meaning of the quantifiers and the truth values of A. We first define a technical notion. Given a structure $\mathcal{S}$ and two variable assignments v and v' in $\mathcal{S}$, we say that v' is an x-**variant of** v if v' coincides with v on every variable except possibly x. Equivalently, v' is an x-variant of v if there exists an element $a \in \mathcal{S}$ such that $v' = v[x := a]$, where $v[x := a]$ is obtained from v by redefining $v(x)$ to be a.

- $\forall x A(x)$ is true if *every object* a from the domain of $\mathcal{S}$, assigned as a value of x, *satisfies* (i.e., renders true) the formula A.
  Formally, $\mathcal{S}, v \models \forall x A(x)$ if $\mathcal{S}, v' \models A(x)$ for every variable assignment v' that is an x-variant of v.

- $\exists x\, A(x)$ is true if *there is an object* **a** from the domain of $\mathcal{S}$ which, assigned as a value of x, satisfies the formula A.

  Formally, $\mathcal{S}, v \models \forall x\, A(x)$ if $\mathcal{S}, v' \models A(x)$ for some variable assignment v' that is an x-variant of v.

3.2.4.4   Computing the truth of first-order formulae

We can now (at least theoretically) compute the truth of any first-order formula in a given structure for a given variable assignment, step-by-step, following the logical structure of the formula and applying recursively the respective truth condition for the main connective of the currently evaluated subformula.

It is not difficult to show that the truth of a formula in a given structure for a given variable assignment only depends on the assignment of values to the variables occurring in that formula. That is, if we denote the set of variables occurring in the formula A by $\mathrm{VAR}(A)$ and v_1, v_2 are variable assignments in $\mathcal{S}$ such that $v_1 \!\mid_{\mathrm{VAR}(A)} = v_2 \!\mid_{\mathrm{VAR}(A)}$, then

$$\mathcal{S}, v_1 \models A \quad \text{iff} \quad \mathcal{S}, v_2 \models A.$$

Still, note that the truth conditions for the quantifiers given above are not really practically applicable when the structure is infinite because they require taking into account *infinitely many* variable assignments. Evaluating the truth of a first-order formula in an infinite structure is, generally speaking, an infinite procedure. Nevertheless, we can often perform that infinite procedure as finite by applying uniform (yet *ad hoc*) arguments to the infinitely many arising cases.

Example 92 *Consider the structure $\mathcal{N}$ and a variable assignment v such that $v(x) = 0$, $v(y) = 1$, and $v(z) = 2$. The following then holds.*

- $\mathcal{N}, v \models \neg(x > y)$.

- *However, $\mathcal{N}, v \models \exists x(x > y)$. Indeed, $\mathcal{N}, v[x := 2] \models x > y$.*

- *In fact, $\mathcal{N}, v \models \exists x(x > y)$ holds for* any *assignment of value to y, and therefore $\mathcal{N}, v \models \forall y \exists x(x > y)$.*

- *On the other hand, $\mathcal{N}, v \models \exists x(x < y)$, but $\mathcal{N}, v \nvDash \forall y \exists x(x < y)$. Why?*

- *What about $\mathcal{N}, v \models \exists x(x > y \wedge z > x)$? This is false; there is no natural number between 1 and 2.*

- *However, for the same variable assignment in the structure of rationals $\mathcal{Q}$, we have that $\mathcal{Q}, v \models \exists x(x > y \wedge z > x)$.*

  Does this hold for every variable assignment in $\mathcal{Q}$?

3.2.5   Evaluation games

There is an equivalent, but somewhat more intuitive and possibly more entertaining, way to evaluate the truth of first-order formulae in given structures. This is done by playing a special kind of a two-player game, called **(formula) evaluation game**[9]. These games go back to Lorenzen's work in the 1950s (if not much earlier), but were first introduced explicitly for first-order logic by Henkin and Hintikka.

The two players are called the **Verifier** and the **Falsifier**[10].

The game is played on a given first-order structure $\mathcal{S}$, containing a variable assignment v, and a formula A, the truth of which is to be evaluated in the structure $\mathcal{S}$ for the assignment v. As suggested by the names of the players, the objective of Verifier is to defend and demonstrate the claim that $\mathcal{S}, v \models A$, while the objective of Falsifier is to attack and refute that claim.

The game goes in rounds and in each round exactly one of the players, depending on the current "game configuration", has to make a move according to rules specified below until the game ends. The current game configuration $(\mathcal{S}, w, C)$ consists of the structure $\mathcal{S}$, an assignment w in $\mathcal{S}$, and a formula C (the truth of which is to be evaluated in $\mathcal{S}$ for the assignment w). The **initial configuration** is $(\mathcal{S}, v, A)$. We identify every such game with its initial configuration.

At every round, the player to make a move as well as the possible move are determined by the main connective of the formula in the current configuration $(\mathcal{S}, w, C)$, by rules that closely resemble the truth definitions for the logical connectives.

The rules are as follows.

- If the formula C is atomic, the game ends.
  If $\mathcal{S}, w \models C$ then Verifier wins, otherwise Falsifier wins.

- If $C = \neg B$ then Verifier and Falsifier *swap their roles* and the game continues with the configuration $(\mathcal{S}, w, B)$. Swapping the roles means that Verifier wins the game $(\mathcal{S}, w, \neg B)$ iff Falsifier wins the game $(\mathcal{S}, w, B)$, and Falsifier wins the game $(\mathcal{S}, w, \neg B)$ iff Verifier wins the game $(\mathcal{S}, w, B)$.
  Intuition: verifying $\neg B$ is equivalent to falsifying B and *vice versa*.

- If $C = C_1 \wedge C_2$ then Falsifier chooses $i \in \{1, 2\}$ and the game continues with the configuration $(\mathcal{S}, w, C_i)$.
  Intuition: for Verifier to defend the truth of $C_1 \wedge C_2$ he should be able to defend the truth of *any* of the two conjuncts, so it is up to Falsifier to question the truth of either of them.

- If $C = C_1 \vee C_2$ then Verifier chooses $i \in \{1, 2\}$ and the game continues with the configuration $(\mathcal{S}, w, C_i)$.
  Intuition: for Verifier to defend the truth of $C_1 \vee C_2$, it is sufficient to be able to defend the truth of at least one of the two disjuncts; Verifier can choose which one.

- If $C = C_1 \rightarrow C_2$ then Verifier chooses $i \in \{1, 2\}$ and, depending on that choice, the game continues with the configuration $(\mathcal{S}, w, \neg C_1)$ or $(\mathcal{S}, w, C_2)$.
  Intuition: $C_1 \rightarrow C_2 \equiv \neg C_1 \vee C_2$.

- If $C = \exists x B$ then Verifier chooses an element $a \in \mathcal{S}$ and the game continues with the configuration $(\mathcal{S}, w[x := a], B)$.
  Intuition: by the truth definition of $\exists x B$, verifying that $\mathcal{S}, w \models \exists x B$ amounts to verifying that $\mathcal{S}, w[x := a] \models B$ for some suitable element $a \in \mathcal{S}$.

---

[9] Also known as **model checking game**.

[10] Also known by various other names, for example Proponent and Opponent, Eloise and Abelard, Eve and Adam. For the sake of convenience, and without prejudice, here we will assume that both players are male.

- If $C = \forall x B$ then Falsifier chooses an element $a \in S$ and the game continues with the configuration $(S, w[x := a], B)$.
  Intuition: by the truth definition of $\forall x B$, falsifying $S, w \models \forall x B$ amounts to falsifying $S, w[x := a] \models B$ for some suitable element $a \in S$.

It is easy to see that any formula evaluation game always ends after a finite number of steps; this is because the number of logical connectives in the formula in the current configuration strictly decreases after every move until an atomic formula is reached. It is also obvious from the rules that the game always ends with one of the players winning. Clearly the winner of such a game depends not only on the truth or falsity of the claim $S, v \models A$, but also on how well the players play the game; we assume they always play a best possible move. The game will therefore be won by the player who has a **winning strategy** for that game, that is, a rule that, for every possible configuration from which that player is to move assigns such a move, that he is guaranteed to eventually win the game, no matter how the other player plays. It is not quite obvious that one of the players is sure to have a winning strategy in every such game, but it follows from a more general result in game theory. That also follows from the following claim, which we state here without proof[11], relating the existence of a winning strategy to the truth of the formula in the initial configuration.

Theorem 93 *For every configuration (S, v, A):*

1. $S, v \models A$ *iff Verifier has a winning strategy for the evaluation game (S, v, A).*

2. $S, v \nvDash A$ *iff Falsifier has a winning strategy for the evaluation game (S, v, A).*

Example 94 *Consider the structure $\mathcal{N}$ and the variable assignment v such that $v(x) = 0$, $v(y) = 1$, and $v(z) = 2$.*

1. *Verifier has a winning strategy for the game $(\mathcal{N}, v, \forall y \exists x (x > y + z))$.*
   Indeed, the first move of the game is by Falsifier who has to choose an integer n, and the game continues from configuration $(\mathcal{N}, v[y := n], \exists x (x > y + z))$.
   Now, Verifier has to choose an integer m. For every given $n \in \mathcal{N}$ Verifier can choose $m = n + 3$, for example. He then wins the game $(\mathcal{N}, v[y := n][x := m], (x > y + z))$ because $n + 3 > n + 2$. Verifier therefore has a winning strategy for the game $(\mathcal{N}, v[y := n], \exists x (x > y + z))$, for any $n \in \mathcal{N}$.
   Hence, Verifier has a winning strategy for the game $(\mathcal{N}, v, \forall y \exists x (x > y + z))$.
   Therefore, $\mathcal{N}, v \models \forall y \exists x (x > y + z)$. In fact, the above winning strategy for Verifier is easy to generalize for any assignment of value to z, demonstrating that $\mathcal{N}, v \models \forall z \forall y \exists x (x > y + z)$.

2. *Verifier has a winning strategy for the game $(\mathcal{N}, v, \forall x (y < x \lor x < z))$.*
   Indeed, the first move in the game is by Falsifier who has to choose an integer n, and the game continues from configuration $(\mathcal{N}, v[x := n], (y < x \lor x < z))$ in which Verifier must choose one of the disjuncts $y < x$ and $x < z$.

---

[11] See references at the end of this section.

The strategy for Verifier is as follows: if Falsifier has chosen $n > 1$ then Verifier chooses the disjunct $y < x$ and wins the game $(\mathcal{N}, v[x := n], y < x)$; if Falsifier has chosen $n \leq 1$ then Verifier chooses the disjunct $x < z$ and wins the game $(\mathcal{N}, v[x := n], x < z)$.
Therefore, $\mathcal{N}, v \models \forall x(y < x \lor x < z)$.

3. *Falsifier has a winning strategy for the game $(\mathcal{N}, v, \forall x(x < z \rightarrow \exists y(y < x)))$.*
   Indeed, let Falsifier choose 0 in the first move. The game then continues from configuration $(\mathcal{N}, v[x := 0], (x < z \rightarrow \exists y(y < x)))$ and now Verifier is to choose the antecedent or the consequent of the implication.
   If Verifier chooses the antecedent, the game continues from configuration $(\mathcal{N}, v[x := 0], \neg(x < z))$ which is won by Falsifier because the game $(\mathcal{N}, v[x := 0], x < z)$ is won by Verifier (since $0 < 2$). If Verifier chooses the consequent of the implication, then the game continues from configuration $(\mathcal{N}, v[x := 0], \exists y(y < x))$ and Verifier chooses a suitable value for y. However, whatever $n \in \mathcal{N}$ Verifier chooses, he loses the game $(\mathcal{N}, v[x := 0][y := n], y < x)$ because $n < 0$ is false for every $n \in \mathcal{N}$.
   Thus, Verifier has no winning move after the first move of Falsifier choosing 0 as a value for x. Therefore $\mathcal{N}, v \not\models \forall x(x < z \rightarrow \exists y(y < x))$. Furthermore, the winning strategy for Falsifier in the game $(\mathcal{N}, v, \forall x(x < z \rightarrow \exists y(y < x)))$ is a winning strategy for Verifier in the game $(\mathcal{N}, v, \neg\forall x(x < z \rightarrow \exists y(y < x)))$. Therefore $\mathcal{N}, v \models \neg\forall x(x < z \rightarrow \exists y(y < x))$.

3.2.6   *Translating first-order formulae to natural language*

First-order formulae formalize statements about first-order structures, and these statements can be translated back into natural language. A formula is of course just a string of symbols, so a formal translation could simply consist of writing all symbols in words. However, that would be of no use for understanding the meaning of the formula. While evaluating the truth of the formula in a given structure formally does not require translating that formula to natural language and understanding its intuitive meaning, that is essential for practically carrying out the truth evaluation procedure. Indeed, the meaning of (the statement expressed by) a first-order formula in a given interpretation is closely related to its truth in that interpretation. In fact, it could be argued that to understand the logical meaning of a formula, spelled out in a natural language, and to determine its truth are the two sides of the same coin. (This is not really true; recall for instance Goldbach's conjecture.) In any case, a sensible translation of the formula would surely help the player who has a winning strategy in the evaluation game to come up with the right strategy, and therefore to establish the truth of the formula in the given interpretation.

Example 95 *Let us look at some examples of translating to natural language and evaluating the truth of first-order formulae, now interpreted in the structure of real numbers $\mathcal{R}$. Note that a good translation is usually not word-for-word, but it takes some polishing and rephrasing in the target language so that it eventually sounds natural and makes good sense.*

1.

$$\exists x(x = x^2)$$

—*"There is a real number which equals its square." (True, take $x = 0$.)*

2.

$$\forall x(x < \mathbf{0} \to x^3 < \mathbf{0})$$

—*"Every negative real number has a negative cube." (True.)*

3.

$$\forall x \forall y(xy > \mathbf{0} \to (x > \mathbf{0} \lor y > \mathbf{0}))$$

—*"If the product of (any) two real numbers is positive, then at least one of them is positive." (False: take $x = y = -1$.)*

4.

$$\forall x(x > \mathbf{0} \to \exists y(y^2 = x))$$

—*"Every positive real number is a square of a real number." (True: algebraic fact.)*

5.

$$\exists x \forall y(xy < \mathbf{0} \to y = \mathbf{0})$$

—*"There is a real number x such that, for every real number y, if xy is negative, then y is 0." (True or false?)*

Now, some examples of formulae in $\mathcal{L}_\mathcal{H}$:

1.

John $=$ f(Mary) $\land$ L(John, Mary)

—*"John is the father of Mary and he loves her."*

2.

$(\text{John} = \text{f(Mary)}) \to \exists x \text{L}(x, \text{Mary})$

—*"If John is the father of Mary then (there is) someone (who) loves Mary."*

3.

$\exists x \forall z(\neg \text{L}(z, y) \to \text{L}(x, z))$

—*"There is someone (x) who loves everyone who does not love y."*
(Note that y stands for an unspecified person here, so this is not a proposition.)

4.

$$\forall x(\exists y(x = \mathsf{m}(y)) \to \forall z(x = \mathsf{m}(z) \to \mathsf{L}(x, z)))$$

—*"Every mother loves all her children."*

References for further reading
For further discussion and more details on semantics of first-order logic, see Tarski (1965), Kalish and Montague (1980), Hamilton (1988), Ebbinghaus *et al.* (1996), Barwise and Echemendy (1999), Enderton (2001), Hodges (2001), Smith (2003), Hedman (2004), Bornat (2005), Chiswell and Hodges (2007), Ben-Ari (2012), and van Benthem *et al.* (2014).

For more on evaluation games, see Ebbinghaus *et al.* (1996), Hodges (2001), and van Benthem *et al.* (2014).

Exercises

3.2.1 Evaluate the occurring terms and determine the truth of the following atomic formulae in $\mathcal{R}$:
  (a) $\times(s(s(x)), \mathbf{2}) = \times(x, s(s(\mathbf{2})))$, where x is assigned value 3.
  (b) $+(\mathbf{5}, \times(x, +(\mathbf{3}, y))) > s(\times(\mathbf{3}, \times(\mathbf{2}, x)))$, where x is assigned the value $-1/2$ and y is assigned the value -5.
  (c) $(s(\times(\times(x, x), x))) = \times(s(x), +(\times(x, x), -(\mathbf{1}, x)))$, for each of the following values of x: 2, $\sqrt{2}$, π.

3.2.2 Show by induction on the definition of terms in any given first-order language $\mathcal{L}$ that if v_1, v_2 are variable assignments in $\mathcal{S}$ such that $v_1 \!\mid_{\mathrm{VAR}(t)} = v_2 \!\mid_{\mathrm{VAR}(t)}$, then $v_1^{\mathcal{S}}(t) = v_2^{\mathcal{S}}(t)$. The value of any term therefore only depends on the assignment of values to the variables occurring in that term.

3.2.3 Show by induction on the definition of formulae in any given first-order language $\mathcal{L}$ that, if v_1, v_2 are variable assignments in $\mathcal{S}$ such that $v_1 \!\mid_{\mathrm{VAR}(A)} = v_2 \!\mid_{\mathrm{VAR}(A)}$, then $\mathcal{S}, v_1 \models A$ iff $\mathcal{S}, v_2 \models A$. The truth value of any formula therefore only depends on the assignment of values to the variables occurring in that formula. (We will see later, in Proposition 98, that not all of these variables matter.)

3.2.4 Consider the structure $\mathcal{N}$ and a variable assignment v such that $v(x) = 1$, $v(y) = 2$, $v(z) = 3$. Determine the truth in $\mathcal{N}$ for v of each of the following formulae, using either the semantic definition or evaluation games. Many of these look similar, but you will see that they say different things:

  (a) $z > y \to x > y$
  (b) $\exists x(z > y \to x > y)$
  (c) $\exists y(z > y \to x > y)$
  (d) $\exists z(z > y \to x > y)$
  (e) $\forall y(z > y \to x > y)$
  (f) $\exists x \forall y(z > y \to x > y)$
  (g) $\exists x \forall z(z > y \to x > y)$
  (h) $\exists z \forall y(z > y \to x > y)$
  (i) $\exists z \forall x(z > y \to x > y)$
  (j) $\exists y \forall x(z > y \to x > y)$

(k) $\exists y \forall z(z > y \rightarrow x > y)$

(l) $\forall x \exists y(z > y \rightarrow x > y)$

(m) $\forall x \exists z(z > y \rightarrow x > y)$

(n) $\forall y \exists x(z > y \rightarrow x > y)$

(o) $\forall y \exists z(z > y \rightarrow x > y)$

(p) $\forall z \exists x(z > y \rightarrow x > y)$

(q) $\forall z \exists y(z > y \rightarrow x > y)$

(r) $\exists x \forall y \forall z(z > y \rightarrow x > y)$

(s) $\exists y \forall x \forall z(z > y \rightarrow x > y)$

(t) $\exists z \forall x \forall y(z > y \rightarrow x > y)$

(u) $\forall y \exists x \forall z(z > y \rightarrow x > y)$

(v) $\forall x \exists y \forall z(z > y \rightarrow x > y)$

(w) $\forall z \exists x \forall y(z > y \rightarrow x > y)$

(x) $\forall x \forall y \exists z(z > y \rightarrow x > y)$

(y) $\forall y \forall z \exists x(z > y \rightarrow x > y)$

(z) $\forall z \forall x \exists y(z > y \rightarrow x > y)$

3.2.5 Translate into English the following first-order formulae and determine which of them represent true propositions when interpreted in $\mathcal{R}$. Use either the semantic definition or evaluation games.

(a) $\neg \forall x(x \neq 0)$

(b) $\forall x(x^3 \geq x)$

(c) $\forall x(x = x^2 \rightarrow x > 0)$

(d) $\exists x(x = x^2 \wedge x < 0)$

(e) $\exists x(x = x^2 \rightarrow x < 0)$

(f) $\forall x(x > 0 \rightarrow x^2 > x)$

(g) $\forall x(x = 0 \vee \neg x + x = x)$

(h) $\forall x((x = x^2 \wedge x > 1) \rightarrow x^2 < 1)$

(i) $\forall x \forall y(x > y \vee y > x)$

(j) $\forall x \exists y(x > y^2)$

(k) $\forall x \exists y(x > y^2 \vee y > 0)$

(l) $\forall x(x \geq 0 \rightarrow \exists y(y > 0 \wedge x = y^2))$

(m) $\forall x \exists y(x > y \rightarrow x > y^2)$

(n) $\forall x \exists y(\neg x = y \rightarrow x > y^2)$

(o) $\exists x \forall y(x > y)$

(p) $\exists x \forall y(x + y = x)$

(q) $\exists x \forall y(x + y = y)$

(r) $\exists x \forall y(x > y \vee -x > y)$

(s) $\exists x \forall y(x > y \vee \neg x > y)$

(t) $\exists x \forall y(y > x \rightarrow y^2 > x)$

(u) $\exists x \forall y(x > y \rightarrow x > y^2)$

(v) $\exists x \exists y(xy = x + y)$

(w) $\forall x \exists y \forall z(xy = yz)$

(x) $\exists x \forall y \exists z((x + y)z = 1)$

(y) $\forall x \forall y(x > y \rightarrow \exists z(x > z \wedge z > y))$

(z) $\forall x \exists z \forall y(x > z \rightarrow z > y)$

3.2.6 Translate into English the following first-order sentences of the language $\mathcal{L}_\mathcal{N}$ and determine for each of them whether it is true when interpreted in the structure of the natural numbers $\mathcal{N}$. Justify your answers. Use either the semantic definition or evaluation games. (Note that this is an exercise in first-order logic, not in mathematics. No special knowledge of arithmetic, going beyond standard high school curriculum, is needed.)

(a) $\exists x(x > 5 \rightarrow \neg x > 6)$

(b) $\forall x \neg(x > 6 \vee \neg x > 5)$

(c) $\neg \exists x(x > 5 \wedge \neg x > 6)$

(d) $\forall x(\neg x > 4 \rightarrow \neg x > 3)$

(e) $\forall x(\neg x < 7 \rightarrow \neg x < 8)$

(f) $\forall x(((\neg x = 1) \wedge (\neg x = 31)) \rightarrow \forall y(\neg x \times y = 31))$

(g) $\forall x((\neg x > 31 \wedge \exists y(x \times y = 31 \wedge y < 31)) \rightarrow x < 31)$

(h) $\forall x \exists z(x < z \wedge \forall y(x < y \rightarrow (z < y \vee z = y)))$

(i) $\exists x(x > 0 \wedge \forall z(z > x \rightarrow \exists y(x < y \wedge y < z)))$

(j) $\neg \forall x(x > 0 \rightarrow \neg \exists z(z > x \wedge \forall y(x < y \rightarrow z \leq y)))$

(k) $\forall x \exists z(x < z \wedge \forall y(x < y \rightarrow z < y))$

(l) $(\forall x(\forall y(y < x \rightarrow P(y)) \rightarrow P(x)) \rightarrow \forall x P(x))$,

   where P is an uninterpreted (arbitrary) unary predicate.

(m) The same sentence as above, but interpreted in the structure of the integers $\mathcal{Z}$.

(n) $\exists x P(x) \rightarrow \exists x (P(x) \wedge \forall y (P(y) \rightarrow (x < y \vee x = y)))$,

 where P is an uninterpreted (arbitrary) unary predicate.

(o) The same sentence as above, but interpreted in the structure of the integers $\mathcal{Z}$.

(p) $(\exists x P(x) \wedge \forall x (P(x) \rightarrow x > 0)) \rightarrow \exists x (P(x) \wedge \forall y (P(y) \rightarrow (x < y \vee x = y)))$,

 where P is an uninterpreted (arbitrary) unary predicate.

3.2.7 Translate into English the following first-order sentences of the language $\mathcal{L}_{\mathcal{H}}$, and determine their truth in the domain of human beings.

(a) $\forall x ((\mathsf{M}(x) \wedge \exists z \mathsf{C}(z, x)) \rightarrow \forall z (\mathsf{C}(z, x) \rightarrow \mathsf{K}(x, z)))$

(b) $\forall x (\mathsf{W}(x) \rightarrow \forall z (\mathsf{C}(z, x) \rightarrow \mathsf{m}(z) = x))$

(c) $\neg \exists x (\mathsf{W}(x) \wedge \forall z (\mathsf{C}(z, x) \rightarrow \neg \mathsf{K}(x, z)))$

(d) $\exists x (\mathsf{W}(x) \wedge \forall z (\mathsf{C}(z, x) \rightarrow \neg \mathsf{m}(z) = x))$

(e) $\forall x (\mathsf{M}(x) \rightarrow \forall z (\mathsf{C}(z, x) \rightarrow \mathsf{f}(z) = x))$

(f) $\forall x (\exists y \, \mathsf{f}(y) = x \rightarrow \neg \exists z \, \mathsf{m}(z) = x)$

(g) $\forall x \exists y (\neg \mathsf{P}(y, x) \wedge \forall z (\mathsf{P}(x, z) \rightarrow \exists u (\mathsf{P}(u, z) \wedge \mathsf{P}(y, u))))$,

 where $\mathsf{P}(x, y)$ means "x is a parent of y."

3.2.8 Formalize and justify the (propositional) reasoning in the cartoon below.

Alfred Tarski (born Alfred Teitelbaum, 14.1.1901–26.10.1983) was a Polish–American logician and mathematician who made seminal contributions to many areas of mathematics including set theory, universal algebra, topology, geometry, universal algebra, and several branches of mathematical logic, including theories of truth, logical consequence and definability, algebraic logic, model theory, and metamathematics (proof theory), as well as in methodology of science.

At age of 22 Tarski became the youngest student awarded a PhD from the University of Warsaw. At that time he changed his religion from Jewish to Roman Catholic and his surname from Teitelbaum to Tarski to reaffirm his Polish identity, but also to make it possible to obtain an academic position which, at that time, was almost impossible for Jews in Poland. Even though Tarski did not have a university position in Warsaw and taught mostly in high schools there until 1939, he made important contributions to logic and set theory which brought him international recognition. In 1933 Tarski published his groundbreaking paper *The concept of truth in formalised languages* where he rigorously defined formal semantics of logical languages and the notion of truth of a logical formula. In 1936 Tarski published another seminal work, *On the concept of logical consequence*, where he precisely defined logical consequence to hold when the conclusion is true in every model in which all premises are true. Yet another work of fundamental importance was *A decision method for elementary algebra and geometry*, which he essentially completed in around 1930 but only published in the USA in 1948. In it Tarski showed, using the method of quantifier elimination, that the first-order theory of the real numbers under addition and multiplication is algorithmically decidable, one of the most important results on decidability of a mathematical theory. In 1936 he also produced his very popular and influential undergraduate textbook on logic, *Introduction to Logic and to the Methodology of Deductive Sciences*.

During Tarski's visit to the USA for a conference in 1939, the Nazi army invaded Poland and World War II began; Tarski could not return so remained in the USA, where he became a naturalized citizen in 1945. Since 1949 until his retirement in 1971 Tarski was a Professor of Mathematics at the University of California at Berkeley, where he organized his famous logic seminar and created a very strong research school. Famous for his extreme productivity and precision, both in his teaching and his scientific writings, he was also an energetic and charismatic teacher whose lectures were meticulously prepared and presented. Tarski was also an inspiring and demanding supervisor of 26 doctoral students, many of whom also became distinguished mathematicians. Tarski was notorious for his all-night working sessions with his students in his smoke-filled study in the basement of his home in Berkeley. He had a strong and colorful personality; he was a heavy smoker and drinker, a womanizer, very ambitious and self-righteous, yet very civilized and polite.

With his enormous legacy in scientific works, students, followers, and general scientific influence, Tarski is widely regarded as one of the greatest logicians of all times, along with Aristotle, Frege, and Gödel.

Wilfrid Augustine Hodges (b. 27.05.1941) is a British logician and mathematician, a prominent expert in classical logic and model theory, an enthusiastic explorer of the history of logic, particularly of Arabic logic, author of some of the most popular logic textbooks, and passionate popularizer of logic in general.

Hodges studied at Oxford during 1959–65, where he received degrees in both Literae Humaniores and Theology. Then he did doctoral studies in logic (model theory) with John Crossley, and in 1970 was awarded a doctorate. He then lectured in both Philosophy and Mathematics at Bedford College, University of London. In 1984 he moved to Queen Mary College, University of London, where he was a Professor of Mathematics until his retirement in 2008; he is now an Emeritus Professor.

Hodges has written several very influential textbooks in logic in a lively, rather informal, yet stimulating and thought-provoking style. His 1997 *Logic: An Introduction to Elementary Logic* is still one of the best popular expositions of the subject, while his 1993 monograph *Model Theory* and the 1997 version *A Shorter Model Theory* are top references in that major field of logic. His 2007 *Mathematical Logic* written with I. Chiswell has become one of the most appreciated textbooks on the subject. Hodges also developed compositional semantics for the so called Independence-Friendly logic.

Since the early 2000s Hodges has also been deeply involved in exploring and interpreting the work in logic and algorithmics of the Persian encyclopedic thinker of the Islamic Golden Age: Ibn Sinā (Avicenna).

Hodges was, *inter alia*, President of the British Logic Colloquium, of the European Association for Logic, Language, and Information, and of the Division of Logic, Methodology, and Philosophy of Science of the International Union of History and Philosophy of Science. In 2009 he was elected a Fellow of the British Academy.

3.3   Basic grammar and use of first-order languages

The best (and only) way to learn a new language is by using it, in particular, by practicing translation between that language and one that you know well. In this section I discuss how to use first-order languages. For that you will have to understand their basic grammar, but first let us look at some examples of translating statements from natural language to first-order languages.

3.3.1   Translation from natural language to first-order languages: warm-up

1. Let us start with translating

   "There is an integer greater than 2 and less than 3."

This is surely false but now we are not concerned with its truth, only with its translation to $\mathcal{L}_{\mathcal{Z}}$. It is immediate:

$$\exists x (x > \mathbf{2} \wedge x < \mathbf{3}).$$

Note that $\mathbf{2}$ and $\mathbf{3}$ here are *terms*, not the numbers 2 and 3.

2. Does the same translation work for $\mathcal{L}_{\mathcal{Q}}$? No, because the quantification is over *all elements of the domain of discourse*, so the same formula would say in $\mathcal{L}_{\mathcal{Q}}$ "There is a rational number greater than 2 and less than 3", which is true, but not what we wanted to say. So, what now? If we want to refer to *integers* while the domain of discourse is a larger set, namely all rational numbers, we need to restrict somehow the scope of quantification. A standard method of doing that is to extend the language with an additional unary predicate, say I, which states that the element of the domain is of the type we want to quantify over, in this case *an integer*. Now, the translation is easy:

$$\exists x (I(x) \wedge x > \mathbf{2} \wedge x < \mathbf{3}).$$

We will come back to this trick later again.

3. Let us now translate to $\mathcal{L}_{\mathcal{R}}$

"There is no real number, the square of which equals -1."

How about

$$\exists x (\neg x^2 = -1)?$$

Although the sentence begins with "There is ...", do not be confused: this is not an existential statement, but a negation of such. It actually states

"*It is not true that* there is a real number the square of which equals -1",

so the correct translation is

$$\neg \exists x (x^2 = -1).$$

3.3.2 Restricted quantification

Take again the sentence

"Every man loves a woman."

In order to formalize it in the language of first-order logic, we need to conduct some preparation. It looks deceptively simple, yet it presents a problem we have seen before: the quantifiers in our formal language range over the *whole* domain of discourse, in our case over all human beings. So, whenever we write $\forall x$, in this context it will mean "every human being", not "every man"; likewise, using $\exists x$ we say "there is a human being", but we cannot say "there is a woman." On the other hand, we usually quantify not over all individuals from the domain, but over a specified family of them, for example "there is a *positive* integer x such that ...", "every *child* ...", etc. To resolve this problem we use (again) a little trick with the predicates, called **restricted quantification**. After thinking a little on the sentence we realize that it can be rephrased in the following way, which is somewhat awkward but more convenient to formalize in first-order logic: "**For every human, if he is a man then there is a human who is a woman and the man loves that woman**". Now the translation into $\mathcal{L}_{\mathcal{H}}$ is immediate:

$$\forall x (\mathsf{M}(x) \rightarrow \exists y (\mathsf{W}(y) \wedge \mathsf{L}(x,y))).$$

In general, we introduce unary predicates for the type of objects we want to quantify over and use logical connectives to express the *restricted quantification schemes*, as we did in the example above. For instance, in order to quantify over positive numbers in the domain of all integers, we can introduce a predicate $P(x)$, stating that "x **is a positive number**", and then write:

$$\exists x (P(x) \wedge \ldots x \ldots)$$

saying that there exists an object x *which is a positive number* and which satisfies $\ldots x \ldots$. Likewise,

$$\forall x (P(x) \rightarrow \ldots x \ldots)$$

says that all objects x, *which are positive numbers*, satisfy $\ldots x \ldots$.

Note, that sometimes we do not really need to introduce a new predicate in the language, if it is already *definable* there. For instance, in the latter case the predicate for (or the set of) positive numbers is definable in the language $\mathcal{L}_{\mathcal{Z}}$ on the structure $\mathcal{Z}$ as $0 < x$. The schemes $\exists x (0 < x \wedge \ldots x \ldots)$ and $\forall x (0 < x \rightarrow \ldots x \ldots)$ will therefore have the same effect as those above.

An alternative way to formalize restricted quantification is to use, as suggested earlier, *many-sorted domains*. For instance, in the domain of real numbers we can have sorts for integers, rational numbers, etc.; in the domain of humans, there are sorts for male, female, child, etc.. Respectively, we would need different and disjoint stocks of variables for each sort[12]. While the former method is more commonly used in informal reasoning, the latter is a more universal method. In usual mathematical practice however, the following convenient combination of the two methods is used. For the most important families of individuals we introduce some standard notation, for example the set of natural numbers is usually denoted **N**, the set of integers **Z**, the set of rational numbers **Q**, and the set of reals **R**. Now, if we want to quantify over integers, we can say: "*for all x in* **Z**", respectively " *there is x in* **Z** *such that*", or symbolically $\forall x \in \mathbf{Z}(\ldots x \ldots)$, and $\exists x \in \mathbf{Z}(\ldots x \ldots)$. For instance, $\forall x \in \mathbf{Z}(x^2 \geq x)$ states (truly) that the square of every *integer* is greater than or equal to that integer, while $\forall x \in \mathbf{Q} \exists z \in \mathbf{Z}(x < z)$ says that for every rational number there is an integer greater than it.

3.3.3 Free and bound variables, and scope of a quantifier

There are essentially two different ways in which we use individual variables in first-order formulae.

---

[12] In fact, this is a common practice in mathematical discourse and writings. For instance, it is a tradition to denote a real number x, y, z etc. while an integer is denoted i, j, k, n, m, a function f, g, h, etc. Of course, there is no inherent reason (apart from the risk of confusion) not to use f for an integer or j for a real number. When working with complex numbers, i of course becomes something quite different.

1. First, we use them to denote *unknown or unspecified objects*, as in

$$(5 < x) \vee (x^2 + x - 2 = 0).$$

We say that the variable x occurs **free** in that formula, or simply that x is a **free variable** in it. As long as a formula contains free variables, it cannot (in general) be assigned a truth value until these free variables are assigned values. A formula containing free variables therefore cannot be regarded as a proposition, for its truth may vary depending on the different possible values of the occurring free variables. For example, assigning value 1 or 6 to x in the formula above turns it into a true proposition in $\mathcal{R}$, while assigning value 2 to x turns it into a false proposition.

2. Second, we use individual variables in order *to quantify* over individuals, as in

$$\exists x((5 < x) \vee (x^2 + x - 2 = 0)) \quad \text{and} \quad \forall x((5 < x) \vee (x^2 + x - 2 = 0)).$$

In these formulae the variable x is said to occur **bound** (or simply, to be a bound variable) by the quantifier $\forall$ (i.e., **existentially bound**) in the formula on the left and, respectively, by $\exists$ (i.e., **universally bound**) in the formula on the right.

Note that the same variable can be *both free and bound* in a formula, for example x in the formula $0 < x \wedge \exists x(5 < x)$. That is why we talk about **free and bound occurrences** of a variable in a formula.

To make the notion of a bound occurrence of a variable more precise, note that every occurrence of a quantifier Q in a formula is the beginning of a *unique* subformula QxA, called the **scope** of that occurrence of the quantifier. For example, in the formula

$$\forall x((x > 5) \rightarrow \forall y(y < 5 \rightarrow (y < x \wedge \exists x(x < 3))))$$

the scope of the first occurrence of $\forall$ is the whole formula, the scope of the second occurrence of $\forall$ is the subformula $\forall y(y < 5 \rightarrow (y < x \wedge \exists x(x < 3)))$, and the scope of the occurrence of $\exists$ is the subformula $\exists x(x < 3)$.

Every bound occurrence of a variable x is bound by the *innermost* occurrence of a quantifier Q over x in which scope that occurrence of x lies. In other words, an occurrence of a variable x is bound by the first occurrence of the quantifier Q in a formula QxA, if and only if that occurrence of x is free in the subformula A. For example, in the formula above the first three occurrences of x are bound by the first occurrence of $\forall$, while the last two are bound by the occurrence of $\exists$.

I now define, for every formula in $A \in \text{FOR}(\mathcal{L})$, the set $\text{FVAR}(A)$ of all individual variables that are free in A, that is, that have a free occurrence in A. The definition is by recursion on the inductive definition of formulae in $\mathcal{L}$ (refer to Section 1.4).

Definition 96 (The set of free variables in a formula) *Let $\mathcal{L}$ be any first-order language. We define a mapping $\text{FVAR} : FOR(\mathcal{L}) \rightarrow \mathcal{P}(VAR)$ recursively on the structure of formulae in $\mathcal{L}$ as follows.*

1. *For every atomic formula $A = p(t_1, \ldots, t_n)$, where $t_1, \ldots, t_n$ are terms in $\mathcal{L}$ and p is an n-ary predicate symbol in $\mathcal{L}$, $\text{FVAR}(A) := VAR(t_1) \cup \ldots \cup VAR(t_n)$.*

2. *If A is a formula in $\mathcal{L}$ then $\text{FVAR}(\neg A) := \text{FVAR}(A)$.*

3. *If A and B are formulae in $\mathcal{L}$ then* $\mathrm{FVAR}(A \wedge B) := \mathrm{FVAR}(A) \cup \mathrm{FVAR}(B)$ *and the same holds for* $\mathrm{FVAR}(A \vee B), \mathrm{FVAR}(A \to B)$, *and* $\mathrm{FVAR}(A \leftrightarrow B)$.

4. *If A is a formula in $\mathcal{L}$ that has the property $\mathcal{P}$ and x is an individual variable, then* $\mathrm{FVAR}(\forall x A) = \mathrm{FVAR}(A) \setminus \{x\}$ *and* $\mathrm{FVAR}(\exists x A) = \mathrm{FVAR}(A) \setminus \{x\}$.

As an exercise, define in a similar way the set of *bound variables*, that is, variables with a bound occurrence $\mathrm{BVAR}(A)$ for every formula A.

Definition 97 *A formula with no bound variables is called an **open formula**. A formula with no free variables is called a **closed formula** or a **(first-order) sentence**.*

Once interpreted in a structure, first-order sentences have a determined meaning and represent propositions about that structure.

The following important observation allows us to only take into account the assignment of values to the free variables occurring in a formula in order to determine the truth of that formula.

Proposition 98 *The truth of a formula in a given first-order structure under a given variable assignment only depends on the assignment of values to the* free *variables occurring in that formula. That is, if v_1, v_2 are variable assignments in $\mathcal{S}$ such that $v_1 \,|_{\mathrm{FVAR}(A)} = v_2 \,|_{\mathrm{FVAR}(A)}$, then*

$$\mathcal{S}, v_1 \models A \ \text{ iff } \ \mathcal{S}, v_2 \models A.$$

In particular, if a variable x does not occur free in A, then $\mathcal{S}, v \models A$ if and only if $\mathcal{S}, v' \models A$ for every x-variant v' of v and, therefore, if and only if $\mathcal{S}, v \models \forall x A$.

3.3.4 *Renaming of a bound variable in a formula and clean formulae*

Note that a bound variable in a formula plays an auxiliary rôle and *does not have its own meaning* (it is also called a *dummy* variable) in the sense that we can replace it throughout the formula by another variable, not occurring in the formula, without altering the meaning of that formula. For example, it should be intuitively clear that $\exists x (5 < x \vee x^2 + x - 2 = 0)$ means exactly the same as $\exists y (5 < y \vee y^2 + y - 2 = 0)$; in particular, they are equally true. Likewise, $\forall x (5 < x \vee x^2 + x - 2 = 0)$ means the same as $\forall y (5 < y \vee y^2 + y - 2 = 0)$. On the other hand, $5 < x \vee x^2 + x - 2 = 0$ is *essentially different* from $5 < y \vee y^2 + y - 2 = 0$; depending on the values of the free variables x and y, one of these can be true while the other is false.

We should distinguish very well between the meaning and the use of free and bound variables. For example, the free and the bound occurrences of the variable x in the formula $(x > 5) \wedge \forall x (x < 2x)$ have *nothing* to do with each other.

Furthermore, different occurrences of the same variable can be bound by different quantifiers, and that may also be confusing.

Example 99 *Here are some examples of different uses of the same variable:*

- $\exists x (x > 5) \vee \forall x (2x > x)$. *Clearly, the occurrences of x, bound by the first quantifier, have nothing to do with those bound by the second.*

- $\exists x(x > 5) \wedge \exists x(x < 3)$. *Likewise, the two x's claimed to exist here need not (and, in fact,* cannot*) be the same, so this formula has the same meaning as* $\exists y(y > 5) \wedge \exists x(x < 3)$ *or* $\exists x(x > 5) \wedge \exists z(z < 3)$ *or* $\exists y(y > 5) \wedge \exists z(z < 3)$.
- $\forall x((x > 5) \rightarrow \exists x(x < 3))$. *Again, the occurrences of x in the subformula $\exists x(x < 3)$ are bound by $\exists$ and not related to the first two occurrences of x bound by $\forall$, so this formula has the same meaning as* $\forall x((x>5) \rightarrow \exists y(y<3))$ *or* $\forall z((z>5)\rightarrow\exists x(x<3))$ *or* $\forall z((z > 5) \rightarrow \exists y(y < 3))$.

The best way to avoid confusions like these in formal arguments and proofs is to *always use different variables for different purposes*. In particular, never use the same variable as both free and bound, or as bound by two different quantifiers in the same formula or proof.

Definition 100 *A formula A is **clean** if no variable occurs both free and bound in A and every two occurrences of quantifiers bind different variables.*

Definition 101 *The uniform replacement of all occurrences of a variable x bound by the same occurrence of a quantifier in a formula A with a variable not occurring in A is called a **renaming** of the variable x in A.*

Example 102 *The formula $\exists x(x > 5) \wedge \exists y(y < z)$ is clean, while $\exists x(x > 5) \wedge \exists y(y < x)$ and $\exists x(x > 5) \wedge \exists x(y < x)$ are not.*
The formula $(x > 5) \wedge \forall x(x > 5 \rightarrow \neg\exists x(x < y))$ is not clean, either.
One correct renaming of that formula is $(x > 5) \wedge \forall x(x > 5 \rightarrow \neg\exists z(z < y))$, but neither $(z > 5) \wedge \forall x((x > 5) \rightarrow \neg\exists x(x < y))$, nor $(x > 5) \wedge \forall z((z > 5) \rightarrow \neg\exists z(z < y))$, nor $(x > 5) \wedge \forall x(x > 5 \rightarrow \neg\exists y(y < y))$ is a correct renaming. (Why?)

Every formula can be transformed into a clean formula by means of several consecutive renamings of variables. For example, the formula

$$(x > 5) \wedge \forall x((x > 5) \rightarrow \neg\exists x(x < y))$$

can be transformed into the clean formula

$$(x > 5) \wedge \forall z_1((z_1 > 5) \rightarrow \neg\exists z_2(z_2 < y)).$$

3.3.5 Substitution of a term for a variable in a formula, and capture of a variable

Given a formula A, a variable x, and a term t, we can obtain a formula $A[t/x]$ by substituting *simultaneously* t for all *free* occurrences of x in A. The formula $A[t/x]$ is called **the result of substitution of t for x in A**. The semantic meaning of such substitution is to assign the value of the term t to the variable x when evaluating the truth of A. For instance, given the formula

$$A = \forall x(P(x, y) \rightarrow (\neg Q(y) \vee \exists y P(x, y))),$$

we have

$$A[f(y,z)/y] = \forall x(P(x, f(y,z)) \to (\neg Q(f(y,z)) \vee \exists y P(x,y))),$$

while

$$A[f(y,z)/x] = A,$$

because x does not occur free in A.

Intuitively, what the formula $A[t/x]$ is supposed to say about the individual denoted t is the same as what A says about the individual denoted by x. Is that always the case? No! An unwanted *capture* effect can occur after such a substitution if we substitute a term t for a variable x in a formula A, where x is in the scope of a quantifier over another variable y which occurs in t. For example, substituting $y + 1$ for x in the formula $\exists y(x < y)$, which is true in $\mathcal{N}$ for *every* value assigned to x, will produce the false sentence $\exists y(y + 1 < y)$ since the occurrence of y in the term $y + 1$ is *captured* by the quantifier $\exists y$; this has happened because the substitution mixed free and bound occurrences of y.

Similarly, in the formula $\forall x \exists z(x < z)$, meaning that for every number there is a greater one, renaming x with the variable z will produce $\forall z \exists z(z < z)$, which clearly distorts the meaning of the formula. The reason is that when substituting z for x it is *captured* by the quantifier $\exists z$.

Formally, **capture** happens when new occurrences of some variable, say y, are introduced in the scope of a quantifier Qy in a formula A as a result of substitution of a term t containing y for another variable x in that formula. The following definition is intended to disallow substitutions that cause capture.

Definition 103 *A term t is **free for (substitution for) a variable** x in a formula A, if no variable in t is captured by a quantifier as t is substituted for any free occurrence of x in A.*

Example 104

- *Every ground term (not containing variables), in particular every constant symbol, is free for substitution for any variable in any formula.*

- *More generally, every term that does not contain variables with bound occurrences in a given formula is free for substitution for any variable in that formula.*

- *The term $f(x,y)$ is free for substitution for y in the formula $A = \forall x(P(x,z) \wedge \exists y Q(y)) \to P(y,z)$, resulting in $A[f(x,y)/y] = \forall x(P(x,z) \wedge \exists y Q(y)) \to P(f(x,y),z)$.*

- *However, the same term is not free for substitution for z in A, resulting in $A[f(x,y)/z] = \forall x(P(x,f(x,y)) \wedge \exists y Q(y)) \to P(y, f(x,y))$, because a capture of the variable x occurs in the first occurrence of $f(x,y)$.*

To avoid the problems arising from capture mentioned above, from now on we only allow a substitution t/x in a formula A when t is free for x in A. Hereafter, when t is a constant symbol c, I will often write $A(c)$ instead of $A[c/x]$.

3.3.6 A note on renamings and substitutions in a formula

Note that renaming and substitution are *very different* operations; renaming always acts on *bound* variables, while substitution always acts on *free* variables.

Also, as we will see in Theorem 119, renamings preserve the formula up to logical equivalence while substitutions do not.

On the other hand, a suitable renaming of a formula can prepare it for a substitution by rendering the term to be substituted free for such substitution in the renamed formula. For instance, the term $f(x, y)$ is not free for substitution for y in

$$A = \forall x (P(x, y) \land \exists y Q(y)),$$

but it becomes free for such a substitution after renaming A, for example to:

$$A' = \forall x' (P(x', y) \land \exists y Q(y)).$$

References for further reading

For more examples and further discussion on grammar and use of first-order languages and translations to and from natural language, see Tarski (1965), Kalish and Montague (1980), Barwise and Echemendy (1999), Hodges (2001), Smith (2003), Nederpelt and Kamareddine (2004), Bornat (2005), Chiswell and Hodges (2007), and van Benthem et al. *et al.* (2014).

Exercises

3.3.1 Using the additional predicate $I(x)$ for "x is an integer", formalize the following sentences in the first-order language for the structure of real numbers. $\mathcal{R}$.

   (a) Every square of an integer is greater than 0.
   (b) Every square of a real number which is not an integer is greater than 0.
   (c) Some real numbers are not integers.
   (d) Every integer is even or odd.
   (e) No integer is both even and odd.
   (f) For every integer there is a greater integer.
   (g) Every positive integer is a square of some negative real number.
   (h) Not every real number is greater than an integer.
   (i) There is an integer such that not every real number is greater than it.
   (j) No real number is greater than every integer.
   (k) Every real number which is not zero has a reciprocal.
   (l) There is a real number which, when multiplied by any real number, produces that number.
   (m) Between every two different real numbers there is an integer.

3.3.2 Using unary predicates $T(x)$ for "x talks" and $L(x)$ for "x listens", formalize the following sentences in a first-order language for the domain of all humans.

   (a) Everybody talks or everybody listens.
   (b) Everybody talks or listens.
   (c) If John talks everybody listens.
   (d) If somebody talks everybody listens.

    (e) If somebody talks everybody else listens.

    (f) Nobody listens if everybody talks.

3.3.3 Use binary predicates $T_2(x, y)$ for "x talks to y" and $L_2(x, y)$ for "x listens to y" to formalize the following sentences in a first-order language for the domain of all humans. (Note the ambiguity of some of these.)

    (a) Not everybody talks to somebody.

    (b) Nobody listens to anybody.

    (c) John listens to everybody who talks to somebody.

    (d) John listens to nobody who talks to somebody else.

    (e) John talks to everybody who does not listen to him.

    (f) Somebody does not talk to anybody who does not listen to him.

    (g) Nobody listens to anybody who does not listen to him.

    (h) Not everybody listens if everybody who talks to somebody.

    (i) If everybody talks to somebody then nobody listens to anybody.

3.3.4 Translate the following sentences into the first-order language $\mathcal{L}_{\mathcal{H}}$ for the structure of all humans $\mathcal{H}$. Remember that the language $\mathcal{L}_{\mathcal{H}}$ does not have a unary predicate for " . . . is a child", but a binary predicate " . . . is a child of . . . ."

    (a) Some men love every woman.

    (b) Every woman loves every man who loves her.

    (c) Every man loves some woman who does not love him.

    (d) Some women love no men who love them.

    (e) Some men love only women who do not love them.

    (f) No woman loves a man who loves every woman.

    (g) Every woman loves her children.

    (h) Every child loves his/her mother.

    (i) Everyone loves a child. (*Note the ambiguity here* . . .)

    (j) A mother loves every child. (. . . *and here.*)

    (k) Every child is loved by someone.

    (l) Some woman love every child.

    (m) All children love at least one of their parents.

    (n) Some children love every man who loves their mother.

    (o) No child loves any mother who does not love all her children.

    (p) Not every man loves some woman who loves all his children.

3.3.5 Translate the following sentences into the first-order language $\mathcal{L}_{\mathcal{H}}$ for humans, with functions m for "the mother of" and f for "the father of"; the unary predicates M for "man" and W for "woman"; and the binary predicates P(x,y) meaning "x is a parent of y", K(x,y) meaning "x knows y", L(x,y) meaning "x likes y", and R(x,y) meaning "x respects y". Some of these exercises require the use of equality.

    (a) Every parent is respected by some mother.

    (b) Every human knows his/her mother.

    (c) Some people do not know any of their parents.

    (d) Every woman knows a man who is not her parent.

    (e) Some children do not respect any of their parents.

(f) Some fathers respect every mother whom they know.

(g) No woman likes a man who does not know all of her children.

(h) Not every woman likes no man who does not know some of her children.

(i) No man respects a woman who does not know all of her children.

(j) Nobody respects any man who does not respect both his parents.

(k) If somebody is a child then every mother likes him/her.

(l) No man, except Adam, knows Eve.

(m) Only possibly Adam loves every woman he knows.

(n) Not every woman loves exactly one man.

(o) Only John's mother loves him.

(p) Not only John's mother loves him. (*Note the implicature here!*)

(q) Henry Ford likes every color, as long as it is black.

(r) Any customer can have a car painted any color that he wants, so long as it is black. (Use suitable predicates here.)

3.3.6 Translate the following sentences into the first-order language (with equality) L_R for the structure of real numbers $\langle \mathbb{R}; <, +, \times, 0 \rangle$, with an additional unary functional symbol f and a unary predicate $Q(x)$ for x "*is a rational number.*"

(a) Every rational zero of the function f is positive.

(b) No irrational zero of the function f is non-negative.

(c) No negative real number is greater than every zero of the function f.

(d) Not every negative rational number is greater than every zero of f.

(e) Some positive zero of f is not less than every irrational zero of f.

(f) Every non-positive zero of f is less than some irrational zero of f.

3.3.7 Ⓣ Recall that a natural number is prime if it is greater than 1 and has no other positive integer divisors but 1 and itself. Express the following statements in the language $\mathcal{L}_N$. For some of these you need to use equality (see Section 3.4.4).

(a) x divides y, denoted $x|y$.

(b) The number (denoted) x is prime.

(c) There are infinitely many primes. (Hint: this is equivalent to saying that for every natural number there is a greater prime number.)

(d) (*Goldbach's conjecture*)
     Every even integer greater than 2 equals the sum of two primes.

(e) (*The twin primes conjecture*)
     There are infinitely many pairs of primes that differ by 2.

(f) Every natural number n has a prime divisor not greater than $\sqrt{n}$.
     (NB: the function $\sqrt{\cdot}$ is not in the language.)

(g) The numbers x and y have no common prime divisor.

(h) z is a greatest common divisor of x and y.

(i) Every two natural numbers x, y which are not both 0 have a greatest common divisor, denoted $\gcd(x, y)$.

(j) (*The fundamental theorem about division with remainder*)
     For any natural numbers $x > 0$ and y, there exist unique natural numbers q and r such that $y = q \times x + r$ and $0 \leq r < x$.

3.3.8 (π) Express the following statements in the language of the structure of real numbers $\mathcal{L}_\mathcal{R}$, using the predicate $I(x)$ to mean "x is an integer."

(a) The number (denoted) z is rational.

(b) Every rational number can be represented as an irreducible fraction.

(c) Between every two different real numbers there is a rational number.

(d) $\sqrt{2}$ is not a rational number.

(e) Every quadratic polynomial with real coefficients which has a non-zero value has at most two different real zeros.

3.3.9 Define by recursion on the inductive definition of formulae, for every formula in $A \in \mathrm{FOR}(\mathcal{L})$, the set $\mathrm{BVAR}(A)$ of all individual variables that are bound in A, that is, have a bound occurrence in A.

3.3.10 Determine the scope of each quantifier and the free and bound occurrences of variables in the following formulae where P is a unary and Q a binary predicate.

(a) $\exists x \forall z (Q(z,y) \vee \neg \forall y (Q(y,z) \to P(x)))$,

(b) $\exists x \forall z (Q(z,y) \vee \neg \forall x (Q(z,z) \to P(x)))$,

(c) $\exists x (\forall z Q(z,y) \vee \neg \forall z (Q(y,z) \to P(x)))$,

(d) $\exists x (\forall z Q(z,y) \vee \neg \forall y Q(y,z)) \to P(x)$,

(e) $\exists x \forall z (Q(z,y) \vee \neg \forall y Q(x,z)) \to P(x)$,

(f) $\exists x \forall z Q(z,y) \vee \neg (\forall y Q(y,x) \to P(x))$,

(g) $\exists x (\forall z Q(z,y) \vee \neg (\forall z Q(y,z) \to P(x)))$,

(h) $\exists x (\forall x (Q(x,y) \vee \neg \forall z (Q(y,z) \to P(x))))$,

(i) $\exists x (\forall y (Q(x,y) \vee \neg \forall x Q(x,z))) \to P(x)$.

3.3.11 Rename the bound variables in each formula above to obtain a clean formula.

3.3.12 Show that every formula can be transformed into a clean formula by means of several consecutive renamings of variables.

3.3.13 For each of the following formulae (where P is a unary predicate and Q is a binary predicate), determine if the indicated term is free for substitution for the indicated variable. If so, perform the substitution.

(a) Formula: $\exists x (\forall z P(y) \vee \neg \forall y (Q(y,z) \to P(x)))$; term: $f(x)$; variable: z.

(b) Same formula; term: $f(z)$; variable: y.

(c) Same formula; term: $f(y)$; variable: y.

(d) Formula: $\forall x ((\neg \forall y Q(x,y) \vee P(z)) \to \forall y \neg \exists z \exists x Q(z,y))$; term: $f(y)$; variable: z.

(e) Same formula; term: $g(x, f(z))$; variable: z.

(f) Formula: $\forall y (\neg (\forall x \exists z (\neg P(z) \wedge \exists y Q(z,x))) \wedge (\neg \forall x Q(x,y) \vee P(z)))$; term $f(x)$; variable: z.

(g) Same formula; term: $f(y)$; variable: z.

(h) Formula: $(\forall y \exists z \neg P(z) \wedge \forall x Q(z,x)) \to (\neg \exists y Q(x,y) \vee P(z))$; term: $g(f(z), y)$; variable: z.

(i) Same formula; term: $g(f(z), y)$; variable: x.

 **Bertrand Arthur William Russell** (18.5.1872–2.2.1970) was a famous British philosopher, logician, mathematician, intellectual, and political activist with strong pacifist views, who made seminal contributions to the foundations of mathematics and to the advancement of modern formal logic, also regarded as one of the founders of modern analytic philosophy.

Russell studied mathematics and philosophy at Trinity College in Cambridge, from where he graduated with distinction and later held academic positions there with several short interruptions, during which he worked at several universities in USA.

Russell's first major work was his book *The Principles of Mathematics* published in 1903, where he presented and discussed his famous paradoxical definition of the set of all sets that are not members of themselves, which became known as *Russell's paradox* (see details in Section 5.2.1). That discovery dealt a devastating blow to Cantor's set theory and to Frege's formalization of logic, and played a crucial role in the arising of a major crisis in the foundations of mathematics at the beginning of the 20th century. In this book and in the following three-volume *Principia Mathematica*, written with Whitehead and published during 1910–1913, Russell advanced the theory of *logicism* which aimed to found the entire field of mathematics on purely logical principles (one of the three main philosophical approaches for reforming the foundations of mathematics, prompted by the foundational crisis). According to the logicist theory, first put forward by Frege, all mathematical truths can be translated into logical truths and all mathematical proofs can be transformed into logical proofs; all mathematical results are therefore theorems of logic. In Russell's words, "The fact that all Mathematics is Symbolic Logic is one of the greatest discoveries of our age; and when this fact has been established, the remainder of the principles of mathematics consists in the analysis of Symbolic Logic itself." With these books and later philosophical works, Russell became immensely influential to many prominent philosophers and mathematicians, including his genius student Wittgenstein.

Russell was a staunch pacifist and, shortly after World War I, was convicted and imprisoned for 6 months for his anti-war views and speeches, where he wrote his extremely popular *Introduction to Mathematical Philosophy*. He pursued his anti-war activities and kept a high political profile, which brought him worldwide fame well beyond academic circles until his death in 1970 at age of 97.

Russell also had a very eventful personal life, with four marriages and a number of intermittent affairs. His often revolutionary views and writings on marriage, morals, and society were often controversial, leading, *inter alia*, to his professorship appointment at the City College New York being revoked in 1940. However, 10 years later Russell received the Nobel Prize for literature, in particular for his book *Marriage and Morals*.

In summary, Russell was one of the most influential logicians and philosophers of the 20th century and one of the intellectually strongest personalities of his time. He said of himself: "Three passions, simple but overwhelmingly strong, have governed my life: the longing for love, the search for knowledge, and unbearable pity for the suffering of mankind."

3.4   Logical validity, consequence, and equivalence in first-order logic

Here I introduce and discuss the fundamental logical notions of logical validity, consequence, and equivalence for first-order formulae.

3.4.1   More on truth of sentences in structures: models and counter-models

Recall that a sentence is a formula with no free variables.

As we noted earlier, the truth of a sentence in a given structure *does not depend on the variable assignment*. For a structure S and sentence A we can simply write

$$S \models A$$

if $S, v \models A$ for *any* (hence, *every*) variable assignment v.

We then say that S **is a model of** A and that A **is true in** S, or that S **satisfies** A or A **is satisfied by** S. Otherwise we write $S \nvDash A$ and say that A **is false in** S, S **does not satisfy** A, or S **is a counter-model for** A.

True sentences express properties of structures.

Example 105 (Sentences expressing properties of structures)

- $R \models \forall x \forall y (x + y = y + x)$ *(commutativity of the addition of real numbers),*
  $R \models \forall x \forall y \forall z ((x + y) \times z = (x \times z) + (y \times z))$ *(distributivity of multiplication over addition of real numbers).*

- $N \models \forall x \exists y (x < y)$ *(for every natural number there is a larger one), while*
  $N \nvDash \forall x \exists y (y < x)$ *(not for every natural number is there a smaller one).*
  Hence, $N \models \neg \forall x \exists y (y < x)$*. However,* $Z \models \forall x \exists y (y < x)$*.*

- $Q \models \forall x \forall y (x < y \rightarrow \exists z (x < z \land z < y))$
  (the ordering of the rationals is dense), but
  $Z \nvDash \forall x \forall y (x < y \rightarrow \exists z (x < z \land z < y))$
  (the ordering of the integers is not dense).

- $Q \models \exists x (3x = 1)$*, but*
  $Z \nvDash \exists x (3x = 1)$ *(1/3 is a rational number, but is not an integer).*

- $R \models \forall x (x > 0 \rightarrow \exists y (y^2 = x))$ *(every positive real is a square of a real), but*
  $Q \nvDash \forall x (x > 0 \rightarrow \exists y (y^2 = x))$
  (not every positive rational number is a square of a rational number).

- $R \nvDash \forall x \forall y (xy > 0 \rightarrow (x > 0 \lor y > 0))$. *(For example,* $x = y = -1$*).*

- $R \models \exists x \forall y (xy < 0 \rightarrow y = 0)$*: true or false?*

- $H \models \forall x \forall y (\exists z (x = m(z) \land y = m(z)) \rightarrow x = y)$.
  What does this sentence mean? Is it true only for this interpretation?

- $N \models (\forall x (\forall y (y < x \rightarrow P(y)) \rightarrow P(x)) \rightarrow \forall x P(x))$,
  where P is any (uninterpreted) unary predicate.
  Why is this true? What does this sentence say about N? Is it also true in Z?

3.4.2   Satisfiability and validity of first-order formulae

A first-order formula A is:

- **satisfiable** if $S, v \models A$ for *some* structure S and *some* variable assignment v in S; or
- **(logically) valid**, denoted $\models \mathbf{A}$, if $S, v \models A$ for *every* structure S and *every* variable assignment v in S.

  If A is a sentence, the variable assignment in these definitions is not relevant and can be omitted.

  Validity of any formula can actually be reduced to validity of a sentence, as follows.

Definition 106 *Given a first-order formula A with free variables $x_1, \ldots, x_n$, the sentence $\forall x_1 \ldots \forall x_n A$ is called a **universal closure** of A.*

Note that, depending on the order of the quantified variables, a formula can have several universal closures; as we will see later in Section 3.4.6 they are all essentially equivalent. The following is an easy exercise to prove.

Proposition 107 *A formula A is logically valid if and only if any of its universal closures is logically valid.*

Example 108 *Some examples of logically valid and non-valid first-order formulae include the following.*

- *Every first-order instance of a tautology is logically valid.*
  For instance, $\models \neg\neg(x > 0) \rightarrow (x > 0)$ and $\models P(x) \vee \neg P(x)$, for any unary predicate P.
- *If $\models A$ and x is any variable then we also have $\models \forall x A$.*
  Thus, $\models \forall x(P(x) \vee \neg P(x))$ and $\models \forall x(P(y) \vee \neg P(y))$.
- *$\models \forall x(x = x)$, because of the fixed meaning of the equality symbol $=$.*
- *$\models \forall x \forall y (\exists z(x = f(z) \wedge y = f(z)) \rightarrow x = y)$, where f is any unary functional symbol, because of the very meaning of the notion of a function.*
- *The sentence $\exists x P(x)$ is not valid: take for instance $P(x)$ to be interpreted as the empty set in any non-empty domain. This will give a* counter-model *for that sentence. However, $\exists x P(x)$ is satisfiable; a model is any structure where the interpretation of $P(x)$ is non-empty.*
- *Likewise, the sentence $\forall x P(x) \vee \forall x \neg P(x)$ is not valid, but it is* satisfiable.
  (Find a model and a counter-model.)
- *The sentence $\exists x(P(x) \wedge \neg P(x))$ is not satisfiable. Why?*
- *$\models \exists x \forall y P(x, y) \rightarrow \forall y \exists x P(x, y)$.*
  It is a good exercise to check this from the definition; we will discuss it again later in Section 3.4.6
- *However, $\not\models \forall y \exists x P(x, y) \rightarrow \exists x \forall y P(x, y)$. Find a counter-model!*

3.4.3 Logical consequence in first-order logic

The notion of logical consequence is fundamental in logic. In propositional logic we defined logical consequence $A_1, \ldots, A_n \models B$ as *preservation of the truth from the premises to the conclusion.* The same idea applies in first-order logic but, instead of truth tables, we now use *truth in a structure.*

Definition 109 *A first-order formula B is a **logical consequence** from the formulae $A_1, \ldots, A_n$, denoted $A_1, \ldots, A_n \models B$, if for every structure S and variable assignment v in S that each satisfy the formulae $A_1, \ldots, A_n$, in S by v, the formula B is also satisfied in S by v. Formally:*

$$\text{If } S, v \models A_1, \ldots, S, v \models A_n \text{ then } S, v \models A.$$

In particular, if $A_1, \ldots, A_n, B$ are sentences, then $A_1, \ldots, A_n \models B$ means that B is true in every structure in which all $A_1, \ldots, A_n$ are true, that is, every model of each of $A_1, \ldots, A_n$ is also a model of B.

*If $A_1, \ldots, A_n \models B$, we also say that B **follows logically from** $A_1, \ldots, A_n$, or that $A_1, \ldots, A_n$ **logically imply** B.*

In particular, $\emptyset \models A$ iff $\models A$.

Some easy observations (left as exercises to prove) are listed below. As in propositional logic, the following are equivalent for the first-order logical consequence.

1. $A_1, \ldots, A_n \models B$.
2. $A_1 \wedge \ldots \wedge A_n \models B$.
3. $\models A_1 \wedge \cdots \wedge A_n \to B$.
4. $\models A_1 \to (A_2 \to \cdots (A_n \to B) \ldots)$.
   Logical consequence in first-order logic is therefore reduced to logical validity.

Example 110 *Here are some examples and non-examples of first-order logical consequences.*

1. *If $A_1, \ldots, A_n, B$ are propositional formulae such that $A_1, \ldots, A_n \models B$, and $A'_1, \ldots, A'_n, B'$ are first-order instances of $A_1, \ldots, A_n, B$ obtained by the same uniform substitution of first-order formulae for propositional variables, then $A'_1, \ldots, A'_n \models B'$.*
   For instance, $\exists x A, \exists x A \to \forall y B \models \forall y B$.

2. $\forall x A(x) \models A[t/x]$ *for any formula A and term t free for x in A.*
   Indeed, if $\forall x A(x)$ is true in a structure S for some assignment, then A is true for every possible value of x in S, including the value of t for that assignment.
   To put it in simple words: if everybody is mortal then, in particular, John is mortal, and the mother of Mary is mortal, etc.

3. $\forall x(P(x) \to Q(x)), \forall x P(x) \models \forall x Q(x)$. *Why?*
   (Note that this is not *an instance of a propositional logical consequence.)*

4. $\exists x P(x) \wedge \exists x Q(x) \not\models \exists x (P(x) \wedge Q(x))$.
   *Indeed, the structure $\mathcal{N}'$ obtained from $\mathcal{N}$ where $P(x)$ is interpreted as 'x is even' and $Q(x)$ is interpreted as 'x is odd' is a **counter-model**:*
   $\mathcal{N}' \models \exists x P(x) \wedge \exists x Q(x)$, *but* $\mathcal{N}' \not\models \exists x (P(x) \wedge Q(x))$.

Example 111 *Here we illustrate "semantic reasoning", based on the formal semantics of first-order formulae, for proving or disproving first-order logical consequences.*

1. *Show the logical validity of the following argument:*
   "If Tinkerbell is a Disney fairy and every Disney fairy has blue eyes, then some-one has blue eyes."

   Proof. Let us first formalize the argument in first-order logic by introducing predicates: $P(x)$ meaning "x is a Disney fairy", $Q(x)$ meaning "x has blue eyes", and a constant symbol c interpreted as "Tinkerbell."

   The argument can now be formalized as follows:

   1. $P(c) \wedge \forall x (P(x) \rightarrow Q(x)) \models \exists y Q(y)$.
      To show its validity take any structure $\mathcal{S}$ for the first-order language introduced above and any variable assignment v in $\mathcal{S}$. Now, suppose
   2. $\mathcal{S}, v \models (P(c) \wedge \forall x (P(x) \rightarrow Q(x))$.
      We have to show that:
   3. $\mathcal{S}, v \models \exists y Q(y)$.
      From the assumption (2) we have, by the truth definition of $\wedge$, that:
   4. $\mathcal{S}, v \models P(c)$ *and*
   5. $\mathcal{S}, v \models \forall x (P(x) \rightarrow Q(x))$.
      Let v' be a variable assignment obtained from v by redefining it on x as follows: $v'(x) = c^{\mathcal{S}}$. (Recall that $c^{\mathcal{S}}$ is the interpretation of the constant symbol c in $\mathcal{S}$.) Then:
   6. $\mathcal{S}, v' \models P(x)$.
      According to the truth definition of $\forall$, it follows from (5) that:
   7. $\mathcal{S}, v' \models P(x) \rightarrow Q(x)$.
      From (6) and (7) it follows that:
   8. $\mathcal{S}, v' \models Q(x)$.
      Now, let v'' be a variable assignment obtained from v by redefining it on y as fol-lows: $v''(y) = v'(x) = c^{\mathcal{S}}$. We then have:
   9. $\mathcal{S}, v'' \models Q(y)$.

   According to the truth definition of $\exists$, it follows from (9) that:

   (3) $\mathcal{S}, v \models \exists y Q(y)$, so we are done. ∎

2. *Prove that the following argument is not logically valid:*
   "If everything is black or white then everything is black or everything is white."

Proof. Again, we first formalize the argument in first-order logic by introducing predicates:

$B(x)$ *meaning "x* **is black***" and* $W(x)$ *meaning "x* **is white***."*

Now the argument can be formalized as follows:

$$\forall x(B(x) \lor W(x)) \models \forall x B(x) \lor \forall x W(x).$$

To falsify this logical consequence it is sufficient to find any counter-model, *that is, a structure* $\mathcal{S}$ *and a variable assignment* v *such that*
$\mathcal{S}, v \models \forall x(B(x) \lor W(x))$ *and* $\mathcal{S}, v \not\models \forall x B(x) \lor \forall x W(x).$

Note that the interpretations of the predicates B *and* W *need not have anything to do with black, white, or any colors in that structure. We therefore choose to take the structure* $\mathcal{S}$ *with domain the set* **Z** *of all integers, where the interpretation of* B *is the predicate* **Even***, where* **Even**(n) *means "n is even", and the interpretation of* W *is the predicate* **Odd***, where* **Odd**(n) *means "n is odd."*

We already know that the variable assignment is irrelevant here because there are no free variables in this argument, but we will nevertheless need it in order to process the truth definitions of the quantifiers. Consider any *variable assignment* v *in* $\mathcal{S}$. *Then take* any *variable assignment* v' *that differs from* v *possibly only on* x. *The integer* $v'(x)$ *is even or odd, therefore* $\mathcal{S}, v' \models B(x)$ *or* $\mathcal{S}, v' \models W(x)$, *hence (by the truth definition of* $\lor$):

1. $\mathcal{S}, v' \models B(x) \lor W(x)$.
   Thus, by the truth definition of $\forall$, *it follows that:*
2. $\mathcal{S}, v \models \forall x(B(x) \lor W(x))$.
   On the other hand, consider the variable assignment v_1 *obtained from* v *by redefining it on* x *as follows:* $v_1(x) = 1$. *Then:*
3. $\mathcal{S}, v_1 \not\models B(x)$.
   By the truth definition of $\forall$, *it follows that:*
4. $\mathcal{S}, v \not\models \forall x B(x)$.

Likewise, consider the variable assignment v_2 *obtained from* v *by redefining it on* x *as follows:* $v_2(x) = 2$. *Then:*

(3) $\mathcal{S}, v_2 \not\models W(x)$.

By the truth definition of $\forall$, *it follows that:*

(4) $\mathcal{S}, v \not\models \forall x W(x)$.

By the truth definition of $\lor$, *it follows that*

(5) $\mathcal{S}, v \not\models \forall x B(x) \lor \forall x W(x)$.

Therefore, by the truth definition of $\rightarrow$, *it follows from (2) and (5) that*

(6) $\mathcal{S}, v \not\models \forall x(B(x) \lor W(x)) \rightarrow (\forall x B(x) \lor \forall x W(x))$. ∎

As noted earlier, logical consequence in first-order logic satisfies all basic properties of propositional logical consequence. Further, some important additional properties related to the quantifiers hold. They will be used as rules of inference in deductive systems for first-order logic, so I give them more prominence here.

Theorem 112 *For any first-order formulae $A_1, \ldots, A_n, A, B$, the following hold.*

1. *If $A_1, \ldots, A_n \models B$ then $\forall x A_1, \ldots, \forall x A_n \models \forall x B$.*
2. *If $A_1, \ldots, A_n \models B$ and $A_1, \ldots, A_n$ are sentences, then $A_1, \ldots, A_n \models \forall x B$, and hence $A_1, \ldots, A_n \models \bar{B}$, where $\bar{B}$ is any universal closure of B.*
3. *If $A_1, \ldots, A_n \models B[c/x]$ where c is a constant symbol not occurring in $A_1, \ldots, A_n$, then $A_1, \ldots, A_n \models \forall x B(x)$.*
4. *If $A_1, \ldots, A_n, A[c/x] \models B$ where c is a constant symbol not occurring in $A_1, \ldots, A_n, A$, or B, then $A_1, \ldots, A_n, \exists x A \models B$.*
5. *For any term t free for substitution for x in A:*
   (a) $\forall x A \models A[t/x]$.
   (b) $A[t/x] \models \exists x A$.

The proofs of these are easy exercises using the truth definition.

3.4.4   Using equality in first-order logic

First-order languages (especially those used in mathematics) usually contain the **equality** symbol $=$, sometimes also called **identity**. This is regarded as a special binary relational symbol, and is always meant to be interpreted as the identity of objects in the domain of discourse.

Example 113 *The equality is a very useful relation to specify constraints on the size of the model, as the following examples in the first-order language with $=$ show.*

1. *The sentence*
$$\lambda_n = \exists x_1 \cdots \exists x_n \left(\bigwedge_{1 \leq i \neq j \leq n} \neg x_i = x_j \right)$$

   states that the domain has at least n elements.
2. *The sentence $\mu_n = \neg \lambda_{n+1}$ or, equivalently,*
$$\mu_n = \forall x_1 \cdots \forall x_{n+1} \left(\bigvee_{1 \leq i \neq j \leq n+1} x_i = x_j \right)$$

   states that the domain has at most n elements.
3. *The sentence $\sigma_n = \lambda_n \wedge \mu_n$ states that the domain has exactly n elements.*

The proofs of these claims are left as an exercise (Exercise 9 in Section 3.4.8).

Example 114 *The equality is often an indispensable relation to express important mathematical properties as shown in the following examples.*

1. *The sentences in Example 113 are easily relativized for every formula $A(x, \bar{z})$ containing, among others, the free variable x, to the subset of the domain consisting of the elements satisfying A. For instance, we can combine these to say things like "of all the students in the class, at most two scored distinctions in at least five exams each."*

2. *In particular, for any formula $A(x, \bar{z})$, the formula*

$$\exists! x A(x, \bar{z}) = (A(x, \bar{z}) \wedge \forall y (A(y, \bar{z}) \rightarrow x = y))$$

   states that there is a unique element in the domain of discourse satisfying A, for the current values of the parameters $\bar{z}$.

3. *The sentence $\forall x \exists! y (R(x, y))$ in the language with $=$ and a binary relational symbol R therefore states that the relation R is **functional**, that is, every element of the domain is R-related to a unique element.*

4. *The sentence $\forall x \forall y (f(x) = f(y) \rightarrow x = y))$ in the language with $=$ and a unary functional symbol f states that the function f is **injective**, that is (by contraposition), assigns different values to different arguments.*

Sometimes the equality is implicit or even hidden in natural language expressions such as "Everyone, except possibly John, understood the joke." It becomes explicit and readily translatable (exercise) to first-order logic when rephrased as "Everyone, who is not (equal to) John, understood the joke." Likewise, "No-one, but John and Mary, enjoyed the party" can be rephrased as "Everyone, who is not (equal to) John and not (equal to) Mary, enjoyed the party."

The following proposition captures the characteristic properties of the equality expressible in a first-order language.

Proposition 115 *The following sentences in an arbitrary first-order language $\mathcal{L}$ are logically valid:*

(Eq1) $x = x$;

(Eq2) $x = y \rightarrow y = x$;

(Eq3) $x = y \wedge y = z \rightarrow x = z$;

(Eq$_f$) $x_1 = y_1 \wedge \dots \wedge x_n = y_n \rightarrow f(x_1, \dots, x_n) = f(y_1, \dots, y_n)$ for n-ary functional symbol f in $\mathcal{L}$;

(Eq$_r$) $x_1 = y_1 \wedge \dots \wedge x_n = y_n \rightarrow (p(x_1, \dots, x_n) \rightarrow p(y_1, \dots, y_n))$ for n-ary predicate symbol p in $\mathcal{L}$.

However, these axioms cannot guarantee that the interpretation of any binary relational symbol $=$ satisfying them is equality, but only that it is a *congruence* (see Section 4.6.3) in the given structure.

Lastly, the following theorem states an important generalization of the equality axioms listed above stating that equal terms can be equivalently replaced for each other in any formula.

Theorem 116 (Equivalent replacement) *For any formula $A(x)$ and terms s, t free for x in A, the following holds:*

$$\models s = t \rightarrow A[s/x] \leftrightarrow A[t/x].$$

The proof is by structural induction on the formula $A(x)$, and is left as an exercise for the reader.

3.4.5 Logical equivalence in first-order logic

Logical equivalence in first-order logic is based on the same idea as in propositional logic: the first-order formulae A and B are logically equivalent if always one of them is true if and only if the other is true. This is defined formally as follows.

Definition 117 *The first-order formulae A and B are **logically equivalent**, denoted $A \equiv B$, if for every structure $\mathcal{S}$ and variable assignment v in $\mathcal{S}$:*

$$\mathcal{S}, v \models A \text{ if and only if } \mathcal{S}, v \models B.$$

In particular, if A and B are sentences, $A \equiv B$ means that every model of A is a model of B, and every model of B is a model of A.

The following theorem summarizes some basic properties of logical equivalence.

Theorem 118
1. $A \equiv A$.
2. *If $A \equiv B$ then $B \equiv A$.*
3. *If $A \equiv B$ and $B \equiv C$ then $A \equiv C$.*
4. *If $A \equiv B$ then $\neg A \equiv \neg B$, $\forall x A \equiv \forall x B$, and $\exists x A \equiv \exists x B$.*
5. *If $A_1 \equiv B_1$ and $A_2 \equiv B_2$ then $A_1 \circ A_2 \equiv B_1 \circ B_2$ where $\circ$ is any of $\wedge, \vee, \rightarrow, \leftrightarrow$.*
6. *The following are equivalent:*
   (a) $A \equiv B$;
   (b) $\models A \leftrightarrow B$; and
   (c) $A \models B$ and $B \models A$.

Theorem 119 *The result of renaming of any variable in any formula A is logically equivalent to A. Consequently, every formula can be transformed into a logically equivalent clean formula.*

Example 120 *Some examples of logical equivalences between first-order formulae are as follows.*

- *Any first-order instance of a pair of equivalent propositional formulae is a pair of logically equivalent first-order formulae. For example:*
  $\neg\neg\exists x Q(x, y) \equiv \exists x Q(x, y)$ *(being an instance of $\neg\neg p \equiv p$);*
  $\exists x P(x) \rightarrow Q(x, y) \equiv \neg\exists x P(x) \vee Q(x, y)$ *(being an instance of $p \rightarrow q \equiv \neg p \vee q$).*

- $\exists x(5< x \vee x^2 + x - \mathbf{2} =0) \equiv \exists y(\mathbf{5} < y \vee y^2 + y - \mathbf{2} = \mathbf{0})$, *as the formula on the right is the result of renaming of the variable x in the formula on the left.*
- $\neg\exists x P(x) \not\equiv \exists x \neg P(x)$. *For example,* "No student passed the exam" *should not be equivalent to* "There is a student who did not pass the exam."
- "The integer x is not less than 0" *is* not logically equivalent *to* "The integer x is greater than or equal to 0." *Mathematically these mean the same (due of the mathematical property of the ordering of integers called* trichotomy, *which arranges all integers in a line) but* not *because of logical reasons. Likewise, 2+2=4 is a mathematical but not a logical truth.*

    To put it simply: Logic does not know any mathematics. *It is important to distinguish logical from non-logical truths, and logical from non-logical equivalences.*

3.4.6 Logical equivalences involving quantifiers

Here is a summary of the most important logical properties relating the quantifiers.

1. To begin with, the negation swaps the quantifiers as follows:

$$\neg\forall x A \equiv \exists x \neg A;$$

$$\neg\exists x A \equiv \forall x \neg A.$$

    For example:

    - "Not every student wrote the test"
      means the same as
      "There is a student who did not write the test."
    - "There is no natural number less than 0" means
      "Every natural number is not less than 0."

2. By negating both sides of the equivalences above we find that each of the universal and existential quantifier is *definable* in terms of the other:

$$\forall x A \equiv \neg\exists x \neg A;$$

$$\exists x A \equiv \neg\forall x \neg A.$$

3. Universal quantifiers distribute over conjunctions: $\forall x P \wedge \forall x Q \equiv \forall x(P \wedge Q)$.

4. Existential quantifiers distribute over disjunctions: $\exists x P \vee \exists x Q \equiv \exists x(P \vee Q)$.

5. Assuming that x does not occur free in Q, the following distributive equivalences hold:

    (a) $\forall x P \wedge Q \equiv Q \wedge \forall x P \equiv \forall x(P \wedge Q)$;
    (b) $\forall x P \vee Q \equiv Q \vee \forall x P \equiv \forall x(P \vee Q)$;
    (c) $\exists x P \vee Q \equiv Q \vee \exists x P \equiv \exists x(P \vee Q)$;
    (d) $\exists x P \wedge Q \equiv Q \wedge \exists x P \equiv \exists x(P \wedge Q)$.

6. Two nested quantifiers of the same type commute:

$$\forall x \forall y A \equiv \forall y \forall x A;$$

$$\exists x \exists y A \equiv \exists y \exists x A.$$

7. Consequently, every two universal closures of a formula are logically equivalent.

8. The case of two different nested quantifiers is more delicate: while

$$\exists x \forall y A \rightarrow \forall y \exists x A$$

is valid (check that yourself!), the converse implication

$$\forall y \exists x A \rightarrow \exists x \forall y A$$

is not. For instance,

"For every integer x there is an integer y such that $x + y = 0$",

which is true, certainly does not imply that

"There is an integer y such that for every integer x it holds that $x + y = 0$",

which is false.

Likewise,

"Every man loves a woman"

does not imply that

"There is a woman whom every man loves."

Therefore, $\exists x \forall y A$ and $\forall y \exists x A$ are *not equivalent*. It is important to remember which of these imply the other, and why.

3.4.7 Negating first-order formulae: negation normal form

Using the first pair of equivalences listed in Section 3.4.6, in addition to those from propositional logic, we can now systematically negate not only propositions but also any first-order sentences. We can transform any first-order sentence into **negation normal form**, where negation only occurs in front of atomic formulae. For example, the sentence

"For every car, there is a driver who, if (s)he can start it, then (s)he can stop it"

can be formalized in a suitable first-order language as

$$\forall x (\mathsf{Car}(x) \rightarrow \exists y (\mathsf{Driver}(y) \wedge (\mathsf{Start}(x, y) \rightarrow \mathsf{Stop}(x, y)))).$$

Now, we negate:

$$\neg\forall x(\mathsf{Car}(x) \rightarrow \exists y(\mathsf{Driver}(y) \wedge (\mathsf{Start}(x,y) \rightarrow \mathsf{Stop}(x,y))))$$
$$\equiv \exists x\neg(\mathsf{Car}(x) \rightarrow \exists y(\mathsf{Driver}(y) \wedge (\mathsf{Start}(x,y) \rightarrow \mathsf{Stop}(x,y))))$$
$$\equiv \exists x(\mathsf{Car}(x) \wedge \neg\exists y(\mathsf{Driver}(y) \wedge (\mathsf{Start}(x,y) \rightarrow \mathsf{Stop}(x,y))))$$
$$\equiv \exists x(\mathsf{Car}(x) \wedge \forall y\neg(\mathsf{Driver}(y) \wedge (\mathsf{Start}(x,y) \rightarrow \mathsf{Stop}(x,y))))$$
$$\equiv \exists x(\mathsf{Car}(x) \wedge \forall y(\neg\mathsf{Driver}(y) \vee \neg(\mathsf{Start}(x,y) \rightarrow \mathsf{Stop}(x,y))))$$
$$\equiv \exists x(\mathsf{Car}(x) \wedge \forall y(\neg\mathsf{Driver}(y) \vee (\mathsf{Start}(x,y) \wedge \neg\mathsf{Stop}(x,y)))).$$

Since $\neg A \vee B \equiv A \rightarrow B$, the last formula is equivalent to

$$\exists x(\mathsf{Car}(x) \wedge \forall y(\mathsf{Driver}(y) \rightarrow (\mathsf{Start}(x,y) \wedge \neg\mathsf{Stop}(x,y)))).$$

The negation of the sentence above is therefore equivalent to:

"There is a car such that every driver can start it and cannot stop it."

Note that the restricted quantifiers satisfy the same equivalences (and non-equivalences) mentioned in Section 3.4.6. In particular,

$$\neg\forall x(P(x) \rightarrow A) \equiv \exists x(P(x) \wedge \neg A),$$
$$\neg\exists x(P(x) \wedge A) \equiv \forall x(P(x) \rightarrow \neg A),$$

and hence:

$$\neg\forall x \in \mathbf{X}(A) \equiv \exists x \in \mathbf{X}(\neg A),$$
$$\neg\exists x \in \mathbf{X}(A) \equiv \forall x \in \mathbf{X}(\neg A).$$

For instance, the negation of the definition of a limit of a function:

$$\neg(\forall \epsilon > \mathbf{0}\exists \delta > \mathbf{0}(0 < |x - c| < \delta \rightarrow |f(x) - L| < \epsilon))$$

is logically equivalent to

$$\exists \epsilon > \mathbf{0}\forall \delta > \mathbf{0}(0 < |x - c| < \delta \wedge \neg|f(x) - L| < \epsilon).$$

References for further reading
For more examples, discussion and properties of truth, logical validity, consequence, and equivalence in first-order logic, see Tarski (1965), Kalish and Montague (1980), van Dalen (1983), Hamilton (1988), Ebbinghaus *et al.* (1996), Barwise and Echemendy (1999), Hodges (2001), Smith (2003), Nederpelt and Kamareddine (2004), Chiswell and Hodges (2007), Ben-Ari (2012), and van Benthem *et al.* (2014). More specifically, on first-order logic with equality see van Dalen (1983) and Hamilton (1988).

Exercises

In all exercises below, "semantic reasoning" refers to reasoning based on the formal semantics presented in the advanced track. For the basic track, read "informal reasoning" instead.

3.4.1 Use semantic reasoning to decide which of the following first-order formulae are logically valid. For each of those which are not, give a *counter-model*, that is, a structure which falsifies it.

(a) $\forall x(P(x) \vee \neg P(x))$

(b) $\forall x P(x) \vee \forall x \neg P(x)$

(c) $\forall x P(x) \vee \neg \forall x P(x)$

(d) $\exists x(P(x) \vee \exists x \neg P(x))$

(e) $\exists x P(x) \rightarrow \forall x P(x)$

(f) $\exists x P(x) \rightarrow \exists y P(y)$

(g) $\exists x(P(x) \rightarrow \forall y P(y))$

(h) $\exists x(P(x) \rightarrow \forall x P(x))$

(i) $\forall x(P(x) \rightarrow \exists y P(y))$

(j) $\forall x(P(x) \rightarrow \forall y P(y))$

(k) $\forall x \exists y Q(x, y) \rightarrow \forall y \exists x Q(y, x)$

(l) $\forall x \exists y Q(x, y) \rightarrow \forall y \exists x Q(x, y)$

(m) $\forall x(\exists y Q(x, y) \rightarrow \exists y Q(y, x))$

(n) $\forall x(\forall y Q(x, y) \rightarrow \exists y Q(y, x))$

(o) $\exists x \neg \exists y P(x, y) \rightarrow \forall y \neg \forall x P(x, y)$

(p) $\forall y \neg \forall x P(x, y) \rightarrow \exists x \neg \exists y P(x, y)$

(q) $(\forall x \exists y P(x, y) \wedge \forall x \forall y(P(x, y) \rightarrow P(y, x))) \rightarrow \exists x P(x, x)$.

(r) $(\forall x \exists y P(x, y) \wedge \forall x \forall y \forall z((P(x, y) \wedge P(y, z)) \rightarrow P(x, z))) \rightarrow \exists x P(x, x)$.

3.4.2 Show that for any first-order formulae $A_1, \ldots, A_n, B$, the following are equivalent:

(a) $A_1, \ldots, A_n \models B$.

(b) $A_1 \wedge \ldots \wedge A_n \models B$.

(c) $\models A_1 \wedge \cdots \wedge A_n \rightarrow B$.

(d) $\models A_1 \rightarrow (A_2 \rightarrow \cdots (A_n \rightarrow B) \ldots)$.

3.4.3 Prove Theorem 112 using the truth definition.

3.4.4 Prove Theorem 118 using the truth definition.

3.4.5 Prove Theorem 119.

3.4.6 Let A, B be any first-order formulae. Using semantic arguments show that the following logical consequences hold.

(a) $\forall x A(x) \models \neg \exists x \neg A(x)$

(b) $\neg \exists x \neg A(x) \models \forall x A(x)$

(c) $\exists x A(x) \models \neg \forall x \neg A(x)$

(d) $\neg \forall x \neg A(x) \models \exists x A(x)$

(e) $\exists x \exists y B(x, y) \models \exists y \exists x B(x, y)$

(f) $\exists x \forall y B(x, y) \models \forall y \exists x B(x, y)$.

3.4.7 Let A, B be any first-order formulae and assume x is not free in B. Show that the following logical consequences hold by giving semantic arguments.

(a) $\forall x(A(x) \vee B) \models \forall x A(x) \vee B$

(b) $\forall x A(x) \vee B \models \forall x(A(x) \vee B)$

(c) $\exists x(A(x) \wedge B) \models \exists x A(x) \wedge B$

(d) $\exists x A(x) \wedge B \models \exists x(A(x) \wedge B)$

(e) $\forall x(B \rightarrow A(x)), B \models \forall x A(x)$

(f) $\exists x A(x) \rightarrow B \models \forall x(A(x) \rightarrow B)$

(g) $\exists x(A(x) \rightarrow B), \forall x A(x) \models B$

(h) $\exists x(B \rightarrow A(x)), B \models \exists x A(x)$.

3.4.8 Using semantic reasoning, determine which of the following logical consequences hold. For those that do, try to give a semantic argument. For those that
do not, construct a *counter-model*, that is, a structure in which all premises are
true, while the conclusion is false.

(a) $\forall x A(x), \forall x B(x) \models \forall x (A(x) \wedge B(x))$

(b) $\forall x (A(x) \wedge B(x)) \models \forall x A(x) \wedge \forall x B(x)$

(c) $\forall x A(x) \vee \forall x B(x) \models \forall x (A(x) \vee B(x))$

(d) $\forall x (A(x) \vee B(x)) \models \forall x A(x) \vee \forall x B(x)$

(e) $\forall x A(x) \rightarrow \forall x B(x) \models \forall x (A(x) \rightarrow B(x))$

(f) $\forall x (A(x) \rightarrow B(x)) \models \forall x A(x) \rightarrow \forall x B(x)$

(g) $\exists x (A(x) \wedge B(x)) \models \exists x A(x) \wedge \exists x B(x)$

(h) $\exists x A(x), \exists x B(x) \models \exists x (A(x) \wedge B(x))$

(i) $\exists x (A(x) \vee B(x)) \models \exists x A(x) \vee \exists x B(x)$

(j) $\exists x A(x) \vee \exists x B(x) \models \exists x (A(x) \vee B(x))$

(k) $\exists x (A(x) \rightarrow B(x)) \models \exists x A(x) \rightarrow \exists x B(x)$

(l) $\exists x A(x) \rightarrow \exists x B(x) \models \exists x (A(x) \rightarrow B(x))$.

3.4.9 Prove the claims in Example 113, Section 3.4.4, by semantic reasoning:

(a) The sentence $\lambda_n = \exists x_1 \cdots \exists x_n (\wedge_{1 \le i \ne j \le n} \neg x_i = x_j)$ is true in a structure iff
its domain has at least n elements.

(b) The sentence $\mu_n = \neg \lambda_{n+1} \equiv \mu_n = \forall x_1 \cdots \forall x_{n+1} (\vee_{1 \le i \ne j \le n+1} x_i = x_j)$ is
true in a structure iff its domain has at most n elements.

(c) The sentence $\sigma_n = \lambda_n \wedge \mu_n$ is true in a structure iff its domain has exactly
n elements.

3.4.10 Prove that for any formula $A(x)$ containing, among others, the free variable x, the
formula $\exists! x A(x) = (A(x) \wedge \forall y (A(y) \rightarrow x = y))$ states that there is a unique
element in the domain of discourse satisfying A.

3.4.11 Prove Theorem 116 for equivalent replacement by structural induction on the
formula $A(x)$.

3.4.12 Prove each of the following equivalences (listed in Section 3.4.6).

(a) $\forall x A \equiv \neg \exists x \neg A$

(b) $\neg \forall x A \equiv \exists x \neg A$

(c) $\exists x A \equiv \neg \forall x \neg A$

(d) $\neg \exists x A \equiv \forall x \neg A$

(e) $\forall x P \wedge \forall x Q \equiv \forall x (P \wedge Q)$

(f) $\exists x P \vee \exists x Q \equiv \exists x (P \vee Q)$

(g) $\forall x \forall y A \equiv \forall y \forall x A$

(h) $\exists x \exists y A \equiv \exists y \exists x A$.

3.4.13 Use semantic arguments to show that the following logical equivalences hold,
given that x does not occur free in Q.

(a) $\forall x P \wedge Q \equiv Q \wedge \forall x P \equiv \forall x (P \wedge Q)$

(b) $\forall x P \vee Q \equiv Q \vee \forall x P \equiv \forall x (P \vee Q)$

(c) $\exists x P \vee Q \equiv Q \vee \exists x P \equiv \exists x (P \vee Q)$

(d) $\exists x P \wedge Q \equiv Q \wedge \exists x P \equiv \exists x (P \wedge Q)$

(e) $\forall x (Q \rightarrow P) \equiv Q \rightarrow \forall x P$

(f) $\forall x (P \rightarrow Q) \equiv \exists x P \rightarrow Q$

(g) $\exists x (P \rightarrow Q) \equiv \forall x P \rightarrow Q$

(h) $\exists x (Q \rightarrow P) \equiv Q \rightarrow \exists x P$.

3.4.14 Use semantic reasoning to decide which of the following logical equivalences hold. For each of those that do not hold, give a counter-model, that is, a structure in which one formula is true while the other false.

(a) $\forall x(P(x) \vee Q(x)) \equiv \forall x P(x) \vee \forall x Q(x)$

(b) $\exists x(P(x) \wedge Q(x)) \equiv \exists x P(x) \wedge \exists x Q(x)$

(c) $\forall x(P(x) \rightarrow Q(x)) \equiv \forall x P(x) \rightarrow \forall x Q(x)$

(d) $\exists x(P(x) \rightarrow Q(x)) \equiv \exists x P(x) \rightarrow \exists x Q(x)$

(e) $\forall x Q(x, x) \equiv \forall x \exists y Q(x, y)$

(f) $\forall x \exists y Q(x, y) \equiv \exists x Q(x, x)$

(g) $\exists x \forall y Q(x, y) \equiv \exists y Q(y, y)$

(h) $\forall x \forall y Q(x, y) \equiv \forall x \forall y Q(y, x)$

(i) $\exists x P(x) \rightarrow \forall x P(x) \equiv \exists x \neg P(x) \rightarrow \forall x \neg P(x)$.

3.4.15 Negate each of the following formulae and import the negations inside all other logical connectives.

(a) $\forall x(x = x^2 \rightarrow x > 0)$

(b) $\forall x((x = x^2 \wedge x > 1) \rightarrow x^2 < 1)$

(c) $\forall x \exists y(x > y \rightarrow x > y^2)$

(d) $\forall x(x = 0 \vee \exists y \neg(xy = x))$

(e) $\forall x \exists y(\neg x = y \rightarrow x > y)$

(f) $\exists x \exists y(x > y \vee -x > y)$

(g) $\exists x(P(x) \rightarrow \forall y P(y))$

(h) $\forall x(P(x) \rightarrow Q(x)) \rightarrow (\forall x P(x) \rightarrow \forall x Q(x))$

(i) $(\exists x P(x) \rightarrow \exists x Q(x)) \rightarrow \exists x(P(x) \rightarrow Q(x))$.

3.4.16 For each of the following pairs of statements, check whether either of them logically implies the other by formalizing them in first-order logic and using semantic reasoning. If either of the logical consequences does not hold, give an appropriate counter-model.

(a) A: "Not everybody who is a lawyer is greedy."

   B: "There exists a lawyer and not everybody is greedy."

(b) A: "No man is happy only when he is drunk."

   B: "There is a happy man and there is a man who is not drunk."

3.4.17 Formalize the following arguments in FOL by introducing suitable predicates. Using semantic reasoning, then determine which of them are logically correct.

(a)

All philosophers are humans.
All humans are mortal.
____
All philosophers are mortal.

(c)

Some negative numbers are rationals.
All integers are rationals.
____
Some integers are negative numbers.

(b)

No work is fun.
Some entertainment is work.
____
Some entertainment is not fun.

(d)

All bank managers are rich.
No bank managers are teachers.
____
No teachers are rich.

(e)
> No mathematicians are stupid.
> Some mathematicians are bachelors.
> _____
> Some bachelors are not stupid.

(f)
> No poor people are politicians.
> Some politicians are crooks.
> _____
> Some crooks are not poor people.

(g)

Some penguins are white.
All penguins are birds.
_____
Therefore, some white bird is a penguin.

(h)

> Every mother loves some child.
> Every mother is a woman.
> Some mothers hate watching TV.
> _____
> Some woman loves some child.

(m)
> No food is better than a good steak.
> Stale bread is better than no food.
> _____
> Stale bread is better than a good steak.

(n) As above, but adding a new premise:
For any x, y, z, if x is better than y and y is better than z then x is better than z.
(*Watch out for the quantifiers!*)

(i) As in (m), but replacing
"Some mothers hate watching TV"
with "All mothers like watching TV."

(j)

No successful politicians are poor.
Someone is poor only if he is a politician.
_____
Therefore, nobody who is poor is successful.

(k)
> No penguins are white.
> Some penguins are fat.
> All penguins are birds.
> _____
> Therefore, no fat bird is white.

(l)

Every bachelor loves a woman he knows.
Every bachelor is a man.
Eve is a woman whom every man knows.
_____
Some bachelor loves some woman.

Alonzo Church (14.6.1903–11.8.1995) was an American mathematician and one of the leading logicians of the 20th century, who made profound contributions to mathematical logic and the theoretical foundations of computer science.

Church studied and had his academic career at the Mathematics Department of Princeton University where he worked until 1967. He then formally retired and moved to the University of California, Los Angeles, as Flint Professor of Philosophy and Mathematics for a further 23 years, until his final retirement at age 87 in 1990.

In the early 1930s Church created **lambda-calculus** as a formal system of defining functions that are effectively computable. In connection with his two-part work *A Set of Postulates for the Foundations of Logic*, published in 1932–1933, he intended to develop a set of axioms that "would lead to a system of mathematical logic free of some of the complications entailed by Bertrand Russell's theory of types, and would at the same time avoid the well-known paradoxes, in particular the Russell paradox." While it later turned out that the original system of lambda-calculus did contain an inconsistency, the idea flourished and became very influential in computer science as the basis of functional programming.

In 1936 Church published two papers of fundamental importance to logic and computability theory. In the first, *An Unsolvable Problem of Elementary Number Theory*, he defined the notion of "effective calculability" by identifying it with the notion of recursive function, or – as it had turned out to be equivalent – of the lambda-definable function of natural numbers. He then proved that the problem of deciding whether a given first-order formula of the language with addition and multiplication is true of the arithmetic of natural numbers is not effectively calculable. By equating effective calculability with algorithmic solvability, Church put forward what was later called by Kleene **Church's Thesis**. Church essentially proved that the truth in the elementary arithmetic is algorithmically unsolvable.

In his other historic 1936 paper, *A Note on the Entschiedungsproblem* (just two pages long, followed by a two-page correction), Church proved what is now called **Church's Undecidability Theorem**: the problem of deciding validity of formulas in first-order logic is not effectively calculable, which means – assuming Church's Thesis – it is algorithmically unsolvable.

Church had an extremely long and fruitful scientific career, spanning over more than 70 years since his first publication as an undergraduate student in 1924 until his last paper, *A Theory of the Meaning of Names*, published in 1995. In 1956 he published his classic textbook *Introduction to Mathematical Logic*, which educated and influenced generations of logicians. Church was also an extremely popular and successful supervisor and had 31 doctoral students, many of whom became distinguished logicians and computer scientists, including Martin Davis, Leon Henkin, John Kemeny, Steven Kleene, Michael O. Rabin, Nicholas Rescher, Hartley Rogers, J. Barkley Rosser, Dana Scott, Reymond Smullyan, and Alan Turing.

Stephen Cole Kleene (05.01.1909–25.01.1994) was an American mathematician and logician, the founder (along with Church, Gödel, Turing, and Post) of **recursion theory**, one of the main branches of mathematical logic, providing the logical foundations of the theory of algorithms and computability and of theoretical computer science in general.

Kleene received a PhD in mathematics from Princeton University in 1934, for the thesis *A Theory of Positive Integers in Formal Logic* supervised by Alonzo Church. He also made important contributions to Church's lambda-calculus. In 1935, Kleene joined the Mathematics Department at the University of Wisconsin-Madison, where he spent most of his academic career. In 1939–40, while he was a visiting scholar at the Institute for

Advanced Study in Princeton, he laid the foundation for recursion theory. During WW II, Kleene served in the United States Navy as a lieutenant commander. He was later an instructor of navigation and a project director at the Naval Research Laboratory.

Several fundamental concepts in the theories of computability and recursion were invented by Kleene and named after him, including **Kleene's recursion theorems**, **Kleene–Mostowski's arithmetical hierarchy** of sets of natural numbers, **Kleene fixpoint theorem**, the **Kleene star** operation, the **Kleene algebra of regular expressions**, and **Kleene's theorem** of the equivalence between regular expressions and finite automata.

Kleene wrote, *inter alia*, two very popular and influential books from which generations of logicians learned mathematical logic: *Introduction to Metamathematics* (1952) and *Mathematical Logic* (1967). He also made significant contributions to the foundations of mathematical intuitionism (founded by L. Brouwer) and wrote, together with Vesley, the book *The Foundations of Intuitionistic Mathematics* in 1965.

3.5 Syllogisms

Formal logic was founded by the famous Greek philosopher Aristotle some 2400 years ago. (See more about Aristotle in the biographical box devoted to him at the end of Section 2.1.) Aristotle's logical system was not based on propositional logic, which was originally developed a little later, again in ancient Greece, mainly by Chrysippus from the Stoic school of logic. Instead, Aristotle's system of logic was based on a fragment of *monadic first-order logic*, consisting of a special kind of logical arguments called **syllogisms** (from the Greek word *syllogismos*, meaning conclusion, inference). Aristotle developed a very systematic logical theory of syllogisms, now called the **Syllogistic**. This earliest known formal logical system remained the gold standard for logical reasoning until the mid-19th century when the modern algebraic treatment of logic began with the works of Boole, de Morgan, Schröder, and others. Even though it was later subsumed by first-order logic, the syllogistic has nevertheless remained an important and useful logical system, not only for historical but also for methodological and practical computational reasons; I therefore present it briefly here.

3.5.0.1 Basics of categorical syllogisms

The best-known syllogism is probably:

> All men are mortal.
> Socrates is a man.
> ───────────────────
> (Therefore) Socrates is mortal.

Even though the notion of syllogism introduced by Aristotle was more general, he mostly focused on the so-called **categorical syllogisms** that formally exclude that above (but can still accommodate it). Categorical syllogisms consist of three propositions, each

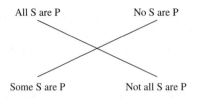

Figure 3.1 The Square of Opposition

of them cast in one of the following four patterns, respectively denoted by the letters A, E, I, and O, where S and P are unary predicates:

A: "All S are P", or, in first-order logic: $\forall x(S(x) \to P(x))$.

E: "No S are P". Equivalently, "All S are not P", that is, $\forall x \neg (S(x) \wedge P(x))$.

I: "Some S are P", that is, $\exists x(S(x) \wedge P(x))$.

O: "Not all S are P". Equivalently, "Some S are not P", that is, $\exists x(S(x) \wedge \neg P(x))$.

These four patterns can be arranged in terms of their logical relationships in a so-called **Square of Opposition**, presented in Figure 3.1. For further discussion and explanation, see the references on philosophical logic.

Example 121 *For some examples of categorical syllogisms see Exercise 17(a–f) from Section 3.4.8. Some other examples include:*

1.

   All humans are mortal.
   All logicians are humans.
   _____
   All logicians are mortal.

3.

   No philosophers are crooks.
   Some philosophers are politicians.
   _____
   No politicians are crooks.

2.

   No philosophers are crooks.
   Some politicians are crooks.
   _____
   No politicians are philosophers.

4.

   All philosophers are clever.
   Some clever people are not happy.
   _____
   Some happy people are not philosophers.

Up to logical equivalence, every categorical syllogistic proposition therefore has the form:

"Some/All S are/are_not P."

Syllogistic propositions are therefore simply quantified sentences involving only unary predicates (and possibly constant symbols), but no functions.

Some standard terminology and conditions of categorical syllogisms are defined in the following.

- The S and P in any syllogistic proposition are called **terms**; S is called its **subject** and P is called its **predicate**.

- A categorical syllogism consists of three syllogistic propositions, the fist two of which are **premises** and the third which is the **conclusion** of the syllogism.

- The predicate of the conclusion is called the **major term** of the syllogism.

- The subject of the conclusion is called the **minor term** of the syllogism.

- One of the premises contains the major term and is called the **major premise**. Usually this is the first premise.

- The other premise contains the minor term and is called the **minor premise**. Usually this is the second premise.

- The two premises must share a third, common term called the **middle term**, which does not appear in the conclusion.

- Any combination of the three propositions, each of one of the four types above, is called a **syllogistic form**. These forms are named by listing the type letters of the major premise, the minor premise, and the conclusion. For instance: AAA denotes the form where each of these is of type A; EIO is the form where the major premise is of type E, the minor premise is of type I, and the conclusion is of type O, etc.

Furthermore, all syllogistic forms are grouped in four **figures**, each referring to the positions of the middle term in the major premise (**M**) and minor premise (**m**):

- **Figure 1**: the middle term is first (the subject) in **M** and second (the predicate) in **m**.

- **Figure 2**: the middle term is second (the predicate) both in **M** and **m**.

- **Figure 3**: the middle term is first (the subject) both in **M** and **m**.

- **Figure 4**: the middle term is second (the predicate) in **M** and first (the subject) in **m**.

Syllogism type is now specified as XYZ-N, where each of X,Y, Z is one of the letters A,E,I,O and N$\in \{1, 2, 3, 4\}$ indicates the figure. For instance, in each of the examples above the major premise is the first and the minor premise is the second, so they are of the type AAA-1, EIE-2, EIE-3, and AOO-4 (check!).

Since in the categorical syllogism XYZ-N each of X,Y,Z, and N can independently take four different values, there are $4^4 = 256$ possible syllogistic forms. Some of them represent logically valid inferences, while others do not. Aristotle already inspected most of the 256 categorical syllogistic forms, identified and classified all valid forms among them, and constructed an obviously incorrect argument (i.e., a counter-model) for each of the invalid forms.

3.5.0.2 Testing validity of syllogisms by Venn diagrams

How can we determine which of the syllogistic forms represent logically valid arguments? An easy method, much more modern than that of Aristotle, is based on the so-called **Venn diagrams**[13]. Venn diagrams represent sets as intersecting circles in a rectangular box, so the basic operations (union, intersection, complementation) on these sets and the basic relations of equality, inclusion, and disjointness between them can be easily visualized. A three-set Venn diagram is depicted in Figure 3.2, where the three circles represent sets A, B, C and the box represents the whole domain of discourse (the "universal set"). Each

---

[13] Named after the British logician and philosopher John Venn who invented them in around 1880. Read more about him in the biographical box at the end of the section.

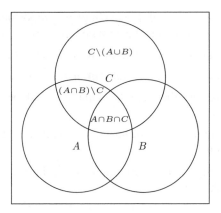

Figure 3.2 Three-set Venn diagram

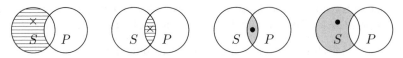

Figure 3.3 Venn diagrams depicting the four types of categorical syllogistic propositions. From left to right: "All S are P", "All S are not P", "Some S are P", and "Some S are not P".

of the eight regions represents a particular set-theoretic combination of the three sets, for example, the region in the middle represents $A \cap B \cap C$ while the region above this represents $C \setminus (A \cup B)$, etc.

Venn diagrams can be used to represent the terms in a categorical syllogism by sets and indicate the relations between them expressed by the premises and the conclusion. In particular, each of the four types of categorical syllogistic propositions can be easily illustrated by Venn diagrams as in Figure 3.3. The circles correspond to the three terms; the barcoded and shaded regions indicate those referred to by the premises; a cross $\times$ in a region means that that region is empty; a bullet $\bullet$ in a region means that that region is non-empty; and no cross or bullet implies that the region could be either empty or non-empty.

Each of the three propositions of a categorical syllogism can therefore be represented on a three-set Venn diagram, corresponding to the three terms. For instance, the syllogisms in Example 121(1) and 121(3) are represented on Figure 3.4 as follows. In the Venn diagram on the left corresponding to Example 121(1), the circle M represents the set of mortals (the major term), the circle L represents the set of all logicians (the minor term), and the circle H represents the set of all humans (the middle term). The diagram depicts the two premises of Example 121(1) by indicating the regions $H \setminus M$ that corresponds to "humans who are not mortals", and $L \setminus H$ that corresponds to "logicians who are not humans", as empty. Likewise, in the Venn diagram on the right corresponding to Example 121(3), C represents the set of crooks (the major term), Po represents the set of all politicians (the minor term), and Ph represents the set of all philosophers (the middle term). The diagram depicts the two premises of Example 121(3) by indicating the region $Ph \cap C$ corresponding to all "philosophers who are crooks" as empty, while indicating the region $Ph \cap Po$ corresponding to all "philosophers who are politicians" as non-empty (the bullet).

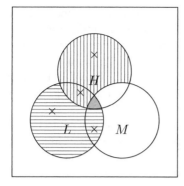 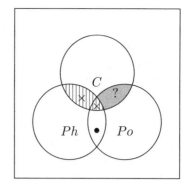

Figure 3.4   Venn diagrams for the syllogisms in Exercise 121(1) (left) and 121(3) (right)

Once the Venn diagram representing the relations stated by the two premises of the syllogism is constructed, the testing for validity is immediate: check, by inspection of the diagram, whether the relation stated by the conclusion must also hold. If so, the syllogism is valid; otherwise, it is not. It is as simple as that!

For example, in Example 121(1), the conclusion claims that in the diagram on the left on Figure 3.4 it must be the case that $L \subseteq M$, that is, the non-barcoded part of L (where all elements of L must live), which is the shaded triangle in the middle, must be included in M. It is included in M, and so this syllogism is valid.

On the other hand, in Example 121(3) the conclusion claims that $Po \cap C = \emptyset$ but in the diagram the right (shaded) part of that region is not depicted as empty and therefore it *need not* be empty. Hence, this syllogism is invalid.

The proof of correctness of the Venn diagrams method is not very difficult and is left as an exercise.

Non-categorical syllogisms of the "Socrates is mortal" type mentioned earlier can also be easily treated here. By regarding proper names, such as Socrates, as singleton predicates/sets, the individual Socrates can be identified with the property of "being Socrates." I leave the easy details of that to the reader.

It turns out that of the 256 categorical syllogistic forms, there are 24 valid forms under the natural assumption of the meaning of the types A and E discussed further. In the Middle Ages these were given special Latin names (some of which appear in the literature with slight variations), each containing three vowels which were used as mnemonics of the types of the respective forms. (Some of the other letters in these mnemonic words also contain some information.) The 24 valid forms are as follows.

Figure 1	Figure 2	Figure 3	Figure 4
1. *AAA-1 Barbara*	1. *AEE-2 Camestres*	1. *AII-3 Datisi*	1. *AAI-4 Bramantip*
2. *EAE-1 Celarent*	2. *EAE-2 Cesare*	2. *AAI-3 Darapti*	2. *AEE-4 Camenes*
3. *AII-1 Darii*	3. *AOO-2 Baroko*	3. *IAI-3 Disamis*	3. *IAI-4 Dimaris*
4. *EIO-1 Ferio*	4. *EIO-2 Festino*	4. *EAO-3 Felapton*	4. *EAO-4 Fesapo*
5. *AAI-1 Barbari*	5. *EAO-2 Cesaro*	5. *EIO-3 Ferison*	5. *EIO-4 Fresison*
6. *EAO-1 Celaront*	6. *AEO-2 Camestrop*	6. *OAO-3 Bocardo*	6. *AEO-4 Camenop*

It is important to note that the validity of some of these forms depends on an assumption which Aristotle made about the interpretation of the quantifiers in the types A and E, known as **Existential Import**: in the propositions "All S are P" and "No S are P" the predicate S is non-empty, that is, there are S objects. Under this assumption, these universal proposition types, which are at the top of the Square of Opposition, logically imply respectively their existential counterparts at the bottom, "Some S are P" and "Some S are not P."

The Existential Import assumption is compliant with the normal natural language usage of such expressions, but is not formally justified by the semantics of first-order logic that we have assumed; we must therefore handle it with special care and will only apply it in the context of syllogisms. Without this assumption, nine of the syllogistic forms listed above cease to be valid. I leave it as an exercise for the reader to identify these.

Note that the syllogistic forms are not formulae but inference rules. It is however possible to develop a deductive system for deriving some valid syllogisms from others; see the references for further details.

To summarize, categorical syllogisms capture a natural and important fragment of logical reasoning in first-order logic involving only unary predicates and constant symbols. They do not capture simple sentences involving binary predicates, such as "Every man loves a woman", but have one great technical advantage: the simple *decision procedure* for testing valid syllogisms given by the method of Venn diagrams. As I explain in the next chapter, *no* such decision procedure exists for first-order logic in general, even when the language involves a single binary predicate.

References for further reading

For more details on syllogisms see Carroll (1897), Barwise and Echemendy (1999), and van Benthem *et al.* (2014), plus many books on philosophical logic.

Exercises

3.5.1   For each of the syllogisms listed in Example 121, verify its type and figure as claimed in the text.

3.5.2   For each of the following syllogisms (taken from Section 3.4.8, Exercise 17(a–f)) identify its major, minor, and middle terms and its type and figure.

(a)

All philosophers are humans.
All humans are mortal.

All philosophers are mortal.

(b)

No work is fun.
Some entertainment is work.

Some entertainment is not fun.

(c)

Some negative numbers are rationals.
All integers are rationals.

Some integers are negative numbers.

(d)

All bank managers are rich.
No bank managers are teachers.

No teachers are rich.

(e)   (f)

No mathematicians are stupid.   No poor people are politicians.
Some mathematicians are bachelors.   Some politicians are crooks.
――――――――――――――――――   ――――――――――――――――
Some bachelors are not stupid.   Some crooks are not poor people.

3.5.3   Using Venn diagrams, show that each of the 24 syllogistic forms listed in Section 3.5 represent logically valid arguments.

3.5.4   Identify the 9 syllogistic forms among those 24 which are only valid conditionally on the assumption of Existential Import.

3.5.5   Prove the correctness of the method of Venn diagrams for testing validity of categorical syllogisms.

3.5.6   Using Venn diagrams, determine whether each the syllogisms listed in Section 3.5.1, Exercise 2 is valid.

John Venn (4.08.1834–4.04.1923) was an English logician and philosopher, mostly known for introducing the **Venn diagrams** used in set theory, probability, logic, statistics, and computer science.

Venn studied at Gonville and Caius College, Cambridge and worked there as a lecturer, and later professor, of moral science, logic, and probability theory during most of his career. In 1903 he was elected President of the College and held that post until his death.

In 1866 Venn published his book *The Logic of Chance*, in which he developed the frequency theory of probability. He developed further George Boole's theory in his 1881 book *Symbolic Logic*, where he applied his diagrams which he had invented for illustration in his teaching.

Lewis Carroll (Charles Lutwidge Dodgson, 7.1.1832–14.1.1898) was an English writer, mathematician, logician, and photographer, most famous for his children's books *Alice's Adventures in Wonderland* and *Through the Looking-Glass*, published under the pen name Lewis Carroll.

Lewis Carroll was talented and creative since his early childhood in many areas, including mathematics, poetry, storytelling, and designing various new games. After completing his university studies in Oxford he became a lecturer at Christ Church College which was his main job for life, although he did not like it much.

Stammering and other medical problems since early age made him feel less at ease among adults, but he was quite relaxed with children for whom he enjoyed creating and telling stories, including his famous books about Alice written for the daughter of the dean of his college.

Lewis Carroll did some research in geometry, algebra, logic, and recreational mathematics, producing several books and articles under his real name. In *The Alphabet-Cipher* (1868) he described a well-known cryptographic scheme, and in a paper on *The Principles of Parliamentary Representation* (1884) he proposed a voting system, still known as Dodgson's method. He was also an avid popularizer of logic, on which he wrote the book *The Game of Logic* (1887, still popular today) and the textbook *Symbolic Logic* (1896), where he explains the theory of syllogisms with many examples. In the story *What the Tortoise Said to Achilles* (1895), he uses a dialogue between the Tortoise and Achilles to demonstrate the problem with infinite regress in logical deduction.

4

Deductive Reasoning in First-order Logic

Validity, logical consequence, and equivalence in first-order logic can be established by using deductive systems, like for propositional logic.

In this chapter I present and illustrate with examples extensions of the propositional deductive systems introduced in Chapter 2, obtained by adding additional axioms (for Axiomatic Systems) and inference rules (for Propositional Semantic Tableaux and Natural Deduction) for the quantifiers or additional procedures handling them (for Resolution), that are sound and complete for first-order logic.

That such sound and complete deductive systems for first-order logic exist is not *a priori* obvious at all. The first proof of completeness of (an axiomatic) deductive system for first-order logic was first obtained by Kurt Gödel[1] in his doctoral thesis in 1929, and is now known as **Gödel's Completeness Theorem**. In supplementary Section 4.6, generic proofs of soundness and completeness of the deductive systems are sketched for first-order logic introduced here.

Differently from the propositional case, however, none of these deductive systems can be guaranteed to always terminate their search for a derivation, even if such a derivation exists. This can happen, for instance, when the input formula is not valid but can only be falsified in an infinite counter-model.

This is not an accidental shortcoming of the deductive systems. In fact, one of the most profound and important results in mathematical logic, proved in 1935 by Alonso Church and independently by Alan Turing in 1936[2], states that the problem of whether a given first-order sentence is valid is *not algorithmically decidable*. It is now known as **Church's Undecidability Theorem**.

Deciding logical consequence in first-order logic is therefore also not possible by a purely algorithmic procedure. Therefore, no sound, complete, and always terminating deductive system for first-order logic can be designed.

---

[1] See more about Gödel and his groundbreaking results in the biographical box at the end of this chapter.

[2] Read more about Church in the biographical box at the end of Section 3.4 and Turing in the box at the end of Section 5.4.

Logic as a Tool: A Guide to Formal Logical Reasoning, First Edition. Valentin Goranko.
© 2016 John Wiley & Sons, Ltd. Published 2016 by John Wiley & Sons, Ltd.

4.1   Axiomatic system for first-order logic

We now extend the propositional axiomatic system presented in Section 2.2 to first-order logic by adding axioms and rules for the quantifiers. I denote the resulting axiomatic system again by **H**.

In what follows, $\forall$ is regarded as the only quantifier in the language and $\exists$ is definable in terms of it. An equivalent system can be obtained by regarding both quantifiers present in the language and adding the axiom $\exists x A \leftrightarrow \neg \forall x \neg A$.

4.1.1   Axioms and rules for the quantifiers

Additional axiom schemes

(Ax$\forall$1) $\forall x (A(x) \rightarrow B(x)) \rightarrow (\forall x A(x) \rightarrow \forall x B(x))$;

(Ax$\forall$2) $\forall x A(x) \rightarrow A[t/x]$ where t is any term free for substitution for x in A;

(Ax$\forall$3) $A \rightarrow \forall x A$ where x is not free in the formula A.

As an easy exercise, show that all instances of these axiom schemes are valid.

Additional rule

We must also add the following rule of deduction, known as **Generalization**:

$$\frac{A}{\forall x A}$$

where A is any formula and x any variable. Note that x may occur free in A, but need not occur free there in order to apply that rule. Note also that this rule *does not* read as

"Assuming A is true, conclude that $\forall x A$ is also true"

but rather as

"If A is valid in the given model (respectively, logically valid) then conclude that $\forall x A$ is also valid in the given model (respectively, logically valid)"

that is, the rule preserves not truth but validity (in a model, or logical validity).
Respectively, the syntactic/deductive reading of the Generalization rule is:

"If A is derived, then derive $\forall x A$."

4.1.2   Derivations from a set of assumptions

Note that if the Generalization rule is used unrestrictedly in derivations from a set of assumptions, it would derive $A \vdash_{\mathbf{H}} \forall x A$ for example, which should not be derivable because $A \nvDash \forall x A$. To avoid such unsound derivations when adding the Generalization rule to the definition of derivation in **H** from a set of assumptions Γ, we include the proviso: if the formula A is already in the sequence *and x does not occur free in any of the formulae of Γ*, then $\forall x A$ can be added to the sequence.

More generally, we can extend the Generalization rule to work with assumptions on the left, as follows:

$$\frac{A_1, \ldots, A_n \vdash_{\mathbf{H}} A}{\forall x A_1, \ldots, \forall x A_n \vdash_{\mathbf{H}} \forall x A}.$$

In the case where x does not occur free in $A_1, \ldots, A_n$, the vacuous quantification on the left can be achieved by using Axiom (Ax$\forall$3), thus justifying the definition above.

With the refined definition of derivations from a set of assumptions, the Deduction Theorem still holds for the extension of **H** to first-order logic. The proof of this claim is left as an exercise.

Here is an example of derivations in **H**. Check that the rules for the quantifiers have been applied correctly.

Example 122 *Derive* $\forall x(Q \to P(x)), \exists x \neg P(x) \vdash_{\mathbf{H}} \neg Q$, *where x is not free in Q.*
Eliminating $\exists$, *we are to derive* $\forall x(Q \to P(x)), \neg \forall x \neg \neg P(x) \vdash_{\mathbf{H}} \neg Q$.

1. $\vdash_{\mathbf{H}} P(x) \to \neg\neg P(x)$	*derived in the Propositional **H**.*
2. $\vdash_{\mathbf{H}} \forall x(P(x) \to \neg\neg P(x))$	*from 1 and Generalization.*
3. $\forall x(P(x) \to \neg\neg P(x)) \vdash_{\mathbf{H}} \forall x P(x) \to \forall x \neg\neg P(x)$	*by Axiom (Ax$\forall$1).*
4. $\vdash_{\mathbf{H}} \forall x P(x) \to \forall x \neg\neg x P(x)$	*by 2, 3, Deduction Theorem, and Modus Ponens.*
5. $\forall x(Q \to P(x)) \vdash_{\mathbf{H}} \forall x Q \to \forall x P(x)$	*by Axiom (Ax$\forall$1).*
6. $\forall x(Q \to P(x)), \forall x Q \vdash_{\mathbf{H}} \forall x P(x)$	*by 5 and Deduction Theorem.*
7. $Q \vdash_{\mathbf{H}} \forall x Q$	*by Axiom (Ax$\forall$3).*
8. $\forall x(Q \to P(x)), Q \vdash_{\mathbf{H}} \forall x P(x)$	*by 6,7 and Deduction Theorem.*
9. $\vdash_{\mathbf{H}} \neg \forall x \neg \neg P(x) \to \neg \forall x P(x)$	*by 4 and contraposition (derived in the propositional **H**).*
10. $\forall x(Q \to P(x)), \neg \forall x P(x) \vdash_{\mathbf{H}} \neg Q$	*by 8 and contraposition.*
11. $\forall x(Q \to P(x)), \neg \forall x \neg \neg P(x) \vdash_{\mathbf{H}} \neg Q$	*by 9, 10, and Deduction Theorem.*

4.1.3   Extension of the axiomatic system H with equality

Recall that the equality symbol $=$ is a special binary relational symbol, always to be interpreted as the identity of objects in the domain of discourse. For that standard meaning of the equality to be captured in the axiomatic system, the following additional axioms for the equality, mentioned in Section 3.4.4, are needed (though, as noted there, it is not sufficient to express the claim that it is the identity relation), where all variables are implicitly universally quantified.

Axioms for the equality

(Ax$_=$1) $x = x$;

(Ax$_=$2) $x = y \to y = x$;

(Ax$_=$3) $x = y \wedge y = z \to x = z$;

(Ax$_f$) $\quad x_1 = y_1 \wedge \ldots \wedge x_n = y_n \to f(x_1, \ldots, x_n) = f(y_1, \ldots, y_n)$ $\quad$ for $\quad$ n-ary functional symbol f;

(Ax$_r$) $x_1 = y_1 \wedge \ldots \wedge x_n = y_n \to (p(x_1, \ldots, x_n) \to p(y_1, \ldots, y_n))$ for n-ary predicate symbol p.

An example of a derivation in the extension **H** with equality is provided in the following.

Example 123 *Derive* $\forall x \forall y (x = y \to (P(f(x)) \leftrightarrow P(f(y))))$ *where* P *is a unary predicate symbol and* f *is a unary function symbol.*

1. $\vdash_{\mathbf{H}} \forall x \forall y (x = y \to (P(x) \to P(y)))$ $\hfill$ *by Axiom Ax$_r$.*
2. $\vdash_{\mathbf{H}} \forall y (f(x) = y \to (P(f(x)) \to P(y)))$ $\hfill$ *by 1, Axiom Ax$\forall$2, and MP.*
3. $\vdash_{\mathbf{H}} f(x) = f(y) \to (P(f(x)) \to P(f(y)))$ $\hfill$ *by 2 and Ax$\forall$2.*
4. $\vdash_{\mathbf{H}} \forall x \forall y (x = y \to y = x)$ $\hfill$ *by Axiom Ax$_=$2.*
5. $\vdash_{\mathbf{H}} \forall y (f(x) = y \to y = f(x))$ $\hfill$ *by 4 and Axiom Ax$\forall$2.*
6. $\vdash_{\mathbf{H}} f(x) = f(y) \to f(y) = f(x)$ $\hfill$ *by 5 and Axiom Ax$\forall$2.*
7. $\vdash_{\mathbf{H}} \forall y \forall x (y = x \to (P(y) \to P(x)))$ $\hfill$ *by Axiom Ax$_r$.*
8. $\vdash_{\mathbf{H}} \forall x (f(y) = x \to (P(f(y)) \to P(x)))$ $\hfill$ *by 7, Axiom Ax$\forall$2, and MP.*
9. $\vdash_{\mathbf{H}} f(y) = f(x) \to (P(f(y)) \to P(f(x)))$ $\hfill$ *by 8 and Ax$\forall$2.*
10. $f(x) = f(y) \vdash_{\mathbf{H}} (P(f(x)) \to P(f(y)))$ $\hfill$ *by 3 and Deduction Theorem.*
11. $f(x) = f(y) \vdash_{\mathbf{H}} f(y) = f(x)$ $\hfill$ *by 6 and Deduction Theorem.*
12. $f(x) = f(y) \vdash_{\mathbf{H}} (P(f(y)) \to P(f(x)))$ $\hfill$ *by 9, 11, and MP.*
13. $f(x) = f(y) \vdash_{\mathbf{H}} (P(f(x)) \to P(f(y))) \wedge (P(f(y)) \to P(f(x)))$ $\hfill$ *by 10, 12, and Propositional logic.*
14. $f(x) = f(y) \vdash_{\mathbf{H}} (P(f(x)) \leftrightarrow P(f(y)))$ $\hfill$ *by 13 and definition of $\leftrightarrow$.*
15. $\vdash_{\mathbf{H}} f(x) = f(y) \to (P(f(x)) \leftrightarrow P(f(y)))$ $\hfill$ *by 14 and Deduction Theorem.*
16. $\vdash_{\mathbf{H}} \forall y (f(x) = f(y) \to (P(f(x)) \leftrightarrow P(f(y))))$ $\hfill$ *by 15 and Generalization.*
17. $\vdash_{\mathbf{H}} \forall x \forall y (f(x) = f(y) \to (P(f(x)) \leftrightarrow P(f(y))))$ $\hfill$ *by 16 and Generalization.*
    QED.

Example 124 *An old jazz song written by Jack Palmer says*

"Everybody loves my baby, but my baby don't love nobody but me."

Properly formalized (after correcting the grammar) in first-order logic, it states:

$$\textit{Me-and-my-baby} = \forall x L(x, \textit{MyBaby}) \wedge \forall y (\neg y = \textit{Me} \to \neg L(\textit{MyBaby}, y))$$

I will derive in **H** *with equality that the claim above implies "I am my baby," that is,*

$$\forall x L(x, \textit{MyBaby}) \wedge \forall y (\neg y = \textit{Me} \to \neg L(\textit{MyBaby}, y)) \vdash_{\mathbf{H}} \textit{MyBaby} = \textit{Me}.$$

1. *Me-and-my-baby* $\vdash_{\mathbf{H}} \forall x L(x, MyBaby)$ *by Propositional logic.*
2. *Me-and-my-baby* $\vdash_{\mathbf{H}} L(MyBaby, MyBaby)$ *by 1 and Axiom $Ax\forall 2$.*
3. *Me-and-my-baby* $\vdash_{\mathbf{H}} \forall y(\neg y = Me \rightarrow \neg L(MyBaby, y))$ *by Propositional logic.*
4. *Me-and-my-baby* $\vdash_{\mathbf{H}} \neg MyBaby = Me \rightarrow \neg L(MyBaby, MyBaby)$ *by 3 and Axiom $Ax\forall 2$.*
5. *Me-and-my-baby* $\vdash_{\mathbf{H}} L(MyBaby, MyBaby) \rightarrow MyBaby = Me$ *by 4 and Propositional logic.*
6. *Me-and-my-baby* $\vdash_{\mathbf{H}} MyBaby = Me$ QED.   *by 2, 5, and MP.*

The following is a deductive analogue in **H** of the Equivalent replacement theorem 116 and a generalization of the equality axioms, stating that equal terms can provably be replaced for each other in any formula.

Theorem 125 (Equivalent replacement) *For any formula $A(x)$ and terms s, t free for x in A, the following is derivable in **H**:*

$$\vdash_{\mathbf{H}} s = t \rightarrow A[s/x] \leftrightarrow A[t/x].$$

The proof is by structural induction on the formula $A(x)$, left as an exercise.

Theorem 126 (Soundness and completeness of H) *Each of the axiomatic systems **H** for first-order logic with and without equality is sound and complete, that is, for every first-order formula $A_1, \ldots, A_n, C$ of the respective language (respectively, with or without equality):*

$$A_1, \ldots, A_n, \vdash_H C \text{ iff } A_1, \ldots, A_n, \models C.$$

The proof for **H** with equality is outlined in Section 4.6.

References for further reading

For further discussion and examples on derivations in axiomatic systems for first-order logic, see Tarski (1965), Shoenfield (1967), Hamilton (1988), Fitting (1996), Mendelson (1997), Enderton (2001). For discussion and proofs of Church's Undecidability Theorem, see Shoenfield (1967), Jeffrey (1994), Ebbinghaus *et al.* (1996), Enderton (2001), Hedman (2004), and Boolos *et al.* (2007).

Exercises

4.1.1 Show that all instances of the axiom schemes listed in Section 4.1.1 are logically valid, and that the Generalization rule preserves validity in a structure and hence logical validity.

4.1.2 Prove the logical correctness of the inference rule

$$\frac{A_1, \ldots, A_n \vdash A}{\forall x A_1, \ldots, \forall x A_n \vdash \forall x A}$$

by showing that if $A_1, \ldots, A_n \models A$ then $\forall x A_1, \ldots, \forall x A_n \models \forall x A$.

4.1.3 Prove the Deduction Theorem for the extension of **H** to first-order logic, with respect to the definition of derivations from a set of assumptions given in the text.

For each of the following exercises on derivations in the axiomatic system **H**, you may use the Deduction Theorem.

4.1.4 Derive the following deductive consequences in **H**, where P is a unary predicate.
(a) $\vdash_{\mathbf{H}} \forall x P(x) \to \forall y P(y)$
(b) $\vdash_{\mathbf{H}} \exists x P(x) \to \exists y P(y)$
(c) $\forall x A(x) \vdash_{\mathbf{H}} \neg \exists x \neg A(x)$
(d) $\neg \exists x \neg A(x) \vdash_{\mathbf{H}} \forall x A(x)$
(e) $\forall x \forall y A(x, y) \vdash_{\mathbf{H}} \forall y \forall x A(x, y)$
(f) $\exists x \exists y A(x, y) \vdash_{\mathbf{H}} \exists y \exists x A(x, y)$.

4.1.5 Suppose x is not free in Q. Prove the validity of the following logical consequences by deriving them in **H**.
(a) $\forall x (P(x) \vee Q) \vdash_{\mathbf{H}} \forall x P(x) \vee Q$
(b) $\forall x P(x) \vee Q \vdash_{\mathbf{H}} \forall x (P(x) \vee Q)$
(c) $\exists x (P(x) \wedge Q) \vdash_{\mathbf{H}} \exists x P(x) \wedge Q$
(d) $\exists x P(x) \wedge Q \vdash_{\mathbf{H}} \exists x (P(x) \wedge Q)$
(e) $\exists x P(x) \to Q \vdash_{\mathbf{H}} \forall x (P(x) \to Q)$
(f) $\forall x (Q \to P(x)), Q \vdash_{\mathbf{H}} \forall x P(x)$
(g) $\exists x (Q \to P(x)), Q \vdash_{\mathbf{H}} \exists x P(x)$
(h) $\exists x (P(x) \to Q), \forall x P(x) \vdash_{\mathbf{H}} Q$
(i) $\exists x (\neg P(x) \vee Q) \vdash_{\mathbf{H}} \forall x P(x) \to Q$
(j) $\exists x (P(x) \vee \neg Q) \vdash_{\mathbf{H}} Q \to \exists x P(x)$.

4.1.6 Determine which of the following logical consequences hold by searching for a derivation in **H**. For those that you find not derivable in **H**, consider A and B as unary predicates and look for a counter-model, that is, a structure and assignment in which all premises are true while the conclusion is false.
(a) $\forall x A(x) \wedge \forall x B(x) \models \forall x (A(x) \wedge B(x))$
(b) $\forall x A(x) \vee \forall x B(x) \models \forall x (A(x) \vee B(x))$
(c) $\forall x (A(x) \wedge B(x)) \models \forall x A(x) \wedge \forall x B(x)$
(d) $\forall x (A(x) \vee B(x)) \models \forall x A(x) \vee \forall x B(x)$
(e) $\forall x A(x) \to \forall x B(x) \models \forall x (A(x) \to B(x))$

(f) $\forall x(A(x) \to B(x)) \models \forall x A(x) \to \forall x B(x)$
(g) $\forall x(A(x) \to B(x)) \models \exists x A(x) \to \exists x B(x)$
(h) $\exists x A(x) \land \exists x B(x) \models \exists x(A(x) \land B(x))$
(i) $\exists x(A(x) \land B(x)) \models \exists x A(x) \land \exists x B(x)$
(j) $\exists x(A(x) \lor B(x)) \models \exists x A(x) \lor \exists x B(x)$
(k) $\exists x A(x) \lor \exists x B(x) \models \exists x(A(x) \lor B(x))$
(l) $\exists x A(x) \to \exists x B(x) \models \exists x(A(x) \to B(x))$
(m) $\exists x(A(x) \to B(x)) \models \exists x A(x) \to \exists x B(x)$
(n) $\forall y \exists x(P(x) \to Q(y)) \models \forall x P(x) \to \forall z Q(z)$
(o) $\exists y \forall x(P(x) \to Q(y)) \models \exists x P(x) \to \exists z Q(z)$
(p) $\models \exists x(P(x) \to \forall y P(y))$.

4.1.7 Formalize the following arguments in first-order logic and try to prove their logical correctness by deriving them in **H**. For those that you find not derivable in **H**, look for a counter-model.

(a)

> All logicians are clever.
> All clever people are rich.
> _____
> All logicians are rich.

(b)

> All penguins are birds.
> No penguins can fly.
> Every white bird can fly.
> _____
> Therefore, no penguin is white.

(c)

> No lion eats birds.
> Every penguin is a bird.
> Simba is a lion.
> _____
> Therefore some lion eats no penguins.

4.1.8 Prove the validity of all axioms for equality.

4.1.9 Derive the following in the extension of **H** with equality (universal quantification is assumed wherever omitted):
(a) $x_1 = y_1 \land x_2 = y_2 \to g(f(x_1, f(x_2))) = g(f(y_1, f(y_2)))$;
(b) $\forall x(x = f(x) \to (P(f(f(x))) \to P(x)))$; and
(c) $\forall x \forall y(f(x) = y \to g(y) = x) \to \forall z(g(f(z)) = z)$.
For more exercises on derivations with equality in set theory, see Section 5.2.7.

David Hilbert (23.01.1862–14.02.1943) was one of the most prominent and influential mathematicians of the late 19th and early 20th centuries.

Hilbert studied at the University of Königsberg and in 1895 became Professor in Mathematics at the University of Göttingen – the world-leading research center for mathematics at that time – where he remained for the rest of his life. He established a famous school there and mentored a great number of students and collaborators who later also became prominent mathematicians, including the chess champion Emanuel Lasker, Ernst Zermelo, Otto Blumenthal, Felix Bernstein, Hermann Weyl, Richard Courant, Erich Hecke, Hugo Steinhaus, and Wilhelm Ackermann. At some period John von Neumann was his assistant and Alonzo Church joined him as a visiting researcher.

Hilbert's knowledge of mathematics was encyclopedic and his contributions were wide ranging from mathematical logic and foundations of mathematics to the theory of invariants, algebraic number fields, the introduction of (now known as) Hilbert spaces, and the founding of the field of functional analysis as the main instrument of modern mathematical physics.

At the end of the 19th century Hilbert analyzed and revised systematically Euclid's postulates of geometry and their consequences. He modernized and re-cast Euclid's seminal work in his book *Foundations of Geometry* (1899), with which he put the foundations of geometry in a formal axiomatic setting. Starting with this work Hilbert also promoted the formal axiomatic approach to mathematics, essentially proclaimed in the famous *Hilbert's program* for formalizing the entire body of mathematics as the most influential development in the foundations of mathematics throughout the 20th century. By launching and pursuing that program, Hilbert became one of the founders of proof theory, which he called **metamathematics**, a major branch of mathematical logic.

In his historical speech *The Problems of Mathematics* delivered at the Second International Congress of Mathematicians in Paris in 1900, Hilbert presented his famous list of 23 unsolved mathematical problems, the solutions of which would, according to Hilbert, guide and chart the development of mathematics in the forth-coming 20th century. Many of Hilbert's 23 problems were solved during the 20th century and their solutions did influence much of the development of mathematics at the time. Others were foundational problems, such as the Continuum Hypothesis and the existence of well-ordering of the reals, the solutions of which were proved to essentially depend on the axiomatic foundations of set theory. Hilbert's list also included some still famously unsolved problems, such as Goldbach's conjecture and the Riemann hypothesis.

Hilbert was baptized and raised in the Reformed Protestant Church, but he later left the church and became an agnostic. He argued that mathematical truth was independent of the existence of God or other *a priori* assumptions. In response to the Latin maxim *Ignoramus et ignorabimus* (*We do not know, we shall not know*), he stated his mathematician's famous credo in his retirement address in 1930: *Wir müssen wissen. Wir werden wissen.* (*We must know. We will know.*). This credo became the epitaph on his tombstone in Göttingen.

4.2   Semantic Tableaux for first-order logic

We now extend the method of Propositional Semantic Tableaux to first-order logic by adding rules for the quantifiers.

Quantifier rules for Semantic Tableaux

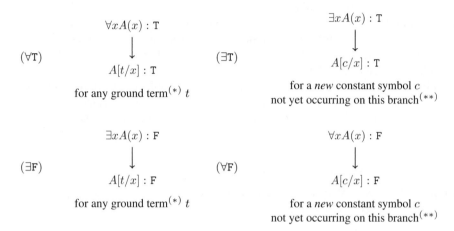

$(\forall T)$

$\forall x A(x) : \text{T}$

$\downarrow$

$A[t/x] : \text{T}$

for any ground term $^{(*)}$ t

$(\exists T)$

$\exists x A(x) : \text{T}$

$\downarrow$

$A[c/x] : \text{T}$

for a *new* constant symbol c
not yet occurring on this branch$^{(**)}$

$(\exists F)$

$\exists x A(x) : \text{F}$

$\downarrow$

$A[t/x] : \text{F}$

for any ground term $^{(*)}$ t

$(\forall F)$

$\forall x A(x) : \text{F}$

$\downarrow$

$A[c/x] : \text{F}$

for a *new* constant symbol c
not yet occurring on this branch$^{(**)}$

(*)*Recall that a ground term is a term with no variables. In order to always enable this rule we assume that there is at least one constant symbol in the language.*
(**)*This rule may only be applied once for the given formula on each branch.*

The rules are quite natural and simple. Their correctness, which will be stated more formally further, follows from some semantic properties of quantifiers and logical consequence that are listed in Theorem 112. However, some discussion is in order. I defer the discussion on the quantifier rules $(\forall T)$ and $(\exists F)$ to the end of this section. Let me first explain the proviso $^{(**)}$. It is only needed to prevent "useless" redundant applications of the rules $(\exists T)$ and $(\forall F)$. Note that without that proviso, each of these rules could be applied in principle infinitely many times: once for every constant symbol in the language that does not yet occur on the current branch. Intuitively, however, such applications should be redundant, because we only need *one* witness of the truth (respectively, of falsity) of the formula assumed to be true in the rule $(\exists T)$ (respectively, false in $(\forall F)$), *if such a witness exists*. We therefore only need to introduce *one name* for such a witness in any given branch of the tableau. If the assumption of the existence of such a witness can lead to a contradiction and close the branch, then one application of the rule should suffice; otherwise, repeated applications of the rule will have no effect, except to make it impossible to declare the branch, and the tableau construction, saturated.

We treat Semantic Tableaux for first-order logic as a deductive system, denoted again as **ST**. Now, the idea of derivation in Semantic Tableaux for first-order logic is essentially the same as in propositional logic: in order to prove that $A_1, \ldots, A_n \models C$, we search systematically for a *counter-model*, that is, a structure where $A_1, \ldots, A_n$ and $\neg C$ are satisfied simultaneously. Closed and open branches and tableaux are defined as in Propositional Semantic Tableaux, but with the more refined notion of saturation of first-order tableaux as discussed above. A closed tableau with the formulae $A_1, \ldots, A_n, \neg C$ at the root certifies that there cannot be such a counter-model, and that constitutes a derivation in Semantic Tableaux of the logical consequence $A_1, \ldots, A_n \models C$, denoted $A_1, \ldots, A_n \vdash_{\text{ST}} C$. In particular, a closed tableau with the formula $\neg C$ at the root, denoted $\vdash_{\text{ST}} C$, certifies that C is not falsifiable; it is therefore valid. On the other hand, a tableau with an open and saturated branch contains the information needed for the construction of a counter-model for the formula or set of formulae in the label of the root.

Unlike propositional logic, where every tableau either closes or else saturates and terminates as open, a third case is also possible where the tableau does not close or produce an open and saturated branch. This case will be illustrated in the last example below and termination is discussed at the end of the section.

Finally, note that the unsigned version of Semantic Tableaux for first-order logic can easily be obtained by modifying the quantifier rules to handle negated quantifiers and adding these to the unsigned version of propositional Semantic Tableaux.

4.2.1   Some derivations in Semantic Tableaux

Example 127 *Using Semantic Tableaux, check if* $\forall x(Q \to P(x)) \models Q \to \forall x P(x)$, *where x does not occur free in Q.*

$$\forall x(Q \to P(x)) : \text{T}, Q \to \forall\, xP(x) : \text{F}$$
$$\downarrow$$
$$\forall x(Q \to P(x)) : \text{T}, Q : \text{T}, \forall x P(x) : \text{F}$$
$$\downarrow$$
$$p : F, \neg q : F$$
$$\downarrow$$
$$P(c) : \text{F}$$
$$\downarrow$$
$$Q \to P(c) : \text{T}$$

$$Q : F \qquad\qquad P(c) : \text{T}$$
$$\times \qquad\qquad\qquad \times$$

The tableau above closes, implying that $\forall x(Q \to P(x)) \vdash_{\text{ST}} Q \to \forall x P(x)$, hence $\forall x(Q \to P(x)) \models Q \to \forall x P(x)$ holds.

Where was the assumption that x is not free in Q used?

Example 128 *Using Semantic Tableaux, show that* $\models \neg\forall x\forall y(P(x) \wedge \neg P(f(y)))$, *where P is a unary predicate symbol and f is a unary functional symbol.*

$$\neg \forall x \forall y (P(x) \wedge \neg P(f(y))) : F$$

$$\downarrow$$

$$\forall x \forall y (P(x) \wedge \neg P(f(y))) : T$$

$$\downarrow$$

$$\forall y (P(f(c)) \wedge \neg P(f(y))) : T$$

$$\downarrow$$

$$P(f(c)) \wedge \neg P(f(c)) : T$$

$$\downarrow$$

$$P(f(c)) : T, \neg P(f(c)) : T$$

$$\downarrow$$

$$P(f(c)) : T, \, P(f(c)) : F$$

$$\times$$

The tableau closes, so $\vdash_{\mathbf{ST}} \neg \forall x \forall y (P(x) \wedge \neg P(f(y)))$, hence $\vDash \neg \forall x \forall y (P(x) \wedge \neg P(f(y)))$. In this example I first applied the rule ($\forall T$) in a somewhat artificial way, by introducing a new constant symbol and guessing the ground term $f(c)$ to substitute for x. Such guessing would be avoided in the Free Variable version of the Semantic Tableaux method, mentioned at the end of this section.

Example 129 *Using Semantic Tableaux, check if*
$\exists x (A(x) \rightarrow B(x)), \exists x A(x) \vDash \exists x B(x).$

$$\exists x (A(x) \rightarrow B(x)) : T, \exists x A(x) : T, \exists x B(x) : F$$

$$\downarrow$$

$$A(c_1) \rightarrow B(c_1) : T$$

$$\downarrow$$

$$A(c_2) : T$$

$$\downarrow$$

$$B(c_1) : F$$

$$\downarrow$$

$$B(c_2) : F$$

$$A(c_1) : F \qquad\qquad B(c_1) : T$$

$$\bigcirc \qquad\qquad\qquad \times$$

Note that, due to the proviso $^{(*)}$, the left-hand branch in this tableau can only be extended further by applying the rule $\exists F$ to the signed formula $\exists x B(x) : F$ to produce instances of the type $B(c) : F$ for new constant symbols c (as there are no function symbols in the formula); these clearly cannot produce a contradictory pair of formulae, so we can treat such applications of the rule as redundant. The left-hand branch therefore remains open, and the tableau does not close; $\exists x(A(x) \rightarrow B(x)), \exists x A(x) \models \exists x B(x)$ therefore cannot be derived in **ST**, and this consequence is not valid.

Indeed, the necessary information for building a counter-model can be collected from the open branch of the tableau. It must have at least two elements, say a_1 and a_2, interpreting c_1 and c_2, respectively, and these should suffice because no other names of elements are mentioned in the tableau. The predicate A must then be false for a_1 and true for a_2, that is, its interpretation in the domain $\{a_1, a_2\}$ is $\{a_2\}$; the predicate B must be false for both a_1 and a_2, that is, its interpretation in the domain $\{a_1, a_2\}$ is $\emptyset$. Indeed, it is easy to check that the resulting model falsifies the logical consequence $\exists x(A(x) \rightarrow B(x)), \exists x A(x) \models \exists x B(x)$.

Example 130 *Using Semantic Tableaux, check if* $\forall x \exists y A(x, y) \models \exists x A(x, c_0)$, *for a constant symbol* c_0.

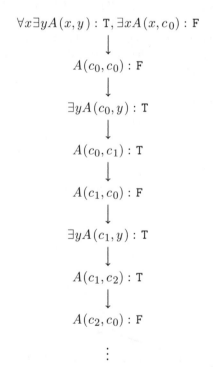

$$\forall x \exists y A(x, y) : T, \exists x A(x, c_0) : F$$
$$\downarrow$$
$$A(c_0, c_0) : F$$
$$\downarrow$$
$$\exists y A(c_0, y) : T$$
$$\downarrow$$
$$A(c_0, c_1) : T$$
$$\downarrow$$
$$A(c_1, c_0) : F$$
$$\downarrow$$
$$\exists y A(c_1, y) : T$$
$$\downarrow$$
$$A(c_1, c_2) : T$$
$$\downarrow$$
$$A(c_2, c_0) : F$$
$$\vdots$$

It should now be evident that this tableau can be extended *forever* without either closing or saturating. That indicates that $\forall x \exists y A(x, y) \models \exists x A(x, c_0)$ cannot be derived in Semantic Tableaux, hence this consequence is not valid. Indeed, a simple counter-model can be defined in the structure of natural numbers $\mathcal{N}$ where $A(x, y)$ is interpreted as $x < y$

and c_0 is interpreted as 0. However, this fact cannot be formally established at any finite stage of the extension of the tableau. We therefore see that the construction of a first-order tableau *does not always terminate*.

4.2.2 *Semantic Tableaux for first-order logic with equality*

The system of Semantic Tableaux presented here does not involve rules for the equality, but such rules can be added to produce a sound and complete system for first-order logic with equality.

Semantic Tableaux rules for the equality

$$A[t/x] : \text{T}$$
$$\downarrow$$
$$(=\text{T}) \qquad\qquad A[s/x] : \text{T}$$

for any terms s, t free for x in A such that
$s = t : \text{T}$ or $t = s : \text{T}$ occurs on the branch

$$A[t/x] : \text{F}$$
$$\downarrow$$
$$(=\text{F}) \qquad\qquad A[s/x] : \text{F}$$

for any terms s, t free for x in A such that
$s = t : \text{T}$ or $t = s : \text{T}$ occurs on the branch

As well as these rules, the following condition for closing a branch is added: a branch closes if for some term t the signed formula $t = t : \text{F}$ occurs on it.

Example 131 *Using Semantic Tableaux with equality, check whether*
$\forall x(P(x) \leftrightarrow x = c) \models P(c)$, *for a constant symbol c.*
The tableau is constructed as follows.

$$\forall x(P(x) \leftrightarrow x = c): \text{ T}, P(c) : \text{F}$$
$$\downarrow$$
$$P(c) \leftrightarrow c = c : \text{T}$$

$$P(c) : \text{T}, c = c : \text{T}, P(c) : \text{F} \qquad\qquad P(c): \text{ F}, c = c : \text{F}, P(c) : \text{F}$$
$$\times \qquad\qquad\qquad\qquad\qquad\qquad \times$$

Note that the right-hand branch closes due to the new closure rule.

Example 132 *Using Semantic Tableaux with equality, derive*
$\vdash_{\mathbf{ST}} \forall x \forall y (x = y \rightarrow (P(f(x)) \leftrightarrow P(f(y)))),$
where P is a unary predicate symbol and f is a unary function symbol:

$$\forall x \forall y (x = y \rightarrow (P(f(x)) \leftrightarrow P(f(y)))) : \mathbf{F}$$
$$\downarrow$$
$$\forall y (c = y \rightarrow (P(f(c)) \leftrightarrow P(f(y)))) : \mathbf{F}$$
$$\downarrow$$
$$(c = d \rightarrow (P(f(c)) \leftrightarrow P(f(d)))) : \mathbf{F}$$
$$\downarrow$$
$$c = d : \mathbf{T}, (P(f(c)) \leftrightarrow P(f(d))) : \mathbf{F}$$

$c = d : \mathbf{T}, (P(f(c)) \rightarrow P(f(d)) : \mathbf{F}$ $c = d : \mathbf{T}, P(f(d)) \rightarrow P(f(c)) : \mathbf{F}$

$\qquad\qquad \downarrow$ $\qquad\qquad \downarrow$

$c = d : \mathbf{T}, P(f(c)) : \mathbf{T}, \ P(f(d)) : \mathbf{F}$ $c = d : \mathbf{T}, P(f(d)) : \mathbf{T}, P(f(c)) : \mathbf{F}$

$=_{\mathbf{T}} \downarrow$ $=_{\mathbf{F}} \downarrow$

$P(f(d)) : \mathbf{T}$ $P(f(d)) : \mathbf{F}$

$\times$ $\times$

The tableaux close. The applications of the rules equality are indicated. In both cases, we had a choice to apply either of these rules.

Example 133 *As a last example we prove once more, now using Semantic Tableaux with equality, that the line in the old jazz song*

> *"Everybody loves my baby, but my baby don't love nobody but me"*
> *implies that "I am my baby," that is,*

$$\forall x L(x, MyBaby) \land \forall y (\neg y = Me \rightarrow \neg L(MyBaby, y)) \vdash_{\mathbf{ST}} MyBaby = Me.$$

Theorem 134 (Soundness and completeness of ST) *Each of the Semantic Tableaux systems* **ST** *for first-order logic, with and without equality, is sound and complete, that is, for all first-order formulae $A_1, \dots, A_n, C$ of the respective language (respectively, with or without equality):*

$$A_1, \dots, A_n, \vdash_{\mathbf{ST}} C \text{ iff } A_1, \dots, A_n, \models C.$$

The proof for **ST** with equality is outlined in Section 4.6.

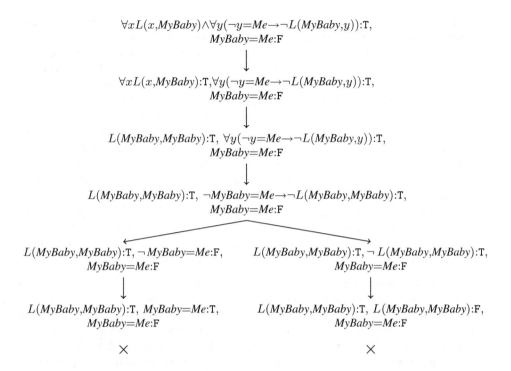

4.2.3   Discussion on the quantifier rules and on termination

We end this section with some discussion on the quantifier rules ($\forall$T) and ($\exists$F) and on the termination of the Semantic Tableaux method for first-order logic.

1. When the language has at least one constant and one functional symbol, there are infinitely many ground terms that can be instantiated in the rules ($\forall$T) and ($\exists$F). All, except for finitely many of these instantiations, would be "useless" for the closure of the branch of the tableau. The rules as they stand therefore appear to be "wasteful", in the sense of allowing many useless applications. As we saw in some of the examples, this could lead to unnecessary non-termination but will not affect the soundness and completeness of the system, as stated in Theorem 134. Still, in automated theorem provers, these rules can be used quite efficiently by employing suitable heuristics and strategies on how to use them.

2. On the other hand, these rules can be strengthened by allowing instantiation with *any* term t that is free for x in A. The resulting rules would still be sound, but even more wasteful in the sense of allowing useless applications.

3. The problem of identifying and preventing useless applications of the rules ($\forall$T) and ($\exists$F) is non-trivial. As we will see soon, this problem is in fact *algorithmically unsolvable* in any sound and complete version of the Semantic Tableaux method. Intuitively, it may seem sufficient to only apply these rules for terms that have already appeared on the branch, if a contradiction is to be reached on that branch. However, such a restriction would be too strong and could prevent the tableau from closing, for example for the unsatisfiable input formula $\forall x \forall y (P(x) \wedge \neg P(f(y)))$.

4. There is a variation of the quantifier rules leading to the so-called **Free Variable Tableaux** version of the method, which defers the term instantiations applied with the rules $(\forall T)$ and $(\exists F)$ to a later stage, where it may become clear exactly what instantiations are needed in order to close the branch, if possible. The idea is (1) to modify the rules $(\forall T)$ and $(\exists F)$ to instantiate not with ground terms but with new free variables, and (2) to modify the rules $(\forall F)$ and $(\exists T)$ to instantiate not with a new constant but with a term $f(v_1, \ldots, v_k)$, where f is a *new* functional symbol and $v_1, \ldots, v_k$ are free variables on the branch that may have appeared there by applications of the rules $(\forall T)$ and $(\exists F)$. The technique behind this modification of the rules is called *Skolemization* and is presented and discussed in Section 4.4.2. In the case when there are no such free variables, the modified rules act like the present rules by instantiating with new constant symbols.

For example, the guessing in Example 128 could be avoided in the Free Variable Tableaux version by instantiating x with a new free variable v_1, then instantiating y with another free variable v_2, and finally instantiating v_1 with $f(v_2)$. This is a simple case of the *term unification* technique, which I discuss in detail in Section 4.5.3.

Finally, some words on termination. We have seen that the system of Semantic Tableaux presented here does not always terminate, even in cases where a suitable application of the rules would close the tableau. The question therefore arises: can the rules of the system be modified so that it remains sound and complete, but also always terminates? The answer is an emphatic *no*, at least not if the rules are to be algorithmically implementable. More precisely, in any sound and complete system of Semantic Tableaux it is not possible to determine algorithmically whether the construction of the tableau for any given first-order formula can terminate (as open or closed) or not. This impossibility is a direct consequence from Church's Undecidability Theorem stated in the introduction to this chapter.

References for further reading
For a comprehensive handbook on tableaux methods, not only in classical logic, see D'Agostino *et al.* (1999). For more details on the theory and examples of derivations in Semantic Tableaux for first-order logic, see Nerode and Shore (1993), Jeffrey (1994, who referred to "analytic trees"), Smullyan (1995, one of the first versions of Semantic Tableaux), Fitting (1996), Smith (2003, who called them "trees"), Ben-Ari (2012), and van Benthem *et al.* (2014).

Exercises

For each of the following exercises use Semantic Tableaux.

4.2.1 Prove the following logical validities and consequences.
(a) $\models \forall x P(x) \rightarrow \forall y P(y)$
(b) $\models \exists x P(x) \rightarrow \exists y P(y)$
(c) $\forall x A(x) \models \neg \exists x \neg A(x)$
(d) $\neg \exists x \neg A(x) \models \forall x A(x)$
(e) $\exists x A(x) \models \neg \forall x \neg A(x)$
(f) $\neg \forall x \neg A(x) \models \exists x A(x)$
(g) $\models \exists x \exists y Q(x, y) \rightarrow \exists y \exists x Q(x, y)$.

4.2.2 Suppose x is not free in Q. Prove the following logical consequences.

(a) $\forall x(P(x) \vee Q) \models \forall x P(x) \vee Q$

(b) $\forall x P(x) \vee Q \models \forall x(P(x) \vee Q)$

(c) $\exists x(P(x) \wedge Q) \models \exists x P(x) \wedge Q$

(d) $\exists x P(x) \wedge Q \models \exists x(P(x) \wedge Q)$

(e) $\forall x(Q \rightarrow P(x)), Q \models \forall x P(x)$

(f) $\exists x P(x) \rightarrow Q \models \forall x(P(x) \rightarrow Q)$

(g) $\exists x(P(x) \rightarrow Q), \forall x P(x) \models Q$

(h) $\exists x(Q \rightarrow P(x)), Q \models \exists x P(x)$.

4.2.3 Check which of the following first-order formulae are logically valid. For each of those which are not, construct a *counter-model*, that is, a structure which falsifies it.

(a) $\exists x(P(x) \rightarrow \forall y P(y))$

(b) $\forall x(P(x) \rightarrow \exists y P(y))$

(c) $\forall x \exists y Q(x, y) \rightarrow \forall y \exists x Q(x, y)$

(d) $\forall x \exists y Q(x, y) \rightarrow \forall y \exists x Q(y, x)$

(e) $\forall x \exists y Q(x, y) \rightarrow \exists y \forall x Q(x, y)$

(f) $\forall x \exists y Q(x, y) \rightarrow \exists x Q(x, x)$

(g) $\exists y \forall x Q(x, y) \rightarrow \exists x Q(x, x)$

(h) $\forall x(\exists y Q(x, y) \rightarrow \exists y Q(y, x))$

(i) $\exists x \neg \exists y P(x, y) \rightarrow \forall y \neg \forall x P(x, y)$

(j) $(\forall x \exists y P(x, y) \wedge \forall x \forall y(P(x, y) \rightarrow P(y, x))) \rightarrow \exists x P(x, x)$.

4.2.4 Check which of the following logical consequences hold. For those which do not, use the tableau to construct a counter-model, that is, a structure and assignment in which all premises are true, while the conclusion is false.

(a) $\forall x A(x), \forall x B(x) \models \forall x(A(x) \wedge B(x))$

(b) $\forall x(A(x) \wedge B(x)) \models \forall x A(x) \wedge \forall x B(x)$

(c) $\forall x A(x) \vee \forall x B(x) \models \forall x(A(x) \vee B(x))$

(d) $\forall x(A(x) \vee B(x)) \models \forall x A(x) \vee \forall x B(x)$

(e) $\forall x A(x) \rightarrow \forall x B(x) \models \forall x(A(x) \rightarrow B(x))$

(f) $\forall x(A(x) \rightarrow B(x)) \models \forall x A(x) \rightarrow \forall x B(x)$

(g) $\forall x(A(x) \rightarrow B(x)) \models \exists x A(x) \rightarrow \exists x B(x)$

(h) $\exists x A(x), \exists x B(x) \models \exists x(A(x) \wedge B(x))$

(i) $\exists x(A(x) \wedge B(x)) \models \exists x A(x) \wedge \exists x B(x)$

(j) $\exists x(A(x) \vee B(x)) \models \exists x A(x) \vee \exists x B(x)$

(k) $\exists x A(x) \vee \exists x B(x) \models \exists x(A(x) \vee B(x))$

(l) $\exists x(A(x) \rightarrow B(x)) \models \exists x A(x) \rightarrow \exists x B(x)$

(m) $\exists x A(x) \rightarrow \exists x B(x) \models \exists x(A(x) \rightarrow B(x))$.

4.2.5 Determine which of the following logical equivalences hold. For each of those that do not hold, construct a counter-model, that is, a structure in which one formula is true while the other is false.

(a) $\forall x(P(x) \wedge Q(x)) \equiv \forall x P(x) \wedge \forall x Q(x)$

(b) $\forall x(P(x) \vee Q(x)) \equiv \forall x P(x) \vee \forall x Q(x)$

(c) $\forall x(P(x) \rightarrow Q(x)) \equiv \forall x P(x) \rightarrow \forall x Q(x)$

(d) $\exists x(P(x) \wedge Q(x)) \equiv \exists x P(x) \wedge \exists x Q(x)$

(e) $\exists x (P(x) \to Q(x)) \equiv \exists x P(x) \to \exists x Q(x)$
(f) $\exists x (P(x) \lor Q(x)) \equiv \exists x P(x) \lor \exists x Q(x)$
(g) $\forall x Q(x, x) \equiv \forall x \exists y Q(x, y)$
(h) $\exists x \forall y Q(x, y) \equiv \exists y Q(y, y)$
(i) $\forall x \exists y Q(x, y) \equiv \exists x Q(x, x)$
(j) $\forall x \forall y Q(x, y) \equiv \forall x \forall y Q(y, x)$
(k) $\exists x P(x) \to \forall x P(x) \equiv \exists x \neg P(x) \to \forall x \neg P(x)$.

4.2.6 In each of the following cases formalize both statements in first-order logic and check which, if any, logically implies the other. If any of the logical consequences do not hold, construct an appropriate counter-model.
   (a) A: "Not every lawyer is greedy."
       B: "There exists a lawyer and not everybody is greedy."
   (b) A: "Some men are happy only when they are drunk."
       B: "There is a drunk man or there is an unhappy man."
   (c) A: " If somebody is talking then everybody is listening."
       B: " There is somebody, such that if he is not listening then nobody is talking."

4.2.7 Formalize the following arguments in FOL by introducing suitable predicates. Then check for each of them whether it is logically valid. If not valid, use the tableau to construct a counter-model.

   (a)

       All logicians are clever.
       All clever people are rich.
       ———————————————
       All logicians are rich.

   (b)

       All natural numbers are integers.
       Some integers are odd numbers.
       ———————————————
       Some natural numbers are odd numbers.

   (c)

       No man loves every woman.
       Some men are bachelors.
       ———————————————
       Therefore, some bachelor does not love some woman.

   (d)

       If everybody is selling then somebody is buying or negotiating.
       No seller is buying.
       ———————————————
       Therefore somebody is selling only if he is negotiating.

(e)

No driver is a pedestrian.
Every child is a pedestrian.
Johnnie is a child.
_____
Therefore somebody is not a driver.

(f)

All penguins are birds.
Some penguins are white.
No white bird eats butterflies.
_____
Therefore, not all penguins eat butterflies.

(g)

Everybody likes some black gmuck.
Nobody likes any white gmuck.
_____
Therefore no gmuck is black and white.

(h)

Some bachelors love every woman they know.
Every bachelor is a man.
Eve is a woman whom every man knows.
_____
Some bachelor loves a woman.

(i) Replace the conclusion in the previous example by
"Every bachelor loves a woman,"
and check whether the resulting argument is logically valid.

4.2.8 Using Semantic Tableaux with equality, derive each of the following (the first five are the axioms of **H** for the equality):
   (a) $(\text{Ax}_=1) \; x = x$
   (b) $(\text{Ax}_=2) \; x = y \rightarrow y = x$
   (c) $(\text{Ax}_=3) \; x = y \wedge y = z \rightarrow x = z$
   (d) $(\text{Ax}_f) \quad x_1 = y_1 \wedge \ldots \wedge x_n = y_n \rightarrow f(x_1, \ldots, x_n) = f(y_1, \ldots, y_n)$ for n-ary functional symbol f
   (e) $(\text{Ax}_r) \; x_1 = y_1 \wedge \ldots \wedge x_n = y_n \rightarrow (p(x_1, \ldots, x_n) \rightarrow p(y_1, \ldots, y_n))$ for n-ary predicate symbol p
   (f) $s = t \rightarrow A[s/x] \leftrightarrow A[t/x]$ for any formula $A(x)$ and terms s, t free for x in A.
   (This is the Theorem 125 for equivalent replacement.)

4.2.9 Using Semantic Tableaux with equality, check whether each of the following validities and logical consequences holds.
   (a) $\models x_1 = y_1 \wedge x_2 = y_2 \rightarrow g(f(x_1, f(x_2))) = g(f(y_1, f(y_2)))$
   (Universal quantification is assumed but omitted.)

(b) $\models \forall x(x = f(x) \rightarrow (P(f(f(x))) \rightarrow P(x)))$
(c) $\forall x \forall y(f(x) = y \rightarrow g(y) = x) \models \forall z(g(f(z)) = z)$
(d) $\forall x \forall y(f(x) = y \rightarrow g(y) = x) \models \forall z(f(g(z)) = z)$
(e) $\forall x \forall y(f(x) = y \rightarrow g(y) = x), \forall x \forall y(g(x) = g(y) \rightarrow x = y) \models$
    $\forall z(f(g(z)) = z)$.

For more exercises on derivations with equality, on sets, functions, and relations, see Section 5.2.7.

Raymond Merrill Smullyan (born 25.5.1919) is an American mathematician, logician, magician, Taoist philosopher, and piano player, best known as an avid popularizer of logic through his series of scientific (mathematical logic) and recreational books with enchanting logical puzzles.

In his childhood Smullyan showed passion and strong talent for music (as a piano player), chess, and mathematics, but was not quite happy at school so he left to study on his own. He was keen on problem solving, both in maths and chess where he composed many problems of his own. He also started learning magic tricks and became a very good magician, earning money by performing magic acts in night-clubs in Greenwich Village. During the early 1950s he studied at the University of Chicago (where one of his teachers was the prominent philosopher and logician Rudolf Carnap), while himself teaching at Dartmouth College. He achieved his PhD at Princeton University under Alonzo Church and published his first paper on logic in 1957, where he showed that Gödel's incompleteness phenomena based on self-reference also apply to much simpler formal systems than Peano arithmetic and gave new, very intuitive explanations of Gödel's theorems.

Smullyan has been for many years a Professor of Mathematics and Philosophy at Yeshiva University, New York, and at Lehman College and the Graduate Center of the City University of New York, where he is still a Professor Emeritus. He is also an Oscar Ewing Professor of Philosophy at Indiana University.

Smullyan has published several well-known books on mathematical logics, including *Theory of Formal Systems* (1959), *First-Order Logic* (1968), *Gödel's Incompleteness Theorems* (1992), *Recursion Theory for Metamathematics* (1993), *Diagonalization and Self-reference* (1994), *Set Theory and the Continuum Problem* (1996, together with Melvin Fitting), and *A Beginner's Guide to Mathematical Logic* (2014). In his book *First-Order Logic* Smullyan developed the method of *analytic tableaux* as a variant of the Semantic Tableaux of Beth and Hintikka, and related it to Gentzen's proofs systems.

Smullyan also wrote a few books on Taoist philosophy, but he is most popular for his fascinating books with logical, mathematical, and chess puzzles, including *What Is the Name of This Book?* (1978), *The Chess Mysteries of Sherlock Holmes* (1979)

The Lady or the Tiger? (1982), *Alice in Puzzle-Land* (1982), *To Mock a Mockingbird* (1985), *Forever Undecided* (1987), *Satan, Cantor, and Infinity* (1992), *The Riddle of Scheherazade* (1997), *King Arthur in Search of his Dog* (2010), *The Gödelian Puzzle Book: Puzzles, Paradoxes and Proofs* (2013), and many others.

As well as teaching and writing about logic, Smullyan is an accomplished piano player and has produced many recordings of his favorite classical piano pieces.

In 2004 the filmmaker Tao Ruspoli produced a documentary film about him called *This Film Needs No Title: A Portrait of Raymond Smullyan.*

Melvin Fitting (born 24.1.1942) is an American logician, mathematician, philosopher, and computer scientist with very broad interests, and has made many contributions to mathematical, philosophical, and computational logic.

Fitting obtained his doctorate in mathematics from Yeshiva University, New York under the supervision of Raymond Smullyan. From 1968 to 2013 he was a professor at City University of New York, Lehman College and at the Graduate Center, in the departments of Computer Science, Philosophy, and Mathematics. He is currently an Emeritus Professor there.

Fitting's philosophy of logic can be formulated succinctly as follows (quoting Wikipedia): "There are many logics. Our principles of reasoning vary with context and subject matter. Multiplicity is one of the glories of modern formal logic. The common thread tying logics together is a concern for what can be said (syntax), what that means (semantics), and relationships between the two. A philosophical position that can be embodied in a formal logic has been shown to be coherent, not correct. Logic is a tool, not a master, but it is an enjoyable tool to use."

Fitting has worked in numerous diverse fields of logic, an impressive testimony of which is the long list of well-known books he has written over the years including *Intuitionistic Logic, Model Theory, and Forcing* (1969), *Fundamentals of Generalized Recursion Theory* (1981), *Proof Methods for Modal and Intuitionistic Logics* (1983), *First-Order Logic and Automated Theorem Proving* (1990), *Set Theory and the Continuum Problem*, with Raymond Smullyan (1996), *First-Order Modal Logic*, with Richard Mendelsohn (1998) *Types, Tableaus, and Gödel's God* (2002), and *Incompleteness in the Land of Sets* (2007). In addition, he has carried out important work on tableaux methods of proof for modal logics, bilattices and the theory of truth, many-valued modal logics, and justification logic.

4.3   Natural Deduction for first-order logic

I now extend propositional Natural Deduction to first-order logic by adding rules for the quantifiers.

4.3.1   Natural Deduction rules for the quantifiers

Introduction rules		Elimination rules	
$(\forall I)^*$	$\dfrac{A[c/x]}{\forall x A(x)}$	$(\forall E)^{**}$	$\dfrac{\forall x A(x)}{A[t/x]}$
$(\exists I)^{**}$	$\dfrac{A[t/x]}{\exists x A(x)}$	$(\exists E)^{***}$	$\dfrac{\exists x A(x) \qquad \begin{array}{c} [A[c/x]] \\ \vdots \\ C \end{array}}{C}$

*where c is a constant symbol, not occurring in $A(x)$ or in any open assumption used in the derivation of $A[c/x]$;
**for any term t free for x in A;
***where c is a constant symbol, not occurring in $A(x)$, C, or in any open assumption in the derivation of C, except for $A[c/x]$.

Let us start with some discussion and explanation of the quantifier rules.

$(\forall I)$: If we are to prove a universally quantified sentence, $\forall x A(x)$, we reason as follows. We say "Let c be any object from the domain (e.g., an arbitrary real number)." However, the name c of that arbitrary object must be new – one that has not yet been used in the proof – to be sure that it is indeed arbitrary. We then try to prove that $A(c)$ holds, without assuming any specific properties of c. If we succeed, then our proof will apply to *any* object c from the structure, so we will have a proof that $A(x)$ holds for *every* object x, that is, a proof of $\forall x A(x)$.

$(\exists I)$: If we are to prove an existentially quantified sentence $\exists x A(x)$, then we try to find an explicit **witness**, an object in the domain, satisfying A. Within the formal system we try to come up with a term t that would name such a witness.

$(\forall E)$: If a premise is a universally quantified sentence $\forall x A(x)$, we can assume $A(a)$ for any object a from the structure. Within the formal system, instead of elements of a structure, we must use syntactic objects which represent them, namely, terms[3].

$(\exists E)$: If a premise is an existentially quantified sentence $\exists x A(x)$ then we introduce a name, say c, for an object that satisfies A. That is, we say "if there is an x such

---

[3] Note that this sounds a little weaker because in general not every element of the domain has a name, or is the value of a term. However, it is sufficient because *ad hoc* names can be added whenever necessary.

that $A(x)$, then take one and call it c." We then replace the premise by $A[c/x]$. If we succeed in deriving the desired conclusion C from that new premise, then it can be derived from $\exists x\, A(x)$, provided c does not occur in A or C, or in any other assumption used in the derivation of C from $A[c/x]$. This means that the name c must be a new one that has not yet been used in the proof. At that moment we can discard the assumption $A[c/x]$, as it was only an auxiliary assumption used to justify the derivation of C from $\exists x\, A(x)$.

From this point onward I write $A(t)$ instead of $A[t/x]$ whenever it is clear from the context which variable is substituted by the term t.

Remark 135 *A common mistake is to use this simpler rule for elimination of $\exists$:*

$$(\exists E)\ \ \frac{\exists x\, A(x)}{A(c)}$$

where c is a new constant symbol. Although simple and natural looking, this rule is not logically sound! For instance, using it I can derive the invalid implication $\exists x\, A(x) \to A(c)$ which can be interpreted as, for example, "If anyone will fail the exam then $\langle \texttt{YourName} \rangle$ will fail the exam," which you certainly do not wish to be valid.

4.3.2   Derivations in first-order Natural Deduction

Derivations in first-order Natural Deduction are organized as for derivations in Propositional Natural Deduction: as trees growing upward (but usually constructed downward); with nodes decorated by formulae, where the leaves are decorated with given or introduced assumptions; every node branching upward according to some of the derivation rules; and the formula decorating that node is the conclusion of the rule, while those decorating the successor nodes are the premises. The root of the derivation tree is decorated with the formula to be derived. In particular, the Natural Deduction derivation tree for the logical consequence $A_1, \ldots, A_n \models B$ has leaves decorated with formulae among $A_1, \ldots, A_n$ or with *subsequently canceled* additional assumptions, while the root is decorated with B. If such a derivation tree exists, then we say that $A_1, \ldots, A_n \models B$ is derivable in Natural Deduction and write $A_1, \ldots, A_n \vdash_{\mathbf{ND}} B$.

Here are some examples of derivations in Natural Deduction. Check that the rules for the quantifiers have been applied correctly.

1. $\vdash_{\mathbf{ND}} \forall x \forall y P(x,y) \to \forall y \forall x P(x,y)$:

$$(\to I)\ \cfrac{(\forall I)\ \cfrac{(\forall I)\ \cfrac{(\forall E)\ \cfrac{(\forall E)\ \cfrac{[\forall x \forall y P(x,y)]^1}{\forall y P(c_1, y)}}{P(c_1, c_2)}}{\forall x P(x, c_2)}}{\forall y \forall x P(x, y)}}{\forall x \forall y P(x,y) \to \forall y \forall x P(x,y)}\ 1$$

2. $\forall x(P(x) \wedge Q(x)) \vdash_{\mathbf{ND}} \forall x P(x) \wedge \forall x Q(x)$:

$$
(\wedge I) \dfrac{(\forall I)\dfrac{(\wedge E)\dfrac{(\forall E)\dfrac{\forall x(P(x) \wedge Q(x))}{P(c) \wedge Q(c)}}{P(c)}}{\forall x P(x)} \qquad (\forall I)\dfrac{(\wedge E)\dfrac{(\forall E)\dfrac{\forall x(P(x) \wedge Q(x))}{P(c) \wedge Q(c)}}{Q(c)}}{\forall x Q(x)}}{\forall x P(x) \wedge \forall x Q(x)}
$$

3. $\neg \exists x \neg A(x) \vdash_{\mathbf{ND}} \forall x A(x)$:

$$
\dfrac{\dfrac{\dfrac{\neg \exists x \neg A(x) \quad \dfrac{[\neg A(c)]^1}{\exists x \neg A(x)}}{\bot}}{A(c)}\ {}^1}{\forall x A(x)}
$$

4. $\neg \forall x \neg A(x) \vdash_{\mathbf{ND}} \exists x A(x)$:

$$
\dfrac{\dfrac{\neg \forall x \neg A(x) \qquad \dfrac{\dfrac{[\neg \exists x A(x)]^2 \quad \dfrac{[A(c)]^1}{\exists x A(x)}}{\bot}}{\neg A(c)}\ {}^1}{\forall x \neg A(x)}}{\dfrac{\bot}{\exists x A(x)}}\ {}^2
$$

5. Suppose x is not free in P. Then $\vdash_{\mathbf{ND}} \forall x(P \to Q(x)) \to (P \to \forall x Q(x))$:

$$
(\to I)\dfrac{(\to I)\dfrac{(\forall I)\dfrac{(\to E)\dfrac{(\forall E)\dfrac{[\forall x(P \to Q(x))]^1}{P \to Q(c)} \quad [P]^2}{Q(c)}}{\forall x Q(x)}}{P \to \forall x Q(x)}\ {}^2}{\forall x(P \to Q(x)) \to (P \to \forall x Q(x))}\ {}^1
$$

6. Suppose x is not free in P. Then $\forall x(Q(x) \to P), \exists x Q(x) \vdash_{\mathbf{ND}} P$:

$$
(\exists E)\dfrac{\exists x Q(x) \quad (\to E)\dfrac{(\forall E)\dfrac{\forall x(Q(x) \to P)}{Q(c) \to P} \quad [Q(c)]^1}{P}}{P}\ {}^1
$$

4.3.3   Natural Deduction for first-order logic with equality

The system of Natural Deduction presented so far does not involve rules for the equality. Such rules can be added to produce a sound and complete system of Natural Deduction for first-order logic with equality, as follows. In each of the rules below, t, t_i, s_i are terms.

(Ref)

$$\overline{t = t}$$

(ConFunc)

$$\frac{s_1 = t_1, \ldots, s_n = t_n}{f(s_1, \ldots, s_n) = f(t_1, \ldots, t_n)}$$

for every n-ary functional symbol f.

(Sym)

$$\frac{t_1 = t_2}{t_2 = t_1}$$

(ConRel)

$$\frac{s_1 = t_1, \ldots, s_n = t_n}{p(s_1, \ldots, s_n) \to p(t_1, \ldots, t_n)}$$

for every n-ary predicate symbol p.

(Tran)

$$\frac{t_1 = t_2, t_2 = t_3}{t_1 = t_3}$$

Example 136 *Using Natural Deduction with equality, derive*

$$\forall x \forall y (f(x) = y \to g(y) = x) \vdash_{\textbf{ND}} \forall z(g(f(z)) = z)$$

where f, g are unary function symbols.

$$(\forall I) \cfrac{(\to E) \cfrac{f(x) = f(x) \quad (\forall E) \cfrac{(\forall E) \cfrac{\forall x \forall y (f(x) = y \to g(y) = x)}{\forall y (f(x) = y \to g(y) = x)}}{f(x) = f(x) \to g(f(x)) = x}}{g(f(x)) = x}}{\forall z(g(f(z)) = z)}$$

Finally, two important general results are as follows.

Theorem 137 *[Equivalent replacement] For any formula $A(x)$ and terms s, t free for x in A, the following is derivable in* **ND**:

$$s = t \vdash_{\textbf{ND}} A[s/x] \leftrightarrow A[t/x].$$

The proof can be done by induction on A and is left as an exercise.

The proof of the following fundamental result, for **ND** with equality, will be outlined in Section 4.6.

Theorem 138 (Soundness and completeness of ND) *Each of the systems* **ND** *of Natural Deduction for first-order logic, with and without equality, is sound and complete, that is, for all first-order formulae* $A_1, \dots, A_n, C$ *of the respective language (respectively, with or without equality):*

$$A_1, \dots, A_n, \vdash_{\mathbf{ND}} C \quad \text{iff} \quad A_1, \dots, A_n, \models C.$$

References for further reading
For more details on the theory and examples of derivations in Natural Deduction, see van Dalen (1983), Jeffrey (1994) (who referred to "synthetic trees"), Smullyan (1995), Fitting (1996), Huth and Ryan (2004), Nederpelt and Kamareddine (2004), Prawitz (2006, the original development of the modern version of Natural Deduction), Chiswell and Hodges (2007), and van Benthem *et al.* (2014), as well as Kalish and Montague (1980) and Bornat (2005), where Natural Deduction derivations are presented in a boxed form rather than in tree-like shape.

Exercises

Use Natural Deduction for all exercises below. In all derivations, indicate the rules applied and the succession of all steps by numbering them. Check that the provisos for all applications of rules are satisfied.

4.3.1   Prove the following logical validities and consequences.

(a) $\models \forall x A(x) \rightarrow \forall y A(y)$ (c) $\forall x A(x) \models \neg \exists x \neg A(x)$

(b) $\models \exists x A(x) \rightarrow \exists y A(y)$ (d) $\exists x A(x) \models \neg \forall x \neg A(x)$.

4.3.2   Suppose x is not free in Q. Prove the validity of the following logical consequences by deriving them with Natural Deduction.

(a) $\forall x (P(x) \vee Q) \models \forall x P(x) \vee Q$ (f) $\forall x (Q \rightarrow P(x)), Q \models \forall x P(x)$

(b) $\forall x P(x) \vee Q \models \forall x (P(x) \vee Q)$ (g) $\exists x (Q \rightarrow P(x)), Q \models \exists x P(x)$

(c) $\exists x (P(x) \wedge Q) \models \exists x P(x) \wedge Q$ (h) $\exists x (P(x) \rightarrow Q), \forall x P(x) \models Q$

(d) $\exists x P(x) \wedge Q \models \exists x (P(x) \wedge Q)$ (i) $\exists x (\neg P(x) \vee Q) \models \forall x P(x) \rightarrow Q$

(e) $\exists x P(x) \rightarrow Q \models \forall x (P(x) \rightarrow Q)$ (j) $\exists x (P(x) \vee \neg Q) \models Q \rightarrow \exists x P(x)$.

4.3.3   Determine which of the following logical consequences hold by searching for a derivation in Natural Deduction. For those that you find not derivable in ND, consider A and B as unary predicates and look for a counter-model, that is, a structure and assignment in which all premises are true while the conclusion is false.

(a) $\forall x A(x) \wedge \forall x B(x) \models \forall x (A(x) \wedge B(x))$

(b) $\forall x A(x) \vee \forall x B(x) \models \forall x (A(x) \vee B(x))$

(c) $\forall x (A(x) \wedge B(x)) \models \forall x A(x) \wedge \forall x B(x)$

(d) $\forall x(A(x) \lor B(x)) \models \forall x A(x) \lor \forall x B(x)$
(e) $\forall x A(x) \rightarrow \forall x B(x) \models \forall x(A(x) \rightarrow B(x))$
(f) $\forall x(A(x) \rightarrow B(x)) \models \forall x A(x) \rightarrow \forall x B(x)$
(g) $\exists x A(x) \land \exists x B(x) \models \exists x(A(x) \land B(x))$
(h) $\exists x(A(x) \land B(x)) \models \exists x A(x) \land \exists x B(x)$
(i) $\exists x(A(x) \lor B(x)) \models \exists x A(x) \lor \exists x B(x)$
(j) $\exists x A(x) \lor \exists x B(x) \models \exists x(A(x) \lor B(x))$
(k) $\exists x A(x) \rightarrow \exists x B(x) \models \exists x(A(x) \rightarrow B(x))$
(l) $\exists x(A(x) \rightarrow B(x)) \models \exists x A(x) \rightarrow \exists x B(x)$
(m) $\exists x \exists y A(x,y) \models \exists y \exists x A(x,y)$
(n) $\forall y \exists x(P(x) \rightarrow Q(y)) \models \forall x P(x) \rightarrow \forall z Q(z)$
(o) $\exists y \forall x(P(x) \rightarrow Q(y)) \models \exists x P(x) \rightarrow \exists z Q(z)$
(p) $\forall y(P(y) \rightarrow Q(y)) \models \exists y P(y) \rightarrow \exists y Q(y)$
(q) $\forall y(P(y) \rightarrow Q(y)) \models \exists x \neg Q(x) \rightarrow \exists x \neg P(x)$
(r) $\models \exists x(P(x) \rightarrow \forall y P(y))$.

4.3.4 Formalize the following arguments in first-order logic and try to prove their logical correctness by deriving them in Natural Deduction. For those that you find not derivable in ND, look for a counter-model.

(a)

> No yellow and dangerous reptiles are lizards.
> Some reptile is dangerous or is not a lizard.
> ――――――――――――――――――――――――
> Therefore, not every reptile is a yellow lizard.

(b)

> All penguins are birds.
> No penguin can fly.
> Every white bird can fly.
> ――――――――――――――――――――
> Therefore, no penguin is white.

(c)

> All penguins are birds.
> Some penguins are white.
> All penguins eat some fish.
> ――――――――――――――――――――――――――――――
> Therefore, there is a white bird that eats some fish.

(d)

> No lion eats birds.
> Every penguin is a bird.
> Simba is a lion.
> ――――――――――――――――――――
> Therefore some lion eats no penguins.

4.3.5 Prove Theorem 137 for equivalent replacement in Natural Deduction with equality by structural induction on the formula A.

4.3.6 Derive each of the following in Natural Deduction with equality.
(a) $\vdash_{\mathbf{ND}} x_1 = y_1 \wedge x_2 = y_2 \to g(f(x_1, f(x_2))) = g(f(y_1, f(y_2)))$
(Universal quantification is assumed but omitted.)
(b) $\vdash_{\mathbf{ND}} \forall x(x = f(x) \to (P(f(f(x))) \to P(x)))$
(c) $\forall x \forall y(f(x) = y \to g(y) = x), \forall x \forall y(g(x) = g(y) \to x = y) \vdash_{\mathbf{ND}}$
$\forall z(f(g(z)) = z)$

4.3.7 Prove again, now using Natural Deduction with equality, that the line in the old jazz song *"Everybody loves my baby, but my baby don't love nobody but me"* implies that "I am my baby," that is,

$$\forall x L(x, \mathrm{MyBaby}) \wedge \forall y(\neg y = \mathrm{Me} \to \neg L(\mathrm{MyBaby}, y)) \vdash_{\mathbf{ND}} \mathrm{MyBaby} = \mathrm{Me}.$$

For more exercises on derivations with equality, on sets, functions, and relations, see Section 5.2.7.

Dag Prawitz (born 16.05.1936) is a Swedish philosopher and logician who has made seminal contributions to proof theory as well as to the philosophy of logic and mathematics.

Prawitz was born and brought up in Stockholm. He studied theoretical philosophy at Stockholm University as a student of Anders Wedberg and Stig Kanger, and obtained a PhD in philosophy in 1965. After working for a few years as a docent (associate professor) in Stockholm and in Lund, and as a visiting professor in US at UCLA, Michigan and Stanford, in 1971 Prawitz took the chair of professor of philosophy at Oslo University for 6 years. Prawitz returned to Stockholm University in 1976 as a professor of theoretical philosophy until retirement in 2001, and is now a Professor Emeritus there.

While still a graduate student in the late 1950s, Prawitz developed his algorithm for theorem proving in first-order logic, later implemented on one of the first computers in Sweden (probably the first computer implementation of a complete theorem prover for first-order logic) and published in his 1960 paper with H. Prawitz and N. Voghera *A mechanical proof procedure and its realization in an electronic computer*.

In his doctoral dissertation *Natural deduction: A proof-theoretical study* Prawitz developed the modern treatment of the system of Natural Deduction. In particular, he proved the **Normalization Theorem**, stating that all proofs in Natural deduction can be reduced to a certain normal form, a result that corresponds to Gentzen's celebrated **Hauptsatz** for sequent calculus. Prawitz' Normalization Theorem was later extended to first-order arithmetic as well as to second-order and higher-order logics.

As well as his pioneering technical work in proof theory, Prawitz has conducted important studies on the philosophical aspects of proof theory, on inference and knowledge, and on analyzing the relations between classical and intuitionistic logic. In particular, following Gentzen and Dummett, he has argued that there is a deep relation between proofs rules and the meaning of the logical connectives, and that the meaning of a statement can be seen as determined by what arguments would establish that statement as true.

Prawitz is a member of the Swedish and Norwegian Academies of Science and of several other distinguished scientific organizations.

4.4   Prenex and clausal normal forms

In this section I present the first-order extensions of conjunctive normal form and clausal form of a formula, needed for the extension of the method of Resolution to first-order logic.

4.4.1   Prenex normal forms

Definition 139

- *A **first-order literal** is an atomic formula or a negation of an atomic formula.*

- *A **first-order elementary conjunction/disjunction** is a first-order literal or a conjunction/disjunction of (two or more) first-order literals.*

- *A **first-order disjunctive normal form (DNF)** is a first-order elementary conjunction or a disjunction of (two or more) first-order elementary conjunctions.*

- *Respectively, a **first-order conjunctive normal form (CNF)** is a first-order elementary disjunction or a conjunction of (two or more) first-order elementary disjunctions.*

Definition 140 *A first-order formula $Q_1 x_1 \ldots Q_n x_n A$, where $Q_1, \ldots, Q_n$ are quantifiers and A is an open formula, is said to be in a **prenex form**. The quantifier string $Q_1 x_1 \ldots Q_n x_n$ is called the **prefix**, and the formula A the **matrix** of the prenex form.*
*If A is a first-order DNF (respectively, CNF) then $Q_1 x_1 \ldots Q_n x_n A$ is in a **prenex disjunctive**(respectively, **conjunctive**) **normal form**.*

Example 141

1. *The formula*
$$\forall x \exists y (\neg x > \mathbf{0} \vee (y > \mathbf{0} \wedge \neg x = y^2))$$

   is in a prenex DNF, but not a prenex CNF.

2. *The formula*
$$\forall x \exists y (x > \mathbf{0} \rightarrow (y > \mathbf{0} \wedge x = y^2))$$

is in a prenex form but not in DNF or in CNF.

3. *The formula*
$$\forall x (x > \mathbf{0} \rightarrow \exists y (y > \mathbf{0} \wedge x = y^2))$$

is not in a prenex form.

Theorem 142 *Every first-order formula is equivalent to a formula in a prenex disjunctive normal form (PDNF) and to a formula in a prenex conjunctive normal form (PCNF).*

I only provide a brief outline of an algorithm for construction of these normal forms, for a given any formula A.

1. Eliminate all occurrences of $\rightarrow$ and $\leftrightarrow$ as in the propositional case.

2. Import all negations inside all other logical connectives and transform the formula to negation-normal form.

3. Pull all quantifiers in front and therefore transform the formula into a prenex form. For that, use the equivalences
   (a) $\forall x P \wedge \forall x Q \equiv \forall x (P \wedge Q)$ and
   (b) $\exists x P \vee \exists x Q \equiv \exists x (P \vee Q)$
   to pull some quantifiers outward, after renaming the formula *wherever necessary*. Then use the following equivalences, where x does not occur free in Q, until the formula is transformed to a prenex form:
   (c) $\forall x P \wedge Q \equiv Q \wedge \forall x P \equiv \forall x (P \wedge Q)$
   (d) $\forall x P \vee Q \equiv Q \vee \forall x P \equiv \forall x (P \vee Q)$
   (e) $\exists x P \vee Q \equiv Q \vee \exists x P \equiv \exists x (P \vee Q)$
   (f) $\exists x P \wedge Q \equiv Q \wedge \exists x P \equiv \exists x (P \wedge Q)$.

4. Finally, transform the matrix into a DNF or CNF, as for a propositional formula.

A simplified, but more inefficient procedure for getting the quantifiers out first transforms the formula into an equivalent clean formula by suitable renaming of bound variables. It then pulls out all quantifiers by maintaining the order, tacitly using the last four equivalences (c–f) above. However, this procedure often introduces unnecessary new variables, which makes the result not optimal for the Skolemization step of the procedure, described in the following section.

Example 143 *Let* $A = \exists z (\exists x Q(x, z) \vee \exists x P(x)) \rightarrow \neg(\neg \exists x P(x) \wedge \forall x \exists z Q(z, x))$.

1. *Eliminating* $\rightarrow$: $A \equiv \neg \exists z (\exists x Q(x, z) \vee \exists x P(x)) \vee \neg(\neg \exists x P(x) \wedge \forall x \exists z Q(z, x))$.

2. *Importing the negation:*
   $A \equiv \forall z (\neg \exists x Q(x, z) \wedge \neg \exists x P(x)) \vee (\neg \neg \exists x P(x) \vee \neg \forall x \exists z Q(z, x))$
   $\equiv \forall z (\forall x \neg Q(x, z) \wedge \forall x \neg P(x)) \vee (\exists x P(x) \vee \exists x \forall z \neg Q(z, x))$.

3. *Using the equivalences (a) and (b):*
$A \equiv \forall z \forall x (\neg Q(x, z) \wedge \neg P(x)) \vee \exists x (P(x) \vee \forall z \neg Q(z, x))$.

4. *Renaming:* $A \equiv \forall z \forall x (\neg Q(x, z) \wedge \neg P(x)) \vee \exists y (P(y) \vee \forall w \neg Q(w, y))$.

5. *Using the equivalences (c–f) and pulling the quantifiers in front:*
$A \equiv \forall z \forall x \exists y \forall w ((\neg Q(x, z) \wedge \neg P(x)) \vee P(y) \vee \neg Q(w, y))$.

6. *The resulting formula is in a prenex DNF. For a prenex CNF we have to distribute the*
$\vee$ *over* $\wedge$:
$A \equiv \forall z \forall x \exists y \forall w ((\neg Q(x, z) \vee P(y) \vee \neg Q(w, y)) \wedge (\neg P(x) \vee P(y) \vee \neg Q(w, y)))$.

Note that we could have renamed the formula A into a clean formula right from the beginning. That would have made the resulting prenex formula much larger, however.

4.4.2 Skolemization

Skolemization, named after the logician Thoralf Skolem who first developed it (read more about him in the biographical box at the end of the section), is a procedure for systematic elimination of the existential quantifiers in a first-order formula in a prenex form, by way of uniform replacement of all occurrences of existentially quantified individual variables with terms headed by new functional symbols, called **Skolem functions**. In the traditional version of Skolemization, these Skolem functions take as arguments all variables (if any) which are bound by universal quantifiers in the scope of which the given existential quantifier sits. In particular, existentially quantified variables not in the scope of any universal quantifiers are replaced by constant symbols, called **Skolem constants**.

The Skolemization procedure is explained by the following examples.

Example 144

1. *The result of Skolemization of the formula*

$$\exists x \forall y \forall z (P(x, y) \rightarrow Q(x, z))$$

is

$$\forall y \forall z (P(c, y) \rightarrow Q(c, z)),$$

*where c is a new constant symbol called a **Skolem constant**.*
Intuitively, c names an element of the domain witnessing the truth of $\exists x A(\dots x \dots)$
such that $A(\dots c/x \dots)$ is true.

2. *More generally, the result of Skolemization of the formula*

$$\exists x_1 \cdots \exists x_k \forall y_1 \cdots \forall y_n A(x_1, \dots, x_k, y_1, \dots, y_n)$$

is

$$\forall y_1 \cdots \forall y_n A(c_1, \dots, c_k, y_1, \dots, y_n),$$

where $c_1, \dots, c_k$ are new Skolem constants.

3. *The result of Skolemization of the formula*

$$\exists x \forall y \exists z (P(x,y) \to Q(x,z))$$

is

$$\forall y (P(c,y) \to Q(c,f(y))),$$

*where c is a new Skolem constant and f is a new unary function, called a **Skolem function**.*
Intuitively, assuming the formula $\forall y \exists z A(y,z)$ is true in a given structure, f names a function in its domain of discourse, which for every possible value of y assigns a suitable value for z that makes the formula $A(y,z)$ true.

4. *More generally, the result of Skolemization of the formula*

$$\forall y \exists x_1 \cdots \exists x_k \forall y_1 \cdots \forall y_n A(y, x_1, \ldots, x_k, y_1, \ldots, y_n)$$

is

$$\forall y \forall y_1 \cdots \forall y_n A(y, f_1(y), \ldots, f_k(y), y_1, \ldots, y_n),$$

where $f_1, \ldots, f_k$ are new Skolem functions.

5. *The result of Skolemization of the formula*

$$\forall x \exists y \forall z \exists u A(x,y,z,u)$$

is the formula

$$\forall x \forall z A(x, f(x), z, g(x,z)),$$

where f is a new unary Skolem function and g is a new binary Skolem function.

Proposition 145 *The result of Skolemization of any first-order formula A is a formula which is generally not logically equivalent to A, but is equally satisfiable with A.*

The proof of this claim is not difficult, but it requires use of a special set-theoretic principle called the **Axiom of Choice**.

Thus, for the purposes of checking satisfiability of a formula, Skolemization is an admissible procedure.

4.4.3   Clausal forms

Recall that a **literal** is an atomic formula or a negation of an atomic formula, and a **clause** is a set of literals (representing their disjunction). An example of a clause is as follows:

$$\{P(x), \neg P(f(c,g(y))), \neg Q(g(x),y), Q(f(x,g(c)), g(g(g(y))))\}.$$

Note that all variables in a clause are implicitly assumed universally quantified.
A **clausal form** is a set of clauses (representing their conjunction), for example:

$$\{$$
$$\{P(x)\},$$
$$\{\neg P(f(c)), \neg Q(g(x, x), y)\},$$
$$\{\neg P(f(y)), P(f(c)), Q(y, f(x))\}$$
$$\}.$$

Theorem 146 *Every first-order formula A can be transformed to a clausal form $\{C_1, \ldots, C_k\}$ such that A is equally satisfiable with the universal closure $(C_1 \wedge \cdots \wedge C_k)$ of the conjunction of all clauses, where the clauses are considered as disjunctions.*

The algorithm for transforming a formula into clausal form is as follows.

1. Transform A into a prenex CNF.

2. Skolemize all existential quantifiers.

3. Remove all universal quantifiers.

4. Write the matrix (which is in CNF) as a set of clauses.

Example 147 *Consider the formula*

$$A = \exists z(\exists x Q(x, z) \vee \exists x P(x)) \rightarrow \neg(\neg \exists x P(x) \wedge \forall x \exists z Q(z, x)).$$

from Example 143.

1. *Transforming A to a prenex CNF:*

$$A \equiv \forall z \forall x \exists y \forall w((\neg Q(x, z) \vee P(y) \vee \neg Q(w, y)) \wedge (\neg P(x) \vee P(y) \vee \neg Q(w, y))).$$

2. *Skolemizing all existential quantifiers:*

$$\mathrm{Skolem}(A) = \forall z \forall x \forall w((\neg Q(x, z) \vee P(f(z, x)) \vee \neg Q(w, f(z, x))) \wedge$$
$$(\neg P(x) \vee P(f(z, x)) \vee \neg Q(w, f(z, x)))).$$

3. *Removing all universal quantifiers and writing the matrix as a set of clauses:*

$$\mathrm{Clausal}(A) = \{\{\neg Q(x, z), P(f(z, x)), \neg Q(w, f(z, x))\},$$
$$\{\neg P(x), P(f(z, x)), \neg Q(w, f(z, x))\}\}.$$

References for further reading
For more technical details, discussions, and examples of prenex normal forms, Skolemization, and clausal forms, see van Dalen (1983), Hamilton (1988), Nerode and Shore (1993), Smullyan (1995), Ebbinghaus *et al.* (1996), Fitting (1996), Barwise and Echemendy (1999), Hedman (2004), and Ben-Ari (2012).

Exercises

Transform each of the following formulae into a prenex DNF and a prenex CNF. Then Skolemize the resulting formula and transform it into clausal form.

4.4.1 $\forall x((\exists y P(y) \wedge \forall z Q(z, x)) \to \exists y Q(x, y))$

4.4.2 $\exists z(\exists x Q(x, z) \to (\exists x P(x) \vee \neg \exists z P(z)))$

4.4.3 $\forall y \neg(P(y) \leftrightarrow \exists x Q(y, x))$

4.4.4 $\forall z(\exists y P(y) \leftrightarrow Q(z, y))$

4.4.5 $(\exists x P(x) \to \neg \exists x Q(x)) \to \forall x(P(x) \to \neg Q(x))$

4.4.6 $\forall x(\neg \forall y Q(x, y) \wedge P(z)) \to \exists z(\forall y Q(z, y) \wedge \neg P(x))$

4.4.7 $\neg \exists x(\forall y(\exists z Q(y, z) \leftrightarrow P(z)) \wedge \forall z R(x, y, z))$

4.4.8 $\neg(\forall y(\forall z Q(y, z) \to P(z)) \to \exists z(P(z) \wedge \forall x(Q(z, y) \to Q(x, z))))$

4.4.9 $\neg(\neg \exists z(P(z) \wedge \forall x(Q(z, y) \to Q(x, z))) \to \neg \forall y(\forall z Q(y, z) \to P(z)))$

4.4.10 $\neg(\forall y(\neg \exists z Q(y, z) \to P(z)) \to \exists z((P(z) \to Q(z, y)) \wedge \neg \exists x R(x, y, z)))$.

Thoralf Albert Skolem (23.5.1887–23.3.1963) was a Norwegian mathematician and logician who made important contributions to mathematical logic, set theory, and algebra.

Skolem was extremely productive and published around 180 papers on Diophantine equations, mathematical logic, algebra, and set theory. However, many of his results were not well known until being rediscovered by others, as they were published in Norwegian journals which were not widely accessible.

In 1922 Skolem refined and made precise Zermelo's axioms for set theory, essentially in the form in which they are known today. In 1923 he created a formal theory of primitive recursive arithmetic meant to avoid the paradoxes of the infinite, and formally developed a great deal of number theory in it. His system can now be regarded as a programming language for defining functions on natural numbers, so he made an early contribution to computer science. His other work that later became very important in computer science and instrumental in resolution-based automated theorem proving was on what later became known as **Skolemization** and **Skolem normal form**.

In 1930 Skolem proved that the subsystem of Peano arithmetic involving only axioms for multiplication but not addition, now called **Skolem arithmetic** in his honor, was complete and decidable, unlike the full system of Peano arithmetic (including both addition and multiplication), which was proven to be incomplete by Gödel in 1931 and undecidable by Church and Turing in 1936.

The completeness of first-order logic, first published by Gödel in 1930, follows from results Skolem obtained in the early 1920s and published in 1928. He was apparently not aware of that fact at the time however, because the problem of proving completeness of an axiomatic system for first-order logic was only stated explicitly in 1928 in the first edition of Hilbert–Ackermann's *Principles of Mathematical Logic*.

The most influential and widely recognized work of Skolem was his pioneering research in **model theory**, the study of applications of logic to mathematics. The most well-known result associated with him is the **Löwenheim–Skolem Theorem**, an extension of an earlier work by Löwenheim published in 1920. It states that if a first-order theory, that is, a set of sentences of first-order logic, has a model (a structure that satisfies them all), then it has a countable model. This theorem has some consequences that sound paradoxical, giving rise to what became known as the **Skolem paradox**: assuming that the axiomatic system ZFC of set theory is satisfied in some model, then it must have one with a countable universe, despite the fact that it can prove the existence of uncountable sets. Likewise, the first-order theory of the field of reals, which is uncountable, also has a countable model. (However, according to some sources, Skolem did not believe in the existence of uncountable sets, so some of the results associated with such sets may have been falsely attributed to him.) Skolem also provided some of the first constructions of non-standard models of arithmetic and set theory.

Anatoly Ivanovich Maltsev (14.11.1909–7.06.1967) was a Russian–Soviet mathematician, well known for his work in universal algebra and mathematical logic. He is recognized as one of the founders, along with Löwenheim and Skolem, of **model theory**, one of the main branches of mathematical logic which studies the interaction between mathematics and logic.

Maltsev studied at Moscow University, completed his graduate work there under A. N. Kolmogorov and obtained his PhD in mathematics in 1941 from the Steklov Institute of Mathematics, with a dissertation on the *Structure of isomorphic representable infinite algebras and groups*. During 1932–1960 Maltsev taught mathematics at the Ivanovo Pedagogical Institute near Moscow. In 1960 he moved to Novosibirsk, where he was head of the department of algebra at the Mathematical Institute of the Siberian branch of the academy, as well as head of the chair of algebra and mathematical logic at the University of Novosibirsk.

Maltsev obtained very important results on decidability and undecidability of the first-order theories of various important classes of groups and other algebraic structures, and also made important contributions to the theory of Lie algebras. His most influential legacy, however, was in mathematical logic, where in 1936 and 1941 he proved in full generality the two most fundamental results of model theory: the **compactness theorem**, stating that if every finite subset of a first-order theory (set of sentences) is satisfiable then the entire theory is satisfiable; and the **Löwenheim–Skolem Theorem** (see the box on Skolem).

4.5   Resolution for first-order logic

We are now almost ready to extend the method of Resolution to first-order logic. One more technical topic – unification – is needed for that, described in the following.

4.5.1   Propositional Resolution rule in first-order logic

The Propositional Resolution rule extended to first-order logic reads:

$$\mathbf{Res}_0 : \quad \frac{C \vee Q(s_1, \ldots, s_n), \quad D \vee \neg Q(s_1, \ldots, s_n)}{C \vee D}.$$

This rule, however, is not strong enough. For example, the clause set

$$\{\{P(x)\}, \{\neg P(f(y))\}\}$$

is not satisfiable, as it corresponds to the unsatisfiable formula

$$\forall x \forall y (P(x) \wedge \neg P(f(y))).$$

However, the resolution rule above cannot produce an empty clause, because it *cannot unify* the two clauses in order to resolve them. We therefore need a stronger derivation mechanism that can handle cases like this. There are two natural solutions[4]:

1. **Ground resolution**: generate *sufficiently many ground instances* of every clause (over the so-called **Herbrand universe** of the language) and then apply the standard Resolution rule above.
   In the example, ground resolution would generate the ground clauses $\{P(f(c))\}$ and $\{\neg P(f(c))\}$ for some constant symbol c.
   This method is sound and complete but inefficient, as it leads to the generation of too many unnecessary clauses. It will not be discussed further.

2. **Resolution with unification** introduce a stronger Resolution rule that first tries to match a pair of clauses by applying a suitable substitution that would enable that pair to be resolved. We present this method in further detail in the following.

---

[4] Both proposed by John Alan Robinson; read more at the end of the section.

4.5.2 Substitutions of terms for variables revisited

Recall that substitution of a term t for a variable x in a term s, denoted $s[t/x]$, is the result of *simultaneous* replacements of all occurrences of x in s by t. For instance:

$$f(g(x), f(y, x))[g(a)/x] = f(g(g(a)), f(y, g(a))).$$

This generalizes to simultaneous substitutions of several terms for *different* variables, denoted $s[t_1/x_1, \ldots, t_k/x_k]$. For instance:

$$f(g(x), f(y, x))[g(y)/x, f(x, b)/y] = f(g(g(y)), f(f(x, b), g(y))).$$

Note that the condition of simultaneity is important: if we first substituted $g(y)$ for x and then $f(x, b)$ for y, the result would have been different (and wrong).

We can apply a substitution to several terms simultaneously, for example, when a substitution acts on an atomic formula $P(t_1, \ldots, t_n)$. For instance:

$$P(x, f(x, y), g(y))[g(y)/x, a/y] = P(g(y), f(g(y), a), g(a)).$$

Given the substitution $\sigma = [t_1/x_1, \ldots, t_k/x_k]$, the set of variables $\{x_1, \ldots, x_k\}$ is called the **domain of** σ, denoted $\mathsf{dom}(\sigma)$.

Substitutions can be *composed* by consecutive application: the composition of substitutions τ and σ is the substitution $\tau\sigma$ obtained by applying τ followed by applying σ to the result, that is, $t\tau\sigma = (t\tau)\sigma$ (note the order). For instance:

$$f(g(x), f(y, x))[f(x, y)/x][g(a)/x, x/y] =$$

$$f(g(f(x, y)), f(y, f(x, y)))[g(a)/x, x/y] =$$

$$f(g(f(g(a), x)), f(x, f(g(a), x))).$$

The composition of two substitutions $\tau = [t_1/x_1, \ldots, t_k/x_k]$ and σ can be computed as follows.

1. Extend $\mathsf{dom}(\tau)$ to cover $\mathsf{dom}(\sigma)$ by adding to τ substitutions $[x/x]$ for every variable x in $\mathsf{dom}(\sigma) \setminus \mathsf{dom}(\tau)$.

2. Apply the substitution σ simultaneously to all terms $[t_1, \ldots, t_k]$ to obtain the substitution $[t_1\sigma/x_1, \ldots, t_k\sigma/x_k]$.

3. Remove from the result all cases x_i/x_i, if any.

For instance:

$$[f(x, y)/x, x/y][y/x, a/y, g(y)/z] =$$

$$[f(y, a)/x, y/y, g(y)/z] =$$

$$[f(y, a)/x, g(y)/z].$$

4.5.3   Unification of terms

A substitution σ **unifies** two words s and s' if $\sigma(s) = \sigma(s')$. A substitution σ that unifies the respective arguments of the literals $Q(s_1, \ldots, s_n)$ and $Q(t_1, \ldots, t_n)$, that is, such that $\sigma(s_1) = \sigma(t_1), \ldots, \sigma(s_n) = \sigma(t_n)$, is called a **unifier** of $Q(s_1, \ldots, s_n)$ and $Q(t_1, \ldots, t_n)$. For instance, the substitution $\sigma = [f(c)/x, c/y, c/z]$ unifies the literals $Q(x, c, f(f(z)))$ and $Q(f(y), z, f(x))$.

Two terms are **unifiable** if they have a unifier; otherwise they are **non-unifiable**.

A unifier τ of two terms (or literals) is **more general** than a unifier ρ, if there is a substitution σ such that $\rho = \tau\sigma$. (Note that, in the sense of this definition, every unifier is more general than itself!) For instance, $\rho = [c/x, f(c)/y]$ is a unifier of the literals $P(f(x))$ and $P(y)$, but $\tau = [f(x)/y]$ is a more general unifier because $\rho = \tau\sigma$, where $\sigma = [c/x]$.

A unifier of two terms (or literals) is their **most general unifier** (MGU) if it is more general than any unifier of these terms (literals).

A most general unifier (if it exists) need not be unique. For instance, in the example above, τ is a most general unifier, as well as $[z/x, f(z)/y]$ for any variable z.

We are eventually interested in unifying literals, that is, (possibly negated) atomic formulae. Note that two atomic formulae $P(t_1, \ldots, t_m)$ and $Q(s_1, \ldots, s_n)$ cannot be unified unless $P = Q$ and $m = n$. If these are satisfied, in order to unify the formulae we need to unify the respective pairs of arguments, that is, we are looking for a (most general) unifier of the system

$$t_1 = s_1,$$
$$\vdots$$
$$t_n = s_n.$$

Informally, we do that by computing a most general unifier (if one exists) of the first pair, then applying it to both lists of terms, then computing a most general unifier of the next pair (if there is one) in the resulting lists, and applying it to both resulting lists of terms. In order to unify the current pair of terms we apply the algorithm recursively to the pair of lists of their respective arguments. The composition of all most general unifiers computed as above, if they all exist, is a most general unifier of the two input lists.

A simple recursive algorithm for computing a most general unifier of two lists of terms **p** and **q** is provided in pseudocode as follows.

procedure mgu(p, q)
$\theta := \epsilon$ *(the empty substitution)*.
Scan **p** and **q** simultaneously, left-to-right,
    and search for the first corresponding subterms
    where **p** and **q** disagree (mismatch).
If there is no mismatch, return θ.
Else, let s and t be the first respective subterms
    in **p** and **q** where mismatch occurs.
If variable(s) and s $\notin$ Var(t) then
    $\theta :=$ compose $(\theta, [t/s])$;

$\theta :=$ compose $(\theta, \text{mgu}(\mathbf{p}\theta, \mathbf{q}\theta))$.
   Else, if `variable(t)` and $t \notin \text{Var(s)}$ then
     $\theta :=$ compose $(\theta, [s/t])$;
     $\theta :=$ compose $(\theta, \text{mgu}(\mathbf{p}\theta, \mathbf{q}\theta))$.
   Else, return **failure**.
end

Example 148 *Some examples of unification of atomic formulae using the algorithm above (where* `Bill` *and* `Jane` *are constant symbols,* `father`, `mother` *are unary functional symbols and* `parents` *is a ternary predicate symbol) are as follows.*

1. *Literal 1:* `parents(x, father(x), mother(Bill))`,
   Literal 2: `parents(Bill, father(Bill), y)`,
   Most general unifier: `[Bill/x, mother(Bill)/y]`.

2. *Literal 1:* `parents(x, father(x), mother(Bill))`,
   Literal 2: `parents(Bill, father(y), z)`,
   Most general unifier: `[Bill/x, Bill/y, mother(Bill)/z]`.

3. *Literal 1:* `parents(x, father(x), mother(Jane))`,
   Literal 2: `parents(Bill, father(y), mother(y))`,
   *The procedure **mgu** starts computing:* [`Bill/x`, `Bill/y`, ... *failure*].
   The algorithm fails. Therefore, a unifier does not exist.

4. *Literal 1:* `g(x,f(x))`.
   Literal 2: `g(f(y),y)`.
   *The procedure **mgu** starts computing:* [`f(y)/x`, ... *failure*].
   MGU does not exist. Indeed, any unifier would have to unify $f(f(x))$ *with* x, *which is impossible.*

4.5.4   Resolution with unification in first-order logic

The first-order resolution rule is combined with a preceding unification of the clausal set in order to produce a complementary pair of literals in the resolving clauses:

$$\textbf{Res}: \quad \frac{C \vee Q(s_1, \ldots, s_n), \quad D \vee \neg Q(t_1, \ldots, t_n)}{\sigma(C \vee D)}$$

where:

(i) the two clauses have no variables in common (achieved by renaming of variables); and

(ii) σ is a most general unifier of the literals $Q(s_1, \ldots, s_n)$ and $Q(t_1, \ldots, t_n)$.

Note that we require that the unifier applied in the rule is a *most general unifier* of the respective literals in order not to weaken the resolvent unnecessarily. This is important because that resolvent may have to be used in further applications of resolution. For instance, resolving $\{P(a, y), Q(x)\}$ with $\{\neg P(a, x)\}$ by using the unifier $[a/x, a/y]$

would produce $\{Q(a)\}$ which cannot be resolved with $\{\neg Q(b)\}$, while using a most general unifier, for example $[y/x]$, would produce $\{Q(x)\}$ which can then be further unified and resolved with $\{\neg Q(b)\}$ by using $[b/y]$.

Some examples of resolution with unification include the following.

$$\frac{\{P(x)\}, \{\neg P(f(y))\}}{\{\}} \quad (MGU : [f(y)/x])$$

$$\frac{\{P(a,y), Q(y)\}\{\neg P(x,b), \neg Q(x)\}}{\{Q(b), \neg Q(a)\}} \quad (MGU : [a/x, b/y])$$

$$\frac{\{P(a,y), Q(y)\}\{\neg P(x,f(x)), Q(f(x))\}}{\{Q(f(a))\}} \quad (MGU : [a/x, f(a)/y]).$$

Note that both literals in the resolvent become equal.

$$\frac{\{P(a,y), P(x,f(x))\}\{\neg P(x,f(a))\}}{\{\}} \quad (MGU : [a/x, f(a)/y]).$$

Note that, after unification, both literals in the first clause become $P(a, f(a))$, so they become identified before the resolution rule is applied.

Factoring
In order for the method of Resolution for first-order logic as defined here to be complete, it has to incorporate an auxiliary rule called **Factoring** which unifies two literals in the same clause, whenever possible:

$$\textbf{Fac:} \quad \frac{\{ \ldots P(\bar{s}), P(\bar{t}) \ldots \}}{\sigma\{ \ldots P(\bar{s}) \ldots \}}, \quad (\sigma = MGU(\bar{s}, \bar{t}))$$

For example,

$$\frac{\{P(x), P(c), \neg Q(x,y)\}}{\{P(c), \neg Q(c,y)\}} \quad (MGU : [c/x]).$$

As an exercise, prove that Factoring is a logically sound rule. Moreover, it is sometimes necessary to be applied. For instance, consider the set of clauses

$$\{P(x), P(c)\}, \{\neg P(y), \neg P(c)\}.$$

The empty clause cannot be derived from it without using Factoring, whereas if Factoring is applied first to both clauses the empty clause is immediately derivable.

Hereafter, we assume that Factoring is incorporated into the Resolution method. Note that it extends the deletion of repeated literals from Propositional Resolution.

Definition 149 *A **resolution-based derivation** of a formula C from a list of formulae $A_1, \ldots, A_n$, denoted $A_1, \ldots, A_n, \vdash_{RES} C$, is a derivation of the empty clause $\{\}$ from the set of clauses obtained from $A_1, \ldots, A_n, \neg C$ by successive applications of the rules **Res** and **Fac**.*

Unlike Propositional Resolution, in some cases first-order Resolution may run forever, that is, never terminate, when the conclusion does not follow logically from the premises. It may also run forever even when the conclusion *does* follow logically from the premises, if unnecessary resolvents are produced recurrently in the process. To avoid this, special strategies or additional mechanisms for disposal of used-up clauses need to be applied.

4.5.5   Examples of resolution-based derivations

Example 150 *Using the method of first-order Resolution, prove that:*

$$\forall x(P(x) \rightarrow Q(x)) \models \forall x P(x) \rightarrow \forall x Q(x).$$

1. *Transform*
$$\{\forall x(P(x) \rightarrow Q(x)), \neg(\forall x P(x) \rightarrow \forall x Q(x))\}$$

   to clausal form:
$$\{\neg P(x), Q(x)\}, \{P(y)\}, \{\neg Q(c)\}$$

   for some Skolem constant c.

2. *Successive applications of **Res**:*
   (a) *Unify $P(x)$ and $P(y)$ with MGU $[y/x]$.*
       Then resolve $\{\neg P(x), Q(x)\}$ and $\{P(y)\}$ to obtain $\{Q(y)\}$.
   (b) *Unify $Q(c)$ and $Q(y)$ with MGU $[c/y]$.*
       Then resolve $\{\neg Q(c)\}$ and $\{Q(y)\}$ to obtain $\{\}$.

Example 151 *Show that*
   if Everybody loves somebody
   and Everybody loves a lover
   then Everybody loves everybody.

1. *Formalize the assumptions and the goal conclusion, using the predicate $L(x, y)$ meaning "x loves y."*
   Everybody loves somebody: $\forall x \exists y L(x, y)$.
   Everybody loves a lover: $\forall x(\exists y L(x, y) \rightarrow \forall z L(z, x))$.
   Everybody loves everybody: $\forall x \forall y L(x, y)$.

2. *Transform the set*
$$\{\forall x \exists y L(x, y), \forall x(\exists y L(x, y) \rightarrow \forall z L(z, x)), \neg(\forall x \forall y L(x, y))\}$$

   to a clausal form:
$$\{L(x, f(x))\}, \{\neg L(x_1, y), L(z, x_1)\}, \{\neg L(a, b)\}$$

   for some Skolem constants a, b and a Skolem function f.

3. *Successive applications of **Res**:*
   (a) *Unify $L(z, x_1)$ and $L(a, b)$ with MGU $[b/x_1, a/z]$ and resolve $\{\neg L(x_1, y),$
       $L(z, x_1)\}$ with $\{\neg L(a, b)\}$ to obtain $\{\neg L(b, y)\}$.*
   (b) *Unify $L(x, f(x))$ and $L(b, y)$ with MGU $[b/x, f(b)/y]$ and resolve $\{L(x, f(x))\}$
       with $\{\neg L(b, y)\}$ to obtain $\{\}$.*

Example 152 *Verify the logical consequence*

$$\forall x \exists y R(x, y), \forall x \forall y \forall z ((R(x, y) \wedge R(y, z)) \rightarrow R(x, z)) \models \exists x R(x, x).$$

1. *Transform*

   $$\{\forall x \exists y R(x, y), \forall x \forall y \forall z ((R(x, y) \wedge R(y, z)) \rightarrow R(x, z)), \neg(\exists x R(x, x))\}$$

   to a clausal form:

   $$C_1 = \{R(x, f(x))\}, C_2 = \{\neg R(u, y), \neg R(y, z), R(u, z)\}, C_3 = \{\neg R(v, v)\}$$

   for some unary Skolem function f.

2. *Successive applications of **Res** to the clausal set*

   $$C_1 = \{R(x, f(x))\}, C_2 = \{\neg R(u, y), \neg R(y, z), R(u, z)\}, C_3 = \{\neg R(v, v)\}$$

   *(wherever necessary, we rename variables in the derived clauses to keep all clauses
   with disjoint sets of variables):*
   (a) *Unify $R(u, z)$ in C_2 and C_3, with MGU $[v/u, v/z]$ and resolve: $C_4 =$
       $\{\neg R(v, y), \neg R(y, v)\}$.*
   (b) *Unify $R(v, y)$ in C_4 and $R(x, f(x))$ in C_1 with MGU $[x/v, f(x)/y]$ and resolve:
       $C_5 = \{\neg R(f(x), x)\}$.*
       After renaming x: $C_5 = \{\neg R(f(x_1), x_1)\}$.
   (c) *Unify $R(u, z)$ in C_2 and $R(f(x_1), x_1)$ in C_5 with MGU $[f(x)/u, x/z]$ and
       resolve: $C_6 = \{\neg R(f(x), y), \neg R(y, x)\}$.*
       After renaming: $C_6 = \{\neg R(f(x'), y'), \neg R(y', x')\}$.
   (d) *Unify $R(f(x'), y')$ in C_6 and $R(x, f(x))$ in C_1 with MGU $[f(x')/x, f(f(x'))/y']$
       and resolve: $C_7 = \{\neg R(f(f(x')), f(x'))\}$.*
       After renaming, $C_7 = \{\neg R(f(f(x_2)), f(x_2))\}$.
       Etc.

   Thus, an infinite set of clauses can be generated:

   $$\{\neg R(f(x_1), x_1)\}, \{\neg R(f(f(x_2)), f(x_2))\}, \{\neg R(f(f(f(x_3))), f(f(x_3)))\}, \ldots$$

but the empty clause cannot be derived.
   *In fact, the logical consequence does not hold. A counter-model is the set of natural
numbers with $R(x, y)$ interpreted as $x < y$.*

4.5.6   Resolution for first-order logic with equality

As we have seen in the previous sections on deductive systems, logical deduction in first-order logic with equality requires special attention in the form of either additional axioms or inference rules. The method of Resolution presented so far does not involve rules for the equality. Such rules can be added to produce a sound and complete Resolution-based deductive system for first-order logic with equality. To begin with, we can take all axioms for the equality for the respective first-order language from Section 4.1, transform them all into clausal form and add the resulting clauses to the set of clauses to which we apply Resolution. This approach, however, can be rather inefficient. A more efficient way to extend the method of first-order Resolution to handle equality and equational reasoning is to keep only the identity unit clauses $t = t$ for any term t, and add the following inference rule, called **Paramodulation**[5]:

$$\mathbf{Par}: \quad \frac{C \vee s_1 = s_2, \quad D \vee Q(t, t_1, \dots, t_n)}{\sigma(C \vee D \vee Q(s_2, t_1, \dots, t_n))} \qquad (\sigma = MGU(s_1, t))$$

or

$$\mathbf{Par}: \quad \frac{C \vee s_2 = s_1, \quad D \vee Q(t, t_1, \dots, t_n)}{\sigma(C \vee D \vee Q(s_2, t_1, \dots, t_n))} \qquad (\sigma = MGU(s_1, t))$$

where the parent clauses are assumed to have no variables in common. The resulting clause is called a **paramodulant** of the parent clauses.

This rule enables direct substitution of terms by equal terms deep inside literals, rather than deriving such substitutions every time by using the clauses for equality.

Example 153 *Here are some simple examples of applications of* Paramodulation, *where* a, b, c *are constants and* P, Q *are unary predicates.*

1.
$$\frac{a = b, P(a)}{P(b)}$$

2. *The set of literals* $\{a = b, P(a), \neg P(b)\}$ *is not satisfiable. Indeed, the empty clause is derivable from the clause set* $\{\{a = b\}, \{P(a)\}, \{\neg P(b)\}\}$ *using the previous derivation.*

3.
$$\frac{P(y) \vee a = f(b), Q(f(x), x)}{P(y) \vee Q(a, b)}$$

Here we have used $[b/x] = MGU(f(b), f(x))$.

4.
$$\frac{P(a) \vee f(g(a)) = g(b), Q(f(x), x) \vee \neg P(x)}{P(a) \vee Q(g(b), g(a)) \vee \neg P(g(a))},$$

where we have used $\sigma = [g(a)/x]$, *and*

$$\frac{P(a) \vee f(g(a)) = g(b), Q(f(x), x) \vee \neg P(x)}{P(a) \vee Q(f(g(b)), f(g(a))) \vee \neg P(g(b))},$$

---

[5] Originally introduced by G. Robinson and L. Wos in the 1969 paper *Paramodulation and theorem-proving in first-order theories with equality.*

where we have used $\sigma = [g(b)/x]$,

5. *Applications of* Paramodulation *to equalities of terms sharing a variable may lead to generation of an infinite series of new clauses, for instance:*

$$\frac{f(x) = x, C \vee Q(a)}{C \vee Q(f(a))},$$

$$\frac{f(x) = x, C \vee Q(f(a))}{C \vee Q(f(f(a)))},$$

$$\dots$$

For more examples and some applications of Resolution with equality, see Exercises 10, 11, and 14.

The proof of the following theorem for the case of Resolution with equality is outlined in Section 4.6.

Theorem 154 (Soundness and completeness of RES)

1. *The system* **RES** *of Resolution with Unification and Factoring for first-order logic without equality is sound and complete, that is, for all first-order formulae* $A_1, \dots, A_n, C$ *of the language without equality:*

$$A_1, \dots, A_n, \vdash_{\textbf{RES}} C \ \textit{iff} \ A_1, \dots, A_n, \models C.$$

2. *Likewise, the system* **RES**$^=$ *of Resolution with Unification, Factoring, and Paramodulation for first-order logic with equality is sound and complete, that is, for all first-order formulae* $A_1, \dots, A_n, C$ *of the language with equality:*

$$A_1, \dots, A_n, \vdash_{\textbf{RES}^=} C \ \textit{iff} \ A_1, \dots, A_n, \models C.$$

Sometimes, it is said that the method of Resolution is **refutation-complete**.

4.5.7 Optimizations and strategies for the method of Resolution

Like Propositional Resolution, the method of Resolution for first-order logic is amenable to various optimizations by means of special additional inference rules and derivation strategies that go beyond the scope of this book. Such optimizations are much more important for first-order Resolution because if the search for a derivation of the empty clause is not cleverly organized, it may not terminate even if such a derivation exists. The techniques of *Tautology Deletion*, *Subsumption Deletion*, and *Removal of Clauses with Mono-polar Literals* still apply to first-order Resolution. Moreover, they can be combined with each other and with suitable unification pre-processing, for example to produce more specific optimization rules and techniques.

In particular, widely used additional optimizing techniques are based on special strategies for application of the Resolution rule that, instead of removing redundant clauses, prevent the generation of useless clauses. For instance, the **Set-of-support** resolution strategy is based on the idea that some of the clauses are of higher importance and should be treated with priority. That leads to the idea of designating a subset of the given clause set S as a **support set** T if its complement $S \setminus T$ is known to be satisfiable, and to require that Resolution is only applied when at least one of the resolved clauses comes from T as there is no chance to produce the empty clause from within $S \setminus T$.

Other very successful strategies include **Hyperresolution**, used to reduce the number of intermediate resolvents by combining several resolution steps into a single inference step, and **Linear Resolution**, requiring always that one of the resolved clauses is the most recently derived resolvent, therefore organizing the deduction in a simple linear shape, which is easily and efficiently implementable. All strategies mentioned above are logically sound and preserve the (refutation) completeness of the Resolution method. A special version of the Linear Resolution strategy, called **Selective Linear Definite Clause Resolution**, or just **SLD-resolution**, is the derivation strategy of the logic programming language Prolog, briefly discussed in Section 5.4.

There are also some practically very efficient but generally not completeness-preserving strategies, such as **Unit Resolution**, where one of the resolved clauses must always be a literal, and **Input Resolution**, where one of the resolved clauses is always selected from the set of clauses generated from the initial assumptions (axioms). Other important types of incomplete, yet practically efficient, strategies are the **ordering strategies** based on some partial ordering imposed on the predicate symbols, terms, literals, or clauses used in the deduction, based on various heuristics. Ordered resolution treats clauses not as sets but as (ordered) *lists* of literals and chooses candidate clauses for resolving in a priority order according to these lists.

For more on practical applications of the method of Resolution, see Section 5.4.

References for further reading
For more technical details, discussions, and examples of first-order Resolution with unification, see Gallier (1986), Nerode and Shore (1993), Ebbginhaus *et al.* (1996), Fitting (1996), Chang and Lee (1997), Barwise and Echemendy (1999), Robinson and Voronkov (2001), Hedman (2004), and Ben-Ari (2012).

Exercises

4.5.1   Using the algorithm, compute a most general unifier (MGU) (if one exists) for the following pairs of literals:
(a) $P(f(x), g(y)), P(y, g(x))$; and
(b) $P(x, f(y), y), P(f(y_1), f(g(z)), g(a))$.

For each of the following exercises use first-order Resolution with unification.

4.5.2   Prove the following logical validities and consequences.

    (a) $\models \forall x P(x) \to \forall y P(y)$
    (b) $\models \exists x P(x) \to \exists y P(y)$
    (c) $\models \exists x (P(x) \to \forall y P(y))$
    (d) $\forall x A(x) \models \neg \exists x \neg A(x)$
    (e) $\neg \exists x \neg A(x) \models \forall x A(x)$
    (f) $\exists x A(x) \models \neg \forall x \neg A(x)$
    (g) $\neg \forall x \neg A(x) \models \exists x A(x)$
    (h) $\exists x \exists y A(x, y) \models \exists y \exists x A(x, y)$.

4.5.3 Suppose x is not free in Q. Prove the following logical consequences.
    (a) $\forall x (P(x) \lor Q) \models \forall x P(x) \lor Q$
    (b) $\forall x P(x) \lor Q \models \forall x (P(x) \lor Q)$
    (c) $\exists x (P(x) \land Q) \models \exists x P(x) \land Q$
    (d) $\exists x P(x) \land Q \models \exists x (P(x) \land Q)$
    (e) $\forall x (Q \to P(x)), Q \models \forall x P(x)$
    (f) $\exists x P(x) \to Q \models \forall x (P(x) \to Q)$
    (g) $\exists x (P(x) \to Q), \forall x P(x) \models Q$
    (h) $\exists x (Q \to P(x)), Q \models \exists x P(x)$.

4.5.4 Check which of the following logical consequences hold. For those which do not, construct a counter-model, that is, a structure and assignment in which all premises are true while the conclusion is false.
    (a) $\forall x A(x), \forall x B(x) \models \forall x (A(x) \land B(x))$
    (b) $\forall x (A(x) \land B(x)) \models \forall x A(x) \land \forall x B(x)$
    (c) $\forall x (A(x) \lor B(x)) \models \forall x A(x) \lor \forall x B(x)$
    (d) $\forall x A(x) \lor \forall x B(x) \models \forall x (A(x) \lor B(x))$
    (e) $\forall x A(x) \to \forall x B(x) \models \forall x (A(x) \to B(x))$
    (f) $\exists x (A(x) \land B(x)) \models \exists x A(x) \land \exists x B(x)$
    (g) $\exists x A(x), \exists x B(x) \models \exists x (A(x) \land B(x))$
    (h) $\exists x A(x) \lor \exists x B(x) \models \exists x (A(x) \lor B(x))$
    (i) $\exists x (A(x) \lor B(x)) \models \exists x A(x) \lor \exists x B(x)$
    (j) $\exists x (A(x) \to B(x)) \models \exists x A(x) \to \exists x B(x)$
    (k) $\exists x A(x) \to \exists x B(x) \models \exists x (A(x) \to B(x))$.

4.5.5 Formalize the following arguments in first-order logic by using suitable predicates, interpreted in a suitable domain. Check whether each is logically correct.

    (a)

        All logicians are clever.
        All clever people are rich.

        Therefore all logicians are rich.

    (b)

        All natural numbers are integers.
        Some integers are odd numbers.

        Therefore some natural numbers are odd numbers.

(c)

> Every child is a pedestrian.
> No driver is a pedestrian.
> Johnie is a child.
> _____
> Therefore somebody is not a driver.

(d)

> All penguins are birds.
> Some penguins are white.
> All penguins eat some fish.
> _____
> Therefore, there is a white bird that eats some fish.

(e)

> All penguins are birds.
> No penguins can fly.
> All seagulls can fly.
> All fat birds are penguins.
> Some fat birds are white.
> _____
> Therefore, there is a fat white bird that is not a seagull.

(f)

> No yellow plonks are qlinks.
> Some object is a plonk or not a qlink.
> _____
> Therefore, there is an object which is not a yellow qlink.

(g)

> Victor is a hungry student.
> Every student either studies or parties.
> Every student who is hungry does not study.
> _____
> Therefore, some student parties.

4.5.6 Formalize each of the following statements in first-order logic and check whether it logically implies the other. If any of the logical consequences do not hold, give an appropriate counter-model.
A: "No man is happy only when he is drunk."
B: "There is a happy man and there is a man who is not drunk."

4.5.7 Using the predicates: $D(x)$ for "x is a dragon," $E(x)$ for "x is an elf," $F(x)$ for "x is a fairy," $Fl(x)$ for "x can fly," $G(x)$ for "x is a goblin," $H(x)$ for "x is a hobbit," $T(x)$ for "x is a troll," $U(x)$ for "x is ugly," and $L(x, y)$ for "x likes y," formalize each of the following arguments in first-order logic, in the domain of all "mythic creatures." Check whether each is logically valid. If it is not valid, give an appropriate counter-model.

(a)

> All fairies are elves. Hobbits do not like any elves.
> Kermit likes a goblin or a fairy. Everyone who likes
> any goblins likes no trolls.
> _____
> Therefore, if Kermit likes any trolls then Kermit is not a hobbit.

(b)

> All trolls are elves or goblins. No troll can fly, unless it is an elf.
> All dragons can fly.
> All ugly goblins are trolls. Some goblins are ugly.
> _____
> Therefore, there is an ugly mythic creature that is not a dragon.

(c) Change the first premise in the previous argument to
  "All trolls are either elves or goblins, but not both."
  Formalize it and check if the resulting argument is logically valid.

4.5.8 Formalize the following arguments in first-order logic by using suitable predicates interpreted in suitable domains. Check whether each is logically valid. If it is not valid, give an appropriate counter-model.

(a)

> Some bachelors love every woman they know.
> Every bachelor is a man.
> Eve is a woman whom every man knows.
> _____
> Some bachelor loves a woman.

(b) Replace the conclusion in the previous example by
  "Every bachelor loves a woman"
  and check again whether the resulting argument is logically valid.

(c)

> Every man admires some brave woman.
> Juan is a man who admires every pretty woman.
> Lara is a brave or pretty woman.
> _____
> Some man admires Lara.

4.5.9 Formalize the following arguments in first-order logic by using suitable predicates, interpreted in a suitable domain. Check whether each is logically valid. If it is not valid, give an appropriate counter-model.

(a)

> All talking robots can walk.
> Some walking robot can construct every talking robot.
> _____
> Therefore, some talking robot can construct some walking
> robot and every talking robot.

(b)

>　Some devices are robots.
>　All robots can talk or walk or reason logically.
>　No non-talking robot can reason logically.
>　All talking robots can construct every logically reasoning device.
>　A device can construct a robot only if it is itself a robot.
>　Some logically reasoning devices can construct
>　every non-walking robot.
> _____
>　Therefore, some talking robot can construct itself.

(c) Change the last premise in the argument above
>　　"Some logically reasoning devices can construct every non-walking robot"
>　to:
>　　"All logically reasoning devices can construct every non-walking robot."
>　Formalize it and check if the resulting argument is logically valid.

(d) Add to the original argument in item (a) the premise
>　　"Creepy is a robot that cannot walk."
>　Formalize it and check if the resulting argument is logically valid.

4.5.10　Prove that the following is a logically sound rule:

$$\frac{\neg s = t \vee P(s)}{\sigma(P(t))}$$

where the terms s and t are unifiable and $\sigma = MGU(s,t)$.

4.5.11　Transform all axioms of equality from Section 4.1 for the first-order language with one constant symbol c, one unary predicate symbol P, and one binary functional symbol f into clausal form. Then use Resolution with Paramodulation to prove their validity.

4.5.12　Using Resolution with Paramodulation, derive each of the following (the first five in the list are the axioms of **H** for the equality):

(a) $(\text{Ax}_=1)\ x = x$

(b) $(\text{Ax}_=2)\ x = y \rightarrow y = x$

(c) $(\text{Ax}_=3)\ x = y \wedge y = z \rightarrow x = z$

(d) $(\text{Ax}_f)\ x_1 = y_1 \wedge \ldots \wedge x_n = y_n \rightarrow f(x_1, \ldots, x_n) = f(y_1, \ldots, y_n)$ for n-ary functional symbol f

(e) $(\text{Ax}_r)\ x_1 = y_1 \wedge \ldots \wedge x_n = y_n \rightarrow (p(x_1, \ldots, x_n) \rightarrow p(y_1, \ldots, y_n))$ for n-ary predicate symbol p.

4.5.13　Using Resolution with Paramodulation, check whether each of the following validities and logical consequences holds.

(a) $\models x_1 = y_1 \wedge x_2 = y_2 \rightarrow g(f(x_1, f(x_2))) = g(f(y_1, f(y_2)))$
(Universal quantification is assumed but omitted.)

(b) $\models \forall x(x = f(x) \rightarrow (P(f(f(f(x)))) \rightarrow P(x)))$

(c) $\forall x \forall y(f(x) = y \rightarrow g(y) = x) \models \forall z(g(f(z)) = z)$

(d) $\forall x \forall y(f(x) = y \rightarrow g(y) = x) \models \forall z(f(g(z)) = z)$

(e) $\forall x \forall y(f(x) = y \rightarrow g(y) = x), \forall x \forall y(g(x) = g(y) \rightarrow x = y) \models$
$\forall z(f(g(z)) = z)$

For more exercises on derivations with equality, on sets, functions, and relations, see Section 5.2.7.

4.5.14 ⓟ Consider the first-order theory of groups $\mathcal{G}$, consisting of the following axioms, where e is the constant for group identity, $\circ$ is the (binary) group operation, and $'$ is the (unary) inverse operation, and all variables below are assumed universally quantified.

G1: $(x \circ y) \circ z = x \circ (y \circ z)$ (associativity of $\circ$)

G2: $x \circ e = x$ (e is a right identity)

G3: $x \circ x' = e$ ($'$ is a right inverse operation)

Using Resolution with Paramodulation, derive the following as logical consequences from the axioms **G1–G3**. Again, all variables in the formulae below are assumed universally quantified.

(a) $x' \circ x = e$ ($'$ is a left inverse operation)

(b) $e \circ x = x$ (e is a left identity)

(c) $(x')' = x$

(d) $(x \circ y)' = y' \circ x'$

(e) $x = y \circ z \rightarrow y = x \circ z' \wedge z = y' \circ x$

(f) Adding to $\mathcal{G}$ the axiom $x \circ x = e$, derive the commutativity law $x \circ y = y \circ x$.

(g$**$) Adding to $\mathcal{G}$ the axiom $x \circ x \circ x = e$, derive the identity $((x,y),y) = e$ where $(x,y) = x \circ y \circ x' \circ y'$.

Jacques Herbrand (12.02.1908–27.07.1931) was a French mathematician and logician who made groundbreaking contributions to logic and automated reasoning, even though he died at the age of only 23 years old.

Herbrand worked in mathematical logic in particular and in what would become computability theory, where he essentially introduced recursive functions. He also proved what is now called the **Herbrand Theorem**, informally stating that a first-order formula in a prenex form containing existential quantifiers is only valid (respectively, provable) in first-order logic if and only if some disjunction of ground substitution instances of the quantifier-free subformula of A is a tautology (respectively, derivable in propositional logic). Herbrand's Theorem therefore reduces in a precise sense validity and theoremhood from first-order logic to propositional logic.

Using his theorem, he also contributed to Hilbert's program in the foundations of mathematics by providing a constructive consistency proof for a weak system of arithmetic.

Herbrand completed his doctorate at École Normale Supérieure in Paris under Ernest Vessiot in 1929 and presented his doctoral thesis *Investigations in proof theory*. He then joined the army and only defended his thesis at the Sorbonne in the following year. He was awarded a Rockefeller fellowship that enabled him to study in Germany in 1931, first with John von Neumann in Berlin, then with Emil Artin in Hamburg, and finally with Emmy Noether in Göttingen.

Herbrand submitted his principal study of proof theory and general recursive functions *On the consistency of arithmetic* early in 1931. While the essay was under consideration, Gödel published his celebrated paper announcing the impossibility of formalizing within any sufficiently expressive theory that theory's consistency proof. Herbrand studied Gödel's paper and wrote an appendix to his own study explaining why Gödel's result did not contradict his own. In July of that year he died in a mountain-climbing accident in the French Alps. His other very important work in logic, *On the consistency of arithmetic*, was published posthumously.

In 1992 the International Association for Automated Reasoning's International Conference on Automated Deduction (CADE) established the annual Herbrand Award for Distinguished Contributions to Automated Reasoning.

John Alan Robinson (b. 1928) is a British–American philosopher, mathematician, and computer scientist, best known for developing the method of resolution for first-order logic.

Robinson was born in England and moved to the US in 1952. He obtained a PhD in philosophy from Princeton University in 1956 for a thesis on *Causation, probability and testimony*. In the 1960s he worked as an operations research analyst at Du Pont and as a visiting researcher at the Argonne National Laboratory, while working at Rice University. He joined Syracuse University as a Distinguished Professor of Logic and Computer Science in 1967 and has held the position of a Professor Emeritus there since 1993.

Robinson introduced the method of **resolution with unification for first-order logic** in his seminal 1965 paper *A machine-oriented logic based on the Resolution Principle*. For his pioneering contributions he received the 1996 Herbrand Award for Distinguished Contributions to Automated Reasoning. His work was also instrumental in the development of the concept of logic programming and, in particular, for the programming language Prolog, for which the Association for Logic Programming awarded him the honorary title *Founder of Logic Programming* in 1997.

4.6   Supplementary: Soundness and completeness of the deductive systems for first-order logic

Here I provide a detailed outline of the proof of the most important technical result of this course: the soundness and completeness of each of the deductive systems for first-order logic presented in this book. I outline the proof for a generic deductive system called **D** which will apply, *mutatis mutandis*, to each of the concrete deductive systems for first-order logic introduced here, namely **H**, **ND**, **ST**, and **RES**. The proof is based on several claims which are left as exercises and listed in the exercises section. Some of them are proved generically for any of these deductive systems, while the proofs of others are specific to each of them and sometimes require quite different arguments based on the specific notions of derivation. I will indicate that in the exercises.

The first such completeness result for a certain axiomatic system for first-order logic, presented in Hilbert and Ackermann's 1928 book *Grundzüge der Theoretischen Logik*, was proved by K. Gödel in the late 1920s. The proof presented here is more modern and elegant, based on L. Henkin's completeness proof (around 1950)[6]. The proof will build on the completeness of the propositional deductive systems outlined in Section 2.7. Extra work and several substantial additional steps are, however, needed because, instead of a

---

[6] Read more on Gödel and Henkin in the biographical boxes at the end of this section.

satisfying truth assignment, we now have to build a whole first-order structure as a model for our consistent theory. There is at least one additional problem: the maximal consistent theory constructed by Lindenbaum's Lemma may not be "rich" enough to provide all the information needed for the construction of such a model. In particular, it may occur that a formula $\exists x A(x)$ belongs to the maximal consistent theory, while for every term t in the language that is free for x in A, *not* $A(t/x)$ but its negation is in that theory; the theory therefore contains no "witness" of the truth of $\exists x A(x)$. We will resolve that problem with a few extra technical lemmas, the proofs of which require the deductive power of **D**.

4.6.1 First-order theories

First, we need to adjust and extend our terminology from the propositional case.

A (**first-order**) **theory** is any set of first-order sentences. A theory Γ is **satisfiable** if it has a **model**, that is, a structure S such that $S \models \Gamma$; otherwise it is **unsatisfiable** (or not satisfiable). Proposition 60 also applies to first-order theories.

4.6.1.1 Semantically closed, maximal, and complete theories

The notions of semantically closed, maximal, and complete first-order theories are defined as for propositional logic in Section 2.7.

4.6.1.2 Deductively consistent, maximal, and complete theories

The notions of **D**-consistent, **D**-maximal, deductively closed in **D**, and **D**-complete first-order theories are defined as for propositional logic; see Definitions 61 and 69. In particular, Propositions 62 and 63 also apply here.

Proposition 155 *For any first-order theory Γ and a sentence B, the following hold.*

1. *$\Gamma \cup \{B\}$ is **D**-consistent iff $\Gamma \nvdash_{\mathbf{D}} \neg B$.*
2. *$\Gamma \vdash_{\mathbf{D}} B$ iff $\Gamma \cup \{\neg B\}$ is **D**-inconsistent.*
3. *If $\Gamma \cup \{B\}$ is **D**-inconsistent and $\Gamma \cup \{\neg B\}$ is **D**-inconsistent then Γ is **D**-inconsistent.*

Lemma 156 *A first-order theory Γ is a maximal **D**-consistent theory iff it is deductively closed in **D** and **D**-complete.*

Theorem 157 (Maximal consistent theories) *For every maximal **D**-consistent first-order theory Γ and sentences A, B the following hold.*

1. *$\neg A \in \Gamma$ iff $A \notin \Gamma$.*
2. *$A \wedge B \in \Gamma$ iff $A \in \Gamma$ and $B \in \Gamma$.*
3. *$A \vee B \in \Gamma$ iff $A \in \Gamma$ or $B \in \Gamma$.*
4. *$A \to B \in \Gamma$ iff $A \in \Gamma$ implies $B \in \Gamma$ (i.e., $A \notin \Gamma$ or $B \in \Gamma$).*

5. *If $\forall x A(x) \in \Gamma$ then $A[t/x] \in \Gamma$ for any term t free for x in A.*
6. *If $A[t/x] \in \Gamma$ for any term t free for x in A then $\exists x A(x) \in \Gamma$.*

Proof. The Boolean cases are as for propositional logic. For the quantifier cases, we need to prove that the following hold for each of the deductive systems **H**, **ND**, **ST**, and **RES**:

1. $\forall x A(x) \vdash_{\mathbf{D}} A[t/x]$, and
2. $A[t/x] \vdash_{\mathbf{D}} \exists x A(x)$.

I leave these proofs as exercises for the reader. ∎

4.6.2   Soundness

Hereafter we fix an arbitrary first-order language (with equality) $\mathcal{L}$.

Theorem 158 (Soundness) *For every first-order theory Γ and a formula A:*

$$\text{if } \Gamma \vdash_{\mathbf{D}} A \text{ then } \Gamma \models A.$$

Proof. This is performed by induction on derivations in **D**. Exercise. ∎

Corollary 159 *For any first-order theory Γ, formula A, and structure $\mathcal{S}$: if $\mathcal{S} \models \Gamma$ and $\Gamma \vdash_{\mathbf{D}} A$ then $\mathcal{S} \models A$.*

Corollary 160 *Every satisfiable theory Γ is **D**-consistent.*

4.6.3   Herbrand structures and interpretations

Definition 161 (Herbrand structure) *Let $\mathcal{L}$ be a first-order language (with equality), containing at least one constant symbol.*

1. *The **Herbrand universe** of $\mathcal{L}$ is the set $H_{\mathcal{L}}$ of all ground terms in $\mathcal{L}$.*
2. *The **Herbrand structure** is any $\mathcal{L}$-structure $\mathcal{H}$ with domain $H_{\mathcal{L}}$, where all functional symbols are interpreted as follows. For any ground terms $t_1, \ldots, t_n \in H_{\mathcal{L}}$:*

$$f^{\mathcal{H}}(t_1, \ldots, t_n) := f(t_1, \ldots, t_n).$$

*Such an interpretation is called a **Herbrand interpretation**. See a note on their importance at the end of the section.*

The interpretation of constant and functional symbols in a given structure $\mathcal{S}$ naturally extends to an interpretation $t^{\mathcal{S}}$ of every ground term t. Note that for every such ground term, we have $t^{\mathcal{H}} = t$.

Definition 162 *Let $\mathcal{S}$ be any $\mathcal{L}$-structure with domain X. An equivalence relation $\sim$ on X is a **congruence in $\mathcal{S}$** if the following hold.*

1. *For every n-ary functional symbol f in $\mathcal{L}$ and $a_1, \ldots, a_n, b_1, \ldots, b_n \in X$:*
   if $a_1 \sim b_1, \ldots, a_n \sim b_n$ then $f^{\mathcal{S}}(a_1, \ldots, a_n) \sim f^{\mathcal{S}}(b_1, \ldots, b_n)$.
2. *For every n-ary predicate symbol r in $\mathcal{L}$ and $a_1, \ldots, a_n, b_1, \ldots, b_n \in X$:*
   if $a_1 \sim b_1, \ldots, a_n \sim b_n$ then $r^{\mathcal{S}}(a_1, \ldots, a_n)$ holds iff $r^{\mathcal{S}}(b_1, \ldots, b_n)$ holds.

Definition 163 *Let $\mathcal{S}$ be any $\mathcal{L}$-structure with domain X and $\sim$ be a congruence in $\mathcal{S}$. The **quotient-structure** of $\mathcal{S}$ with respect to $\sim$ is the $\mathcal{L}$-structure $\mathcal{S}_\sim$ with domain*

$$X_\sim = \{[a]_\sim \mid a \in X\}$$

where $[a]_\sim$ is the $\sim$-equivalence class of a and the non-logical symbols are interpreted in $\mathcal{S}_\sim$ as follows.

1. *For every n-ary functional symbol f in $\mathcal{L}$ and $[a_1]_\sim, \ldots, [a_n]_\sim \in X_\sim$:*

$$f^{\mathcal{S}_\sim}([a_1]_\sim, \ldots, [a_n]_\sim) := [f^{\mathcal{S}}(a_1, \ldots, a_n)]_\sim.$$

2. *For every n-ary predicate symbol r in $\mathcal{L}$ and $[a_1]_\sim, \ldots, [a_n]_\sim \in X_\sim$:*

$$r^{\mathcal{S}_\sim}([a_1]_\sim, \ldots, [a_n]_\sim) \quad \text{iff} \quad [r^{\mathcal{S}}(a_1, \ldots, a_n)].$$

Lemma 164 *The definitions above are correct, that is, independent of the representatives of the elements of $X_\sim$.*

Proof. This is immediate from the definition of congruence. Exercise. ∎

4.6.3.1   Canonical structures for first-order theories

From this point onward we assume that the language $\mathcal{L}$ has at least one constant symbol. If not, such can always be added to the language conservatively, that is, without affecting the derivations in the original language.

Definition 165 (Herbrand structure for a theory) *Given a first-order theory Δ, we define the Herbrand structure $\mathcal{H}(\Delta)$ for Δ as follows. For every n-ary predicate symbol r in $\mathcal{L}$, other than the equality, and ground terms $t_1, \ldots, t_n \in H_{\mathcal{L}}$:*

$$r^{\mathcal{H}}(t_1, \ldots, t_n) \quad \text{iff} \quad \Delta \vdash_{\mathbf{D}} r(t_1, \ldots, t_n).$$

We now define the following binary relation in $H_{\mathcal{L}}$:

$$t_1 \sim_\Delta t_2 \quad \text{iff} \quad \Delta \vdash_{\mathbf{D}} t_1 = t_2.$$

Lemma 166 *For any first-order theory Δ, the relation $\sim_\Delta$ is a congruence in $\mathcal{H}(\Delta)$.*

Proof. This follows from the axioms and rules for the equality in **D**. Exercise. ∎

Definition 167 (Canonical structure for a theory) *Given a first-order theory* Δ, *the* *canonical structure for* Δ *is*

$$\mathcal{S}(\Delta) := \mathcal{H}(\Delta)_{\sim_\Delta}.$$

Lemma 168 *For any first-order theory* Δ *and an atomic sentence* A:

$$\Delta \vdash_{\mathbf{D}} A \ \textit{iff} \ \mathcal{S}(\Delta) \models A.$$

Proof. Exercise. ∎

4.6.4   Henkin theories and Henkin extensions

4.6.4.1   Henkin theories

Definition 169 (Henkin theory) *A first-order theory* Δ *is a* **Henkin theory** *if for every formula with one free variable* $B(x)$ *the following holds: if* $\Delta \vdash_{\mathbf{D}} \exists x B(x)$ *then* $\Delta \vdash_{\mathbf{D}} B[t/x]$ *for some ground term* t, *called a* **witness for** $\exists x B(x)$ **in** Δ.

Lemma 170 *For any first-order theory* Δ, *formula* $B = B(x_1, \dots, x_n)$, *ground terms* $t_1, \dots, t_n$, *structure* $\mathcal{S}$, *and a variable assignment* v *in* $\mathcal{S}$ *such that* $v(x_i) = t_i^{\mathcal{S}}$ *for* $i = 1, \dots, n$, *the following holds:*

$$\mathcal{S}, v \models B(x_1, \dots, x_n) \ \textit{iff} \ \mathcal{S} \models B[t_1/x_1, \dots, t_n/x_n].$$

Proof. This is proved by induction on the number of logical connectives in B. Exercise. ∎

Theorem 171 (Henkin models) *For every maximal* **D**-*consistent Henkin theory* Δ,

$$\mathcal{S}(\Delta) \models \Delta.$$

Proof. It is sufficient to prove that, for every sentence A,

$$\Delta \vdash_{\mathbf{D}} A \ \textit{iff} \ \mathcal{S}(\Delta) \models A.$$

The proof is by induction on the number of occurrences of logical connectives in A.

1. The case of atomic sentence follows from Lemma 168.
2. The cases $A = \neg B$, $A = A_1 \wedge A_2$, $A = A_1 \vee A_2$, and $A = A_1 \rightarrow A_2$ are straightforward, using Theorem 157.
3. Let $A = \exists x B$. If x is not free in B then B is also a sentence. In that case, $\models B \leftrightarrow \exists x B$ and $\vdash_{\mathbf{D}} B \leftrightarrow \exists x B$ (exercise), and therefore $\Delta \vdash_{\mathbf{D}} \exists x B$ iff $\Delta \vdash_{\mathbf{D}} B$ iff (by inductive hypothesis) $\mathcal{S}(\Delta) \models B$ iff $\mathcal{S}(\Delta) \models \exists x B$.

Now, suppose x occurs free in B, that is, $B = B(x)$.

If $\mathcal{S}(\Delta) \models \exists x B$ then $\mathcal{S}(\Delta), v \models B$ for some variable assignment v. Let $v(x) = [t]_{\sim_\Delta}$, where t is a ground term. Then, $\mathcal{S}(\Delta) \models B[t/x]$ by Lemma 170, hence, by the inductive hypothesis $\Delta \vdash_{\mathbf{D}} B[t/x]$. Since $\vdash_{\mathbf{D}} B[t/x] \rightarrow \exists x B$, we obtain $\Delta \vdash_{\mathbf{D}} \exists x B$.

Conversely, let $\Delta \vdash_{\mathbf{D}} \exists x B$. Since Δ is a Henkin theory, we have $\Delta \vdash_{\mathbf{D}} B[t/x]$ for some ground term t. Then $\mathcal{S}(\Delta) \models B[t/x]$ by the inductive hypothesis, hence $\mathcal{S}(\Delta), v \models B$ for any variable assignment v such that $v(x) = [t]_{\sim_\Delta}$ by Lemma 170. Therefore, $\mathcal{S}(\Delta) \models \exists x B$.

4. The case $A = \forall x B(x)$ is analogous. It can also be reduced to the previous cases, using $\models \forall x B(x) \leftrightarrow \neg\exists\neg B$ and $\vdash_{\mathbf{D}} \forall x B(x) \leftrightarrow \neg\exists\neg B$. ■

Thus, every maximal Henkin theory has a model. The model $\mathcal{S}(\Delta)$ is called the **Henkin model** of such a theory Δ.

4.6.4.2   Henkin extensions of theories

We now show that every **D**-consistent theory Γ can be extended to a maximal **D**-consistent Henkin theory.

Lemma 172   (Lemma about constants) *For every first-order theory Γ, formula A and a constant symbol c not occurring in $\Gamma \cup A$:*

$$\Gamma \vdash_{\mathbf{D}} A[c/x] \text{ iff } \Gamma \vdash_{\mathbf{D}} \forall x A(x).$$

Proof. The one direction is straightforward, since $\vdash_{\mathbf{D}} \forall A(x) \rightarrow A[c/x]$. The other is specific to each deductive system **D**. Exercise. ■

Corollary 173 *For any theory Γ and a sentence $\exists x A(x)$, if $\Gamma \cup \exists x A(x)$ is **D**-consistent then $\Gamma \cup A[c/x]$ is **D**-consistent for any constant symbol c not occurring in $\Gamma \cup A$.*

Proof. Suppose $\Gamma \cup A[c/x] \vdash_{\mathbf{D}} \bot$. Then $\Gamma \vdash_{\mathbf{D}} \neg A[c/x]$, therefore $\Gamma \vdash_{\mathbf{D}} \forall x \neg A(x)$ by Lemma 172, hence $\Gamma \vdash_{\mathbf{D}} \neg\exists x A(x)$, which contradicts the **D**-consistency of $\Gamma \cup \exists x A(x)$. ■

Lemma 174   (Conservative extensions) *Let Γ be any theory in a first-order language $\mathcal{L}$ and let $\mathcal{L}^+$ be the extension of $\mathcal{L}$ obtained by adding countably many new constants $c_1, \ldots, c_n, \ldots$ Then for any formula A in $\mathcal{L}$, $\Gamma \vdash_{\mathbf{D}} A$ in $\mathcal{L}$ iff $\Gamma \vdash_{\mathbf{D}} A$ in $\mathcal{L}^+$.*

*In particular, Γ is **D**-consistent in $\mathcal{L}$ iff Γ is **D**-consistent in $\mathcal{L}^+$.*

Proof. Clearly, if Γ is **D**-consistent in $\mathcal{L}^+$ then Γ is **D**-consistent in $\mathcal{L}$. The converse direction is specific to each deductive system **D**. See exercises. ■

Lemma 175   (Compactness of the deductive consequence) *The deductive consequence in **D** is compact in the following sense: for any theory Γ and a sentence A, it holds that $\Gamma \vdash_{\mathbf{D}} A$ iff $\Gamma_0 \vdash_{\mathbf{D}} A$ for some finite $\Gamma_0 \subseteq \Gamma$.*

This property is equivalent (exercise) to the following: a first-order theory Γ is **D**-consistent iff every finite subset of Γ is **D**-consistent.

Theorem 176 (Lindenbaum–Henkin Theorem) *Let Γ be* **D***-consistent theory in a countable language $\mathcal{L}$ and let $\mathcal{L}^+$ be the extension of $\mathcal{L}$ obtained by adding countably many new constants $c_1, \ldots, c_n, \ldots$ Then Γ can be extended to a maximal Henkin theory $H(\Gamma)$ in $\mathcal{L}^+$.*

Proof. Let $A_0, A_1, \ldots$ be a list of all sentences in $\mathcal{L}^+$. (NB: these are still only countably many!)

We define a chain by inclusion of theories $\Gamma_0, \Gamma_1, \ldots$ by recursion on n as follows.

- $\Gamma_0 := \Gamma$.
- $\Gamma_{n+1} := \begin{cases} \Gamma_n \cup \{A_n\} & \text{if } \Gamma_n \vdash_{\mathbf{D}} A_n \text{ and } A_n \text{ is not of the type } \exists x B; \\ \Gamma_n \cup \{A_n\} \cup \{B[c/x]\} & \text{if } \Gamma_n \vdash_{\mathbf{D}} A_n \text{ and } A_n = \exists x B, \text{ where } c \text{ is the} \\ & \text{first new constant that does not occur in} \\ & \Gamma_n \cup \{A_n\}; \\ \Gamma_n \cup \{\neg A_n\} & \text{if } \Gamma_n \nvdash_{\mathbf{D}} A_n. \end{cases}$

Note that every Γ_n is an **D**-consistent theory. (Exercise: prove this by induction on n, using Corollary 173.)

We now define

$$H(\Gamma) := \bigcup_{n \in \mathbf{N}} \Gamma_n.$$

Clearly, $\Gamma \subseteq H(\Gamma)$. We claim that $H(\Gamma)$ is a maximal **D**-consistent Henkin theory. Indeed:

- $H(\Gamma)$ is **D**-consistent. To prove this, suppose otherwise. Then $H(\Gamma) \vdash_{\mathbf{D}} A$ and $H(\Gamma) \vdash_{\mathbf{D}} \neg A$. Since the deductive consequence is compact by Lemma 175, it follows that $\Gamma_n \vdash_{\mathbf{D}} A$ and $\Gamma_n \vdash_{\mathbf{D}} \neg A$ for some large enough index n (exercise: complete the details here), which contradicts the consistency of Γ_n.

- $H(\Gamma)$ is maximal **D**-consistent. Indeed, take any sentence A. Let $A = A_m$. Then $A_m \in \Gamma_{m+1}$ or $\neg A_m \in \Gamma_{m+1}$, hence $A \in H(\Gamma)$ or $\neg A \in H(\Gamma)$, hence $H(\Gamma)$ cannot be extended to a larger **D**-consistent theory.

- $H(\Gamma)$ is a Henkin theory. The proof of this is left as an exercise. ∎

The theory $H(\Gamma)$ defined above is called a **maximal Henkin extension of Γ**.

Remark: The result above also applies to uncountable languages, by using Zorn's Lemma.

4.6.5 Completeness theorem

We are now ready to state and prove the completeness theorem generically for any of the deductive systems **D** introduced here.

Theorem 177 (Model existence) *Every* **D**-*consistent first-order theory* Γ *has a model.*

Proof. Given a **D**-consistent theory Γ, construct a maximal Henkin theory $H(\Gamma)$ extending Γ, using Theorem 176. We then take the Henkin model for $H(\Gamma)$. That model, restricted to $\mathcal{L}$, is a model of Γ. ∎

Corollary 178 (Completeness Theorem) *For every first-order theory* Γ *and a formula* A, *if* $\Gamma \models A$ *then* $\Gamma \vdash_{\mathbf{D}} A$.

Proof. If $\Gamma \models A$ then $\Gamma \cup \{\neg A\}$ is unsatisfiable; by the model existence theorem, $\Gamma \cup \{\neg A\}$ is then **D**-inconsistent. Therefore, $\Gamma \vdash_{\mathbf{D}} A$ by Proposition 155. ∎

4.6.6 Semantic compactness of first-order logic

An immediate consequence of the completeness theorem is the compactness of the logical consequence as follows.

Theorem 179 (Semantic Compactness Theorem 1) *For any first-order theory* Γ *and a formula* A, *if* $\Gamma \models A$ *then* $\Gamma_0 \models A$ *for some finite* $\Gamma_0 \subseteq \Gamma$.

Equivalently, we also have the following theorem.

Theorem 180 (Semantic Compactness Theorem 2) *A first-order theory* Γ *is satisfiable iff every finite subset of* Γ *is satisfiable.*

This is a fundamental result about the semantics of first-order logic which can be proved independently of the deductive completeness theorem. It has numerous important applications which we do not discuss here, but see the references.

Finally, we take note of another very important result on first-order logic, known as **Herbrand's Theorem**. First, it follows from the proof of the completeness theorem for **RES** that, if any interpretation satisfies a given set of clauses S, then there is a Herbrand interpretation that satisfies them. Herbrand's Theorem states that, moreover, if a set of clauses S is unsatisfiable then there is a *finite* set of ground instances of these clauses, instantiated with elements from the Herbrand universe, which is unsatisfiable. Since such a set is finite and consists of ground clauses, where no substitutions are possible, its unsatisfiability can be verified in a finite number of steps. This sounds like a way to check any set of first-order clauses for satisfiability; isn't this in a contradiction with Church's

Undecidability Theorem? Alas, no. The catch is that there may be an infinite number of such candidate sets of ground instances to check.

References for further reading
For more detailed proofs of soundness and completeness of deductive systems for first-order logic, see: van Dalen (1983) and Chiswell and Hodges (2007) for a completeness proofs for **ND**; Shoenfield (1967), Hamilton (1988), Mendelson (1997), and Enderton (2001) for completeness of **H**; Nerode and Shore (1993), Fitting (1996), and Ben-Ari (2012) for completeness of **ST** and **RES**; Smullyan (1995) and Smith (2003) for completeness of **ST**; Ebbinghaus *et al.* (1996) and Hedman (2004) for completeness of **RES**; and Boolos *et al.* (2007) for completeness of the Sequent calculus presented there.

For more on Herbrand's Theorem see Shoenfield (1967), Nerode and Shore (1993), and Fitting (1996). For expositions and discussions of Gödel's Incompleteness Theorems, see Jeffrey (1994), Ebbinghaus *et al.* (1996), Enderton (2001), Hedman (2004), and Boolos *et al.* (2007).

Exercises

In the following exercises, FO stands for first-order logic, **H**, **ND**, **ST**, and **RES** refer to the respective deductive systems for FO introduced here, and **D** refers to any of these deductive systems. All references to definitions and results in Section 2.7 now refer to the respective definitions and results for FO.

4.6.1   Prove Proposition 60 for FO.

4.6.2   Prove Proposition 62 for:
    (a) **H**;   (b) **ST**;   (c) **ND**;   (d) **RES**.

4.6.3   Prove Proposition 63 for:
    (a) **H**;   (b) **ST**;   (c) **ND**;   (d) **RES**.

4.6.4   Prove Proposition 68 generically for each deductive system **D**.

4.6.5   Prove Proposition 70 generically for each deductive system **D**.

4.6.6   Prove Theorem 157 for:
    (a) **H**;   (b) **ST**;   (c) **ND**;   (d) **RES**.

4.6.7   Prove Theorem 158 for:
    (a) **H**;   (b) **ST**;   (c) **ND**;   (d) **RES**.

4.6.8   Prove Lemma 164.

4.6.9 Prove Lemma 166 for:
(a) **H**;   (b) **ST**;   (c) **ND**;   (d) **RES**.

4.6.10 Prove Lemma 168 generically for any deductive system **D**.

4.6.11 Prove Lemma 170.

4.6.12 Complete the proof of Lemma 172 for:
(a) **H**;   (b) **ST**;   (c) **ND**;   (d) **RES**.

4.6.13 Prove Lemma 174 for:
(a) **H**;   (b) **ST**;   (c) **ND**;   (d) **RES**.

(Hint for **ND**: suppose $\Gamma \vdash_{\mathbf{D}} \bot$ in $\mathcal{L}^+$ by a derivation Ξ. Let $\widehat{\Xi}$ be the result of replacing each free occurrence of a new constant c_i in Ξ by a new variable x_i not occurring in Ξ. It can be shown, by inspection of the rules of **D**, that $\widehat{\Xi}$ is a valid derivation of $\Gamma \vdash_{\mathbf{D}} \bot$ in $\mathcal{L}$.)

4.6.14 Prove Lemma 175 for:
(a) **H**;   (b) **ST**;   (c) **ND**;   (d) **RES**.

4.6.15 Prove that the deductive compactness property stated in Lemma 175 is equivalent to the following: a first-order theory Γ is **D**-consistent iff every finite subset of Γ is **D**-consistent.

4.6.16 Complete the generic proof details of Theorem 176 for any deductive system **D**.

4.6.17 Assuming soundness and completeness of **H**, prove soundness and completeness of each of **ST**, **ND**, and **RES** by using a first-order analog of Proposition 76.

4.6.18 Assuming soundness and completeness of **ST**, prove soundness and completeness of each of **H**, **ND**, and **RES** by using a first-order analog of Proposition 76.

4.6.19 Assuming soundness and completeness of **ND**, prove soundness and completeness of each of **ST**, **H**, and **RES** by using a first-order analog of Proposition 76.

4.6.20 Assuming soundness and completeness of **RES**, prove soundness and completeness of each of **H**, **ST**, and **ND** by using a first-order analog of Proposition 76.

4.6.21 Prove the equivalence of the semantic compactness theorems 179 and 180.

 **Kurt Friedrich Gödel** (28.4.1906–14.1.1978) was an Austrian–American logician, mathematician, and philosopher, regarded as the most influential logician of the 20th century.

Gödel's first interest in logic was sparked when he attended a seminar on Russell's book *Introduction to Mathematical Philosophy*. Later he attended a lecture by David Hilbert on completeness and consistency of mathematical systems, where Hilbert posed the question of whether there is a consistent formal system of axioms of first-order logic which can derive every valid – that is, true in all models – statement of first-order logic. Gödel chose this problem as the topic of his doctoral work and he completed his doctoral dissertation under the supervision of Hans Hahn in 1929, at the age of 23. In his thesis he proposed an axiomatic system for the first-order predicate logic and established its completeness. This was his first famous result, known as **Gödel's Completeness Theorem**.

In 1931 Gödel published his most important and groundbreaking work, *On Formally Undecidable Propositions of "Principia Mathematica" and Related Systems*, where he proved his famous *incompleteness theorems*.

Gödel's First Incompleteness Theorem states that any axiomatic system of first-order logic which is consistent (i.e., no contradiction can be derived in it), has an effectively recognizable (recursive) set of axioms, and is expressive enough to describe the arithmetic of the natural numbers with addition and multiplication must be incomplete; that is, there are true arithmetical statements that can be stated in the language of that system but cannot be derived from its axioms. In particular, this result applies to the Peano system of axioms of the arithmetic and to Zermelo–Fraenkel set theory. **Gödel's Second Incompleteness Theorem** states that no such consistent system can derive a statement, formalized in the language of arithmetic, claiming its consistency.

Gödel's basic idea of the proof of the incompleteness theorems was conceptually simple but extremely original. It involved using, for any given formal axiomatic system of the arithmetic with an effectively enumerable set of axioms, a specially developed technique of encoding of the notions of formulae, axioms, and derivations in the language of the arithmetic, to eventually construct a formula of that language that claims that it is unprovable in the given formal system. Such a formula cannot be provable in the system, for that would imply its inconsistency, and therefore what it states must be true.

Gödel's incompleteness theorems had a shattering effect on the attempts to find an effectively enumerable set of axioms sufficient to derive all true statements in mathematics, beginning with the work of Frege half a century earlier, and culminating in *Principia Mathematica* and Hilbert's program for formalizing of the mathematics.

Gödel's theorems also had a profound effect on mathematical logic and the foundations and philosophy of mathematics. They are one of the greatest achievements of 20th century mathematics; they are of fundamental importance not only for the foundations of mathematics but also for computer science, as they imply that a computer can never be programmed to derive all true mathematical statements, not even all those about the arithmetic of natural numbers.

Leon Albert Henkin (19.04.1921–1.11.2006) was an American logician and mathematician, best known for the **Henkin Completeness Proof** for axiomatic deductive systems of first-order logic. This proof is much simpler and intuitive than the original Gödel proofs, and has become the standard method for semantic completeness proofs since then.

After studying mathematics and philosophy at Columbia College, Henkin was a doctoral student of Alonzo Church at Princeton University, where he received his PhD in 1947. He then became Professor of Mathematics at the University of California, Berkeley in 1953 where he worked until his retirement in 1991. He was a collaborator of Alfred Tarski and a keen popularizer of logic and of mathematics.

In his doctoral dissertation *The completeness of formal systems*, Henkin originally proved the completeness of Church's higher-order logic using the now-called **general Henkin semantics**, where the higher types are interpreted not by the full space of functions but by a suitably rich subset of it. He then adapted his method to obtain a new completeness proof for first-order logic, introducing the now-called **Henkin constants** and **Henkin models**. His proof method is now used for completeness proofs of a wide variety of logical systems. Henkin also studied extensions of first-order logic with **branching quantifiers** and with infinite strings of quantifiers and proposed **game-theoretic semantics** for them.

Henkin was not only a distinguished academic and promoter of logic and mathematics, but also a passionate social activist, instrumental in designing special programs and scholarships for talented students from disadvantaged minorities.

5

Applications: Mathematical Proofs and Automated Reasoning

Logical systems are used in practice to perform formal reasoning that goes beyond pure logic. Most often they are used for mathematical reasoning, but they have also found numerous other applications in artificial intelligence, especially the field of *knowledge representation and reasoning* (e.g. various *description logics for ontologies*), as well as in many areas of computer science including *database theory, program analysis*, and *deductive verification*. In each of these areas the use of deductive systems is guided by the specific applications in mind, but the underlying methodology is the same: the purely logical engine of the deductive system, which consists of the general, *logical axioms and inference rules* is extended with specific, *non-logical axioms and rules* describing the particular subject area. The logical and non-logical axioms and rules together constitute a formal *theory* in which the formal reasoning is performed by means of derivations from a set of assumptions in the chosen deductive system. These assumptions are usually the relevant non-logical axioms, plus other specific *ad hoc* assumptions applying to the concrete domain or situation for which the reasoning is conducted.

Most typical mathematical theories describe classes of important *relational structures* such as *sets, partial or linear orders, directed graphs, trees, equivalence relations, various geometric structures, etc., algebraic structures*, such as *lattices, Boolean algebras, groups, rings, fields, etc.*, or combined, for example *ordered rings and fields*. Other important mathematical theories are intended to describe *single structures* of special interest, for instance: an axiomatic theory for the *arithmetic of natural numbers*, the most popular one being *Peano Arithmetic*; the subsystem of *Presburger Arithmetic* (involving only addition); or the first-order theory of the field of reals $\mathcal{R}$ or of the field of rational numbers $\mathcal{Q}$, etc.

In this chapter I first discuss generally the logical structure, strategies and tactics of mathematical reasoning and proofs and then illustrate these with several examples in

Logic as a Tool: A Guide to Formal Logical Reasoning, First Edition. Valentin Goranko.
© 2016 John Wiley & Sons, Ltd. Published 2016 by John Wiley & Sons, Ltd.

Section 5.1. I then present and briefly discuss some of the basic theories mentioned above in Section 5.2 where I give just a minimum background on sets, functions, and relations needed for performing meaningful mathematical reasoning in them. The actual proofs will be left as exercises for the reader however, the main purpose of the chapter. Section 5.3.1 is supplementary, intended to provide a basic background on Mathematical Induction to the reader who needs it. Section 5.3.2 focuses on deductive reasoning in the axiomatic system of Peano Arithmetic, again mainly by means of exercises.

Logical deductive reasoning is carried out not only manually but also – with increasing popularity and success – using computers. That use has lead to the active development of *automated reasoning*, including *automated and interactive theorem proving*. On the other hand, logic has made a strong methodological contribution to the theory and practice of programming and computing by suggesting the paradigm of *logic programming*, realized in several programming languages (the most popular of these being *Prolog*). I discuss these topics very briefly in Section 5.4.

5.1 Logical reasoning and mathematical proofs

In mathematics the truth of a statement is established by proving it. A proof may consist of a simple argument or of much complicated calculations, but essentially no logical reasoning. Alternatively, it may consist of a long and intricate argument involving a number of other already-proven statements, conjectures, or logical inferences.

This section is about the *logical* aspects and structure of mathematical reasoning and proofs. I discuss strategies and tactics for proofs, essentially based on the rules of Natural Deduction, and illustrate them with a few examples. I end this section with some brief remarks on Resolution-based automated reasoning.

5.1.1 Proof strategies: direct and indirect proofs

A typical mathematical statement is of the form:

$$\text{if } \mathsf{P}_1, \dots, \mathsf{P}_n \text{ then } \mathsf{C}$$

where $P_1, \dots, P_n$ (if any) are *premises* or *assumptions* and C is a *conclusion*.

While mathematical arguments are very specific to the subject area and the concrete statement, the structure of the logical arguments in a proof only depend on the adopted logical strategy of proof and the logical forms of the premises and the conclusion. Here I provide some *proof tactics*, describing how to go about specific, *local* steps of the proof, but I first discuss possible *proof strategies*, describing how proofs are organized *globally*.

I. Direct proofs With this proof strategy we *assume* that all premises are true, and then try to *deduce* the desired conclusion by applying a sequence of correct – logical or substantially mathematical – inference steps. Schematically, a direct proof takes the following form.

$$\text{Assume } P_1, \ldots, P_n.$$

$$\vdots$$

(a sequence of valid inferences)

$$\vdots$$

Conclude C.

Typical patterns of direct proofs are derivations in Natural Deduction without using the rule of *Reductio ad Absurdum*.

Example 181 *As an example, we provide a direct proof of the statement*

if n is an odd integer, then n^2 is an odd integer.

Proof.

1. *Suppose n is an odd integer.*
2. *Then there is an integer k such that $n = 2k + 1$.*
3. *Therefore, $n^2 = (2k+1)(2k+1) = 4k^2 + 4k + 1 = 2(2k^2 + 2k) + 1$.*
4. *Therefore, n^2 is odd.*
5. *That proves the statement.*

In this proof only the first and last steps are logical, roughly corresponding to the derivation of $A \to B$ by assuming A and deducing B.

II. Indirect proofs Indirect proofs are also known as **proofs by assumption of the contrary** or **proofs by contradiction**. Typical patterns of indirect proofs are derivations in Semantic Tableaux, and also those derivations in Natural Deduction that use an application of the rule of *Reductio ad Absurdum*. The idea of the indirect proof strategy is to assume that all premises are true while the conclusion is false (i.e., the negation of the conclusion is true), and try to reach a contradiction based on these assumptions, again by applying only valid inferences. A contradiction is typically obtained by deducing a statement known to be false (e.g., deducing that $1 + 1 = 3$) or by deducing a statement and its negation. We can often reach a contradiction by deducing the negation of some of the premises (see the example below) which have been assumed to be true. The rationale behind the proof by contradiction is clear: if all our assumptions are true and we only apply valid inferences, then all our conclusions must also be true. By deducing a false conclusion we therefore show that, given that all original assumptions are true, the additional one that we have made (i.e., the negation of the conclusion) must have been wrong, that is, that *the conclusion must be true*.

The scheme of a proof by contradiction is of the following form.

Assume $P_1, \ldots, P_n$.
Assume $\neg C$

$$\vdots$$

(a sequence of valid inferences)

$$\vdots$$

Derive both A and $\neg A$
for some proposition A.

Conclude that $\neg C$ cannot be true.

We therefore conclude C.

Example 182 *Here is an example of a proof by contradiction of the statement:*

If x and y are integers, x is odd and xy is even, then y is even.

Proof.

1. *Suppose x, y are integers, x is odd, and xy is even.*
2. *Suppose also that y is not even.*
3. *Therefore, y is odd.*
4. *Then there is an integer k such that $x = 2k + 1$, since x is odd.*
5. *There is also an integer m such that $y = 2m + 1$, since y is odd.*
6. *Therefore, $xy = (2k + 1)(2m + 1) = \ldots = 2(2km + k + m) + 1$.*
7. *Therefore, xy is odd.*
8. *On the other hand, we have assumed that xy is even and, therefore, xy is not odd.*
9. *xy is therefore odd and xy is not odd.*
10. *This is a contradiction, which completes the proof.*

Example 183 *Euclid's proof by contradiction that*

$\sqrt{2}$ is irrational.

Proof.

1. *Suppose it is not the case that $\sqrt{2}$ is irrational.*
2. *Therefore, $\sqrt{2}$ is rational.*
3. *$\sqrt{2}$ can therefore be represented as an irreducible fraction, that is, there are relatively prime integers m and n such that $\sqrt{2} = \frac{m}{n}$.*
4. *Therefore, $2 = \frac{m^2}{n^2}$, hence $2n^2 = m^2$.*

5. m^2 *is therefore even.*

6. m *must therefore be even since, if it is odd, then m^2 would be odd (see the example of a direct proof).*

7. *There is therefore an integer a such that $m = 2a$.*

8. *Hence, $m^2 = 4a^2$.*

9. *Therefore $2n^2 = 4a^2$, hence $n^2 = 2a^2$.*

10. n^2 *is therefore even.*

11. n *must therefore be even, by the same argument as above.*

12. *There is therefore an integer b such that $n = 2b$.*

13. *The fraction $\frac{m}{n} = \frac{2a}{2b}$ is therefore not irreducible.*

14. *This contradicts the assumption that $\frac{m}{n}$ is irreducible.*

15. *This contradiction completes the proof.*

A variation of proof by contradiction is **proof by contraposition**. This strategy is usually applied when the statement to be proved is of the type "if P, then C," that is, there is only one premise[1]. This proof strategy is based on the fact that the implication "if P, then C" is logically equivalent to its contrapositive "if $\neg C$, then $\neg P$." If we prove the latter, we therefore have a proof of the former. Schematically, the proof by contraposition of "if P, then C" has the following form.

Assume $\neg C$.

⋮

(a sequence of valid inferences)

⋮

Derive $\neg P$.

We can also think of this strategy as obtained from the proof by contradiction strategy, by assuming P together with $\neg C$ at the beginning and then proving $\neg P$ to produce the desired contradiction.

We usually apply the proof by contraposition when it is easier to prove the contrapositive of the original implication.

Example 184 *Here is a proof by contraposition of the statement*

if n is an integer such that n^2 is even, then n is even.

Proof.

1. *Suppose n is not even.*

2. *Therefore, n is odd.*

3. *Then (see the example of a direct proof) n^2 is odd.*

4. *Therefore, n^2 is not even.*

5. *This completes the proof.*

---

[1] It can in fact be applied in the general case, as we can always replace all premises by their conjunction.

Note that while Semantic Tableaux and Resolution are very convenient and conceptually easy-to-use systems of deduction, they are only designed for proofs by contradiction and not for direct proofs. However, reasoning by contradiction is often unnecessary and sometimes unnatural. Proof by contradiction requires that we have a specified goal (conclusion) to prove, and is of no use if we simply want to derive logical consequences from the knowledge that we have and the assumptions that we may have made. In such cases, direct reasoning has no good alternative.

5.1.2 Tactics for logical reasoning

Here we discuss **tactics of proof** that are applicable to any of the methods discussed above. These tactics can be extracted from the rules of Natural Deduction. The choice of tactics depends both on the logical structure of the conclusion and on the logical structure of the premises. Let us look at these separately.

Tactics based on the conclusion to be proved
A specific tactic can be used according to the logical form of the conclusion C to be proved, as follows.

- If the conclusion is a **negation**, $C = \neg A$, then apply a proof by contradiction by assuming A (which is equivalent to the negation of $\neg A$) and trying to reach a contradiction. Alternatively, import the negation in $\neg A$ inside A and try another tactic.

- If the conclusion is a **conjunction**, $C = A \wedge B$, prove each of A and B.

- If the conclusion is a **disjunction**, $C = A \vee B$, try to prove one of A or B. This works sometimes, but not always. If it does not work, then try a proof by contradiction, that is, assume $\neg C$ which is equivalent to $\neg A \wedge \neg B$, hence assume each of $\neg A$ and $\neg B$, and try to reach a contradiction.
  The equivalence $P \vee Q \equiv \neg P \rightarrow Q$ can also be used to transform the disjunction into implication; see below.

- If the conclusion is an **implication**, $C = A \rightarrow B$, assume A in addition to all premises, and try to prove B.
  Alternatively, attempt a proof by contraposition by tackling a conclusion $\neg B \rightarrow \neg A$ instead.
  Lastly, try a proof by contradiction by assuming $\neg(A \rightarrow B)$ or, equivalently, A and $\neg B$.

- If the conclusion is a **biconditional**, $C = A \leftrightarrow B$, then it is equivalent to the conjunction $(A \rightarrow B) \wedge (B \rightarrow A)$ and we follow the tactics mentioned above.

- If the conclusion is a **universally quantified statement**, $C = \forall x A(x)$, typically reason by saying "Let c be a (name of an) arbitrary object from the domain" (e.g., an arbitrary real number). Then you try to prove that $A(c)$ holds with a general argument, without assuming any specific properties of c. In particular, the name c must be a new one which has not yet been mentioned in the proof for the argument to be really general.
  If you have proved that $A(c)$ holds independently of c, then that proof will apply to *any* object c from the domain. That will produce a proof that $A(x)$ holds for *every* object x, that is, a proof of $\forall x A(x)$.

- If the conclusion is an **existentially quantified statement**, $C = \exists x A(x)$, then either try to find an explicit **witness** c (which can be any concrete object in the domain or just a term in the language) such that $A(c)$ holds, or attempt a proof by contradiction, that is, a proof that *it cannot be the case that there does not exist x for which $A(x)$ holds*. The latter is called a **non-constructive proof of existence**, common in mathematics.

Tactics based on the premises
A specific tactic can be used for each premise P, according to its logical form as follows.

- If the premise is a **negation**, $P = \neg A$, then import that negation inside all other logical connectives in A, if any. Alternatively, try to use it for deriving a contradiction if A is assumed or can be derived.

- If the premise is a **conjunction**, $P = A \wedge B$, it can be replaced it by two new premises, namely A and B.

- If the premise is a **disjunction**, $P = A \vee B$, and it is not known which of A and B is true, then reason *per case*, that is, consider separately each of the two possible cases: Case 1: A is true; and Case 2: B is true. If it can be proven that the conclusion C follows in each of these cases, then this provides a proof that C follows from $A \vee B$. (Why?)

- If the premise is an **implication**, $P = A \to B$, assume or try to derive A and then conclude B and use it further.
  Alternatively, replace $A \to B$ by the equivalent disjunction $\neg A \vee B$ and reason as in the previous case.

- If the premise is a **biconditional**, $P = A \leftrightarrow B$, replace it by the pair of premises $(A \to B)$ and $(B \to A)$.

- If the premise is a **universally quantified statement**, $P = \forall x A(x)$, then you can obtain from it new premises (possibly infinitely many) $A(c)$ for each object c from the domain. More generally, deduce $A(t)$ for any term that is free for substitution for x in $A(x)$ and use these further in your reasoning.

- If the premise is an **existentially quantified statement**, $P = \exists x A(x)$, then introduce a new name, say c, for an object that satisfies A ("if there is an object x such that $A(x)$, then let us take one and give it a name, say c"). Then replace the premise by $A(c)$. Note that the name c must be a new one that is not mentioned in the proof or target conclusion, and no assumptions about c can be made other than $A(c)$.

References for further reading
For further discussions and detailed illustrations with many examples of how to perform logically correct and well-structured mathematical reasoning and proofs, see Kalish and Montague (1980, including many formal proofs), Sollow (1990, including many solutions), Garnier and Taylor (1996), Barwise and Echemendy (1999), Devlin (2004, 2012), Nederpelt and Kamareddine (2004), Bornat (2005), Velleman (2006), Makinson (2008), and Day (2012).

Exercises

5.1.1 For each of the examples of proofs given in this section determine which steps in the proof above are based on logical inferences, and which are based on substantial mathematical arguments.

5.1.2 Prove each of the following statements. Clearly indicate the method of proof and explain the steps. Distinguish the logical from the mathematical steps.
  (a) If the sum of two integers is even, then so is their difference.
  (b) If a, b, x are real numbers, $a < b$, and $(x - a)(x - b) < 0$, then $a < x < b$.
  (c) If a and b are rational numbers and $a < b$, then there exists a rational number r such that $a < r < b$.
  (d) For every integers m, n, if mn is even, then m is even or n is even.
  (e) For every natural number n the number $n^2 + n$ is even.
  (f) For every two integers a, b, if ab is odd then $a + b$ is even.
  (g) The number $\sqrt[3]{3}$ is irrational.
  (h) No positive integers m and n exist such that $\frac{1}{m} + \frac{1}{n} > 2$.
  (i) If $x + y$ is an irrational number, then x is an irrational number or y is an irrational number.

Charles Sanders Peirce (10.09.1839–19.04.1914) was an American universal scientist, philosopher, logician, and mathematician. He made pioneering scientific contributions to an extremely wide range of disciplines including logic, mathematics, statistics, philosophy, metrology, chemistry, experimental psychology, economics, linguistics, scientific methodology, and the history of science. In particular, he is regarded as a founding father of **pragmatism**, a philosophical theory and method of scientific research, of modern statistics, and of **semiotics**, the study of the meaning of symbols.

While he trained and worked as a chemist and geodesist, Peirce regarded himself mainly as a logician, philosopher, and mathematician. When Alfred Tarski and his students were developing their theory of **relation algebras** in the 1940s, they rediscovered many ideas of Peirce's pioneering work on the algebraic and logical study of relations, starting with his 1870 paper *Logic of Relatives*. Moreover, Peirce's system of relational logic laid the foundations for the relational model of databases developed by Edgar F. Codd (a doctoral student of a student of Peirce's) for which Codd received a Turing Award in 1981. Peirce developed further Boole's work on algebraic logic and showed, in particular, that a single binary logical operator (logical NOR, now called **Peirce's arrow**) is sufficient to express all other Boolean connectives. He also introduced first-order and second-order quantification independently of Frege and Peano.

Peirce also made contributions to cardinal arithmetic for infinite numbers, years before Cantor's first attempts, as well as to the theory of infinitesimals and mathematics of the continuum. He also conducted important work in several other branches of pure and applied mathematics and statistics. Notably, as early as in 1886 – more than 50 years before the first digital computers were constructed – he showed how Boolean logical operations could be carried out by electrical switching circuits.

Peirce's personal life and professional career were quite troubled. From early age he suffered from a nervous condition now called "facial neuralgia" which made him appear unfriendly and also caused sudden outbursts of bad temper. Together with some other features of his character, this contributed to the fact that he never held a stable employment; the only academic job he had was a non-tenure position as lecturer in logic at the Johns Hopkins University. However, he was fired from this post in 1884 after a scandal following the discovery that, while still legally married pending divorce, he had an affair with another woman (of unknown origin) before they married later and lived together to his end. Peirce spent the last 20–30 years of his life in utmost poverty and constant debt, and died destitute.

While Peirce was largely unrecognized and rejected during his life, gradually he received the recognition he deserved after his death. Russell and Whitehead were not aware of his work when they published *Principia Mathematica* in 1910 and did not mention him there, but in 1959 Russell wrote "Beyond doubt [...] he was one of the most original minds of the later nineteenth century, and certainly the greatest American thinker ever." Much earlier in 1918, the logician C. I. Lewis wrote "The contributions of C.S. Peirce to symbolic logic are more numerous and varied than those of any other writer – at least in the 19th century." In 1934, the philosopher Paul Weiss called Peirce "the most original and versatile of American philosophers and America's greatest logician" and Karl Popper considered him "one of the greatest philosophers of all times."

Clarence Irving Lewis (12.04.1883–3.02.1964) was an American logician, epistemologist, and moral philosopher, one of the founders of modern philosophical logic. Lewis is best known in logic for his work exposing and criticizing the paradoxical features of the truth functional, *material implication*, as used in Russell and Whitehead's *Principia Mathematica*, which declares, for instance, that any true consequent follows from any false antecedent. Lewis proposed to replace that material implication with a **strict implication** which avoids that. Lewis' strict implication was not primitive, but defined as in terms of negation, conjunction, and a prefixed unary intensional **modal operator** $\Diamond$, where $\Diamond A$ reads "*A is possibly true*." Lewis then defined "*A strictly implies B*" as $\neg\Diamond(A \wedge \neg B)$. Lewis later devised the formal systems $S1$ to $S5$ of **modal logic**, adopting different principles for the modal operator $\Diamond$ and its dual $\Box$ (where $\Box A$ read as "*A is necessarily true*"), included in his 1932 book with Langford *Symbolic Logic*, thus laying the foundations of modern, formal modal logic.

5.2 Logical reasoning on sets, functions, and relations

The purpose of this section is *not* to teach the basics of sets, functions, and relations. Here I assume that the reader already has some basic background on these; if necessary, consult the references at the end of the section. In this section I summarize the relevant basic concepts and list a number of simple properties to be proved as exercises on basic logical reasoning in mathematical proofs.

5.2.1 Zermelo–Fraenkel axiomatic theory of sets

The concept of a set is fundamental in contemporary mathematics, but it only developed as a precise mathematical concept in the mid-late 19th century through the work of several mathematicians, most notably Georg Cantor. Soon afterwards, some paradoxes were discovered as inevitably arising in the "naive" theory of sets, where any conceivable collection of objects that can be described in our language was declared as a set. The most famous of these is **Russell's paradox**. Using that unrestricted possibility to define sets, Russell proposed the definition of the set of all sets that are not elements of themselves. Written in a **set-builder notation**, that set can be "defined" as follows:

$$R = \{x \mid x \notin x\}.$$

Now, any yes/no answer to the question of whether R is an element to itself leads to a contradiction, thus creating the paradox. Note the role of self-reference here.

These paradoxes showed that a more conservative and systematic approach was needed when building sets and their theory, and several axiomatic systems were proposed to capture the notion of sets abstractly and consistently. The most popular of them is **Zermelo–Fraenkel axiomatic set theory** (ZF). It is a formal logical theory in a very simple first-order language for sets $\mathcal{L}_{ZF}$, containing only equality plus just one additional binary predicate symbol $\in$ of membership of an object to a set as its element. I will not list all axioms of ZF here, as we will need only very few of them in what follows. The most fundamental axiom that we use very often is the **Extensionality axiom**:

EXT: $\forall x \forall y(x = y \leftrightarrow \forall z(z \in x \leftrightarrow z \in y))$.

This essentially states that two sets are declared equal just in case they have the same elements, so a set is determined completely by the collection of its elements.

We also say that a set A is included in a set B, denoted $A \subseteq B$, just in case every element of A is also an element of B. Formally, we can define the relation inclusion in the language $\mathcal{L}_{ZF}$ as follows:

$$x \subseteq y := \forall z(z \in x \rightarrow z \in y).$$

This definition plus the axiom EXT and a basic logical inference provide a proof of the following very simple claim, which I will present in a very detailed way as an illustration of logic-based mathematical reasoning. The reader may find it *too* detailed, and I agree that it is so, but my purpose here is not just to prove the claim, but also to explain explicitly every single step, logical or mathematical, involved in the proof. You are advised to work through a few more proofs on that level of detail until you feel sufficiently experienced to skip or leave implicit most of these details.

Proposition 185 *For all sets A and B, $A = B$ if and only if $A \subseteq B$ and $B \subseteq A$.*

Proof. We have to prove an equivalence: "... if and only if ... " We do that by proving the implications in both directions.

1. To prove the implication "If $A = B$, then $A \subseteq B$ and $B \subseteq A$" we assume the premise "$A = B$" and try to derive the conclusion "$A \subseteq B$ and $B \subseteq A$." This is a conjunction, so we have to prove both "$A \subseteq B$" and "$B \subseteq A$." To prove that $A \subseteq B$, by definition of set inclusion we must show that every element of A is an element of B. Let x be any (arbitrary) element of A. Then, because $A = B$, we also have $x \in B$. Since $x \in A$ was arbitrary, we conclude that $A \subseteq B$.

Likewise, to show that $B \subseteq A$, let x be any element of B. Because $A = B$, it follows that $x \in A$. Since $x \in B$ was arbitrary, we therefore conclude that $B \subseteq A$.

2. To prove "If $A \subseteq B$ and $B \subseteq A$, then $A = B$," we assume the premise "$A \subseteq B$ and $B \subseteq A$" and try to deduce $A = B$. By the axiom EXT, we have to show that A and B have exactly the same elements. Let x be an arbitrary element. If $x \in A$, then since $A \subseteq B$, we have $x \in B$. Similarly, if $x \in B$, then since $B \subseteq A$, we have $x \in A$. We have therefore shown that, for any x, it holds that $x \in A$ iff $x \in B$, that is, that A and B have the same elements, hence $A = B$, by EXT. ■

Note that the proof above is almost formalized in Natural Deduction. As an exercise, formalize it completely in **ND** or in any deductive system of your choice for first-order logic for $\mathcal{L}_{ZF}$. This can be done in two ways: without explicitly using the symbol $\subseteq$ or using $\subseteq$ and adding its defining equivalence as an assumption:

$$\forall x \forall y (x \subseteq y \leftrightarrow \forall z (z \in x \rightarrow z \in y)).$$

Most of the other axioms of ZF are just statements of existence of some special sets or can be obtained by applying basic operations to already existing sets.

The **empty set axiom** states the existence of an "empty set":

EMPTY: $\exists x \forall y (\neg y \in x)$.

As an exercise, using EXT prove that there is only one empty set, denoted $\emptyset$.

The **pair set axiom** states the existence of a set with elements being a given pair of sets:

PAIR: $\forall x \forall y \exists z \forall u (u \in z \leftrightarrow u = x \lor u = y)$.

As an exercise, using EXT prove that for any sets x, y there is only one pair set for them, denoted $\{x, y\}$.

The **union set axiom** states the existence of the union of any pair of sets:

UNION: $\forall x \forall y \exists z \forall u (u \in z \leftrightarrow u \in x \lor u \in y)$.

As an exercise, using EXT prove that for any sets x, y there is only one union set for them, denoted $x \cup y$.

The **powerset axiom** states the existence of the set of all subsets of any given set:

POWER: $\forall x \exists y \forall z (z \in y \leftrightarrow z \subseteq x)$.

As an exercise, using EXT prove that for any sets x there is only one powerset of x, denoted $\mathcal{P}(x)$.

A natural operation on sets, definable in $\mathcal{L}_{ZF}$, is the **successor set operation** which, applied to any set x, produces the set

$$x' := x \cup \{x\}.$$

The **infinity axiom** states the existence of a set containing the empty set and closed under the successor set operation:

INF: $\exists x(\emptyset \in x \land \forall y(y \in x \to y' \in x))$.

As an exercise, using the other axioms (some are yet to come), prove that $x' \neq x$ for any set x. It then follows that any set x satisfying the formula above must indeed be infinite. "Infinite" here means that it is bijective (see Section 5.2.3) with a proper subset of itself, in this case the subset obtained by removing $\emptyset$.

We next have the **regularity axiom** or the **foundation axiom**:

REG: $\forall x(x \neq \emptyset \to \exists y \forall z(z \in y \to z \notin x))$

which states that every non-empty set x has a disjoint element (i.e., an element having no common elements with x), therefore preventing the existence of an infinite descending chain of set memberships. In particular, this axiom forbids any set to be an element of itself (show this as an exercise).

The next axiom of ZF is actually a scheme, called the **Axiom Scheme of Separation** or **Axiom Scheme of Restricted Comprehension**: for any formula $\phi(\bar{x}, u)$ from $\mathcal{L}_{ZF}$, where $\bar{x}$ is a tuple of free variables $x_1, \dots, x_n$ (to be treated as parameters), it states:

SEP: $\forall y \forall x_1 \dots \forall x_n \exists z \forall u(u \in z \leftrightarrow u \in y \land \phi(\bar{x}, u))$.

Intuitively, this axiom states that, given any set y, for any fixed values of the parameters $\bar{x}$ there exists a set consisting of exactly those elements of u that satisfy the property of sets defined by the formula $\phi(\bar{x}, u)$.

The last axiom in ZF is again a scheme, called the **Axiom Scheme of Replacement**: for any formula $\phi(\bar{x}, y, z)$ from $\mathcal{L}_{ZF}$, where $\bar{x}$ is a tuple of free variables $x_1, \dots, x_n$ (to be treated as parameters), it states:

REP:

$$\forall x_1 \dots \forall x_n \forall u(\forall y(y \in u \to \exists! z \phi(\bar{x}, y, z)) \to$$

$$\exists v \forall z(z \in v \leftrightarrow \exists y \in u \phi(\bar{x}, y, z))).$$

Intuitively, this axiom states that if, for any fixed values of the parameters $\bar{x}$, the formula $\phi(\bar{x}, y, z)$ defines a functional relation $f_{\phi, \bar{x}}(y, z)$, then the image of any set u under that relation is again a set (v).

Another important axiom was added to ZF later, namely the **axiom of choice**, stating that for every set x of non-empty sets there exists a "set of representatives" of these sets, which has exactly one element in common with every element of x:

AC: $\forall x(\forall y(y \in x \to y \neq \emptyset) \to \exists z \forall y(y \in x \to \exists! u(u \in y \land y \in z)))$.

This axiom is sometimes stated only for sets x consisting of pairwise disjoint elements, and then proved for all sets.

The axiomatic system ZF is a very rich theory within which most of the contemporary mathematics can be built. ZF with the added axiom AC is denoted ZFC. While AC looks very natural and almost obvious, it turns out to have some quite unintuitive, even paradoxical, consequences, so it has always been treated with special care. For more on ZF and ZFC, see the references. I do not delve into the foundations of set theory here, but only use that theory to illustrate logical reasoning and mathematical proofs.

5.2.2 Basic operations on sets and their properties

I hereafter denote some sets with capital letters $A, B, C, \ldots, X, Y, Z$ and elements of those sets with lower-case letters $a, b, c, \ldots, x, y, z$. This should hopefully be intuitive and help to avoid confusion.

The existence of the empty set $\emptyset$ and some of the basic operations on sets, namely union $\cup$ and powerset of a set $\mathcal{P}$, are postulated with axioms of ZF. The other basic operations – intersection $\cap$, difference $-$, and Cartesian product $\times$ – can also be defined and justified within ZF.

I list a number of properties of these operations, to be formalized in first-order logic and proved as exercises. First, we make some important general remarks on defining objects in ZF. In order to formally define an operation or relation on sets, we can write its defining equivalence in $\mathcal{L}_{ZF}$ as we did for $\subseteq$ above. For instance, we can define the **proper subset** relation $\subset$ as

$$\forall x \forall y (x \subset y \leftrightarrow x \subseteq y \land x \neq y)$$

and the **intersection** $\cap$ by:

$$\forall x \forall y \forall z (z \in (x \cap y) \leftrightarrow (z \in x \land z \in y)).$$

As an exercise, use the separation axiom SEP to prove the existence of $x \cap y$. For that, write its definition in a set-builder notation and note that it can now be relativized to already "existing" sets.

Likewise, the **Cartesian product** of two sets A, B can formally be defined in set-builder notation as

$$A \times B := \{(a, b) \mid a \in A, b \in B\}$$

where (a, b) is the **ordered pair** consisting of a as a first element and b as a second element. A strange-looking but formally correct (see exercise) and commonly accepted way to define an ordered pair as a set is

$$(a, b) := \{\{a\}, \{a, b\}\}.$$

As an exercise, using axioms of ZF show that the Cartesian product of any two sets exists.

In the next four propositions, note the close parallel between the properties of the logical connectives $\bot, \land, \lor$, and $-$, where $A - B := A \land \neg B$, and of their set-theoretic counterparts $\emptyset, \cap, \cup$, and $-$.

Proposition 186 (Properties of the intersection) *The following hold for any sets* $A, B, C,$ *and* X:

(a) $A \cap \emptyset = \emptyset$

(b) $A \cap A = A$

(c) $A \cap B = B \cap A$

(d) $A \cap (B \cap C) = (A \cap B) \cap C$

(e) $A \subseteq B$ *if and only if* $A \cap B = A$

(f) $A \cap B \subseteq A$ *and* $A \cap B \subseteq B$

(g) *If* $X \subseteq A$ *and* $X \subseteq B$, *then* $X \subseteq A \cap B$.

Proposition 187 (Properties of the union) *The following hold for any sets* $A, B, C,$ *and* X:

(a) $A \cup \emptyset = A$

(b) $A \cup A = A$

(c) $A \cup B = B \cup A$

(d) $A \cup (B \cup C) = (A \cup B) \cup C$

(e) *If* $A \subseteq B$ *if and only if* $A \cup B = B$

(f) $A \subseteq A \cup B$ *and* $B \subseteq A \cup B$

(g) *If* $A \subseteq X$ *and* $B \subseteq X$, *then* $A \cup B \subseteq X$.

Proposition 188 (Properties of set theoretic difference) *The following hold for any sets* A, B, *and* C:

(a) $A - A = \emptyset$

(b) $\emptyset - A = \emptyset$

(c) $A - \emptyset = A$

(d) $A - B \subseteq A$

(e) $(A - B) \cap B = \emptyset$

(f) $A - B = \emptyset$ *if and only if* $A \subseteq B$

(g) $(A - B) \cap (B - A) = \emptyset$

(h) $(A - B) \cup B = A \cup B$

(i) $A - (A \cap B) = A - B$

(j) $A - (B \cap C) = (A - B) \cup (A - C)$

(k) $A - (B \cup C) = (A - B) \cap (A - C)$

(l) $A \cup B = (A - B) \cup (B - A) \cup (A \cap B)$.

Proposition 189 (Interaction between union and intersection) *For all sets* A, B, *and* C *the following hold:*

(a) $A \cap (B \cup C) = (A \cap B) \cup (A \cap C)$ *(distributivity of $\cap$ over $\cup$)*

(b) $A \cup (B \cap C) = (A \cup B) \cap (A \cup C)$ *(distributivity of $\cup$ over $\cap$)*

(c) $A \cap (A \cup B) = A$ *(absorption)*

(d) $A \cup (A \cap B) = A$ *(absorption)*.

Proposition 190 (Properties of the powerset) *The following hold for any sets* A, B, C:

(a) $A \in \mathcal{P}(B)$ *iff* $A \subseteq B$

(b) $\emptyset \in \mathcal{P}(A)$

(c) $A \in \mathcal{P}(A)$

(d) *If* $A \subseteq B$ *then* $\mathcal{P}(A) \subseteq \mathcal{P}(B)$

(e) *If* $A \in \mathcal{P}(X)$ *then* $X - A \in \mathcal{P}(X)$

(f) *If* $A \in \mathcal{P}(X)$ *and* $B \in \mathcal{P}(X)$ *then* $A \cap B$, $A \cup B$, *and* $A - B$ *are in* $\mathcal{P}(X)$

(g) *If* X *has* n *elements then* $\mathcal{P}(X)$ *has* 2^n *elements*.

These properties of the powerset imply that, if U is a **universal set** in which all sets of our interest are included, then $\cup$, $\cap$, and $'$ are operations on $\mathcal{P}(U)$. (We write A' for $U - A$.) Thus, we obtain an algebraic structure $\langle \mathcal{P}(U); \cup, \cap, ', \emptyset, U \rangle$ called the **powerset Boolean algebra** of U.

Proposition 191 (Properties of the Cartesian product) *The following hold for any sets* A, B, C.

(a) $A \times (B \cup C) = (A \times B) \cup (A \times C)$, $(B \cup C) \times A = (B \times A) \cup (C \times A)$.
(b) $A \times (B \cap C) = (A \times B) \cap (A \times C)$, $(B \cap C) \times A = (B \times A) \cap (C \times A)$.
(c) $A \times (B - C) = (A \times B) - (A \times C)$.

5.2.3   Functions

First, we recall the basic terminology: a **function** (or **mapping**) from a set A to a set B is a rule denoted $f : A \to B$ which assigns *to each element* $a \in A$ *a unique element* $f(a) \in B$. The element $f(a)$ is called the **value of** a **under** f, or the **image of** a **under** f. If $f(a) = b$ then a is called a **pre-image of** b **under** f.

The set A is called the **domain** of f, denoted $A = \mathsf{dom}(f)$, and B is called the **co-domain**, or the **target set**, of f, denoted $B = \mathsf{cod}(f)$.

The notion of image can be generalized from elements to subsets of the domain as follows. For any subset $X \subseteq A$ of the domain of f, **the image of** X **under** f is the set $f[X] = \{f(a) \mid a \in X\}$.

The image of the whole domain A under f, that is, the set $f[A] = \{f(a) \mid a \in A\}$ of all values of f, is called the **range** or **image** of f, also denoted $\mathsf{rng}(f)$.

Two functions f and g are **equal** iff $\mathsf{dom}(f) = \mathsf{dom}(g)$, $\mathsf{cod}(f) = \mathsf{cod}(g)$, and $f(a) = g(a)$ for every $a \in \mathsf{dom}(f)$.

The **graph** of a function f is the set of ordered pairs $\{(a, f(a)) \mid a \in \mathsf{dom}(f)\}$.

A function $f : A \to B$ is:

- **injective** or **into** if for every $a_1, a_2 \in A$, $f(a_1) = f(a_2)$ implies $a_1 = a_2$;

- **surjective** or **onto** if $\mathsf{rng}(f) = B$; or

- **bijective** or **one-to-one** if it is both injective and surjective.

If f is injective then the **inverse** of f is the function $f^{-1} : \mathsf{rng}(f) \to \mathsf{dom}(f)$, defined by $f^{-1}(f(a)) = a$ for every $a \in \mathsf{dom}(f)$.

Proposition 192 *For every bijective function* f:

1. f^{-1} *is a bijection; and*   2. $(f^{-1})^{-1} = f$.

If $f : A \to B$ and $g : B \to C$ then the **composition of** f **and** g is the mapping $gf : A \to C$ defined by $gf(a) = g(f(a))$ for each $a \in A$.

Whenever we refer to a composition gf of two mappings f and g, we will assume that $\mathsf{rng}(f) \subseteq \mathsf{dom}(g)$.

Proposition 193 *Composition of mappings is associative, that is, $f(gh) = (fg)h$, whenever either of these is defined.*

Proposition 194 *Let $f : A \to B$, $g : B \to C$ be any mappings. Then the following hold.*

(a) *If f and g are injective then gf is also injective.*
(b) *If f and g are surjective then gf is also surjective.*
(c) *In particular, the composition of bijections is a bijection.*
(d) *If gf is injective then f is injective.*
(e) *If gf is surjective then g is surjective.*

Proposition 195 *If $f : A \to B$, $g : B \to C$ are bijective mappings then gf has an inverse, such that $(gf)^{-1} = f^{-1}g^{-1}$.*

Proposition 196

1. *A mapping f is injective iff for every two mappings g_1 and g_2 with $\mathsf{dom}(g_1) = \mathsf{dom}(g_2)$ and $\mathsf{cod}(g_1) = \mathsf{dom}(f) = \mathsf{cod}(g_2)$, the following **left cancellation property** holds:*

$$\text{if } fg_1 = fg_2 \text{ then } g_1 = g_2.$$

2. *A mapping f is surjective iff for every two mappings g_1 and g_2 with $\mathsf{dom}(g_1) = \mathsf{dom}(g_2) = \mathsf{cod}(f)$, the following **right cancellation property** holds:*

$$\text{if } g_1 f = g_2 f \text{ then } g_1 = g_2.$$

5.2.4 Binary relations and operations on them

We have already discussed relations (predicates) of any number of arguments. Being subsets of a given universal set, sets can be regarded as unary relations. Here we focus on *binary* relations and operations on them.

First, let us summarize the basic terminology. Given sets A and B, a **binary relation between A and B** is any subset of $A \times B$. In particular, a **binary relation on a set** A is any subset of $A^2 = A \times A$. Given a binary relation $R \subseteq A \times B$, if $(a, b) \in R$, we sometimes also write aRb and say that a is R-related to b.

Given sets A and B and a binary relation $R \subseteq A \times B$:

- the **domain of** R is the set $\mathsf{dom}(R) = \{a \in A \mid \exists b \in B(aRb)\}$; and
- the **range of** R is the set $\mathsf{rng}(R) = \{b \in B \mid \exists a \in A(aRb)\}$.
  More generally, given subsets $X \subseteq A$ and $Y \subseteq B$, we define
- the **image of** X **under** R: $R[X] = \{b \in B \mid \exists x \in X(xRb)\}$; and
- the **inverse image of** Y **under** R: $R^{-1}[Y] = \{a \in A \mid \exists y \in Y(aRy)\}$.

Notice that $\mathsf{dom}(R) = R^{-1}[B]$ and $\mathsf{rng}(R) = R[A]$.

Some special relations on any given set A include:

- the **empty relation** $\emptyset$;
- the **equality(identity, diagonal)** relation $E_A = \{(a, a) \mid a \in A\}$; and
- the **universal relation** A^2.

We can also restrict the domain of a relation as for that of a function as follows.

If R is a binary relation on a set A and $B \subseteq A$, then the **restriction of R to B** is the relation $R\mid_B = R \cap (B \times B) = \{(x, y) \mid (x, y) \in R \text{ and } x, y \in B\}$.

5.2.4.1   Operations on binary relations

Boolean operations

Since binary relations are sets themselves, the set operations $\cup, \cap$, and $-$ also apply to them. Besides, we define the **complementation** of a relation $R \subseteq A \times B$ as $R' = (A \times B) - R$.

Proposition 197 *For any relations $R, S \subseteq A \times B$ the following hold:*

(a) $\mathsf{dom}(R \cup S) = \mathsf{dom}(R) \cup \mathsf{dom}(S)$;   (c) $\mathsf{dom}(R \cap S) \subseteq \mathsf{dom}(R) \cap \mathsf{dom}(S)$;

(b) $\mathsf{rng}(R \cup S) = \mathsf{rng}(R) \cup \mathsf{rng}(S)$;   (d) $\mathsf{rng}(R \cap S) \subseteq \mathsf{rng}(R) \cap \mathsf{rng}(S)$.

Inverse of a relation

Let $R \subseteq A \times B$. The **inverse of** R is the relation $R^{-1} \subseteq B \times A$ defined as: $R^{-1} = \{(b, a) \mid (a, b) \in R\}$. Thus, aRb if and only if $bR^{-1}a$.

Proposition 198 *For any relations $R, S \subseteq A \times B$ the following hold:*

(a) $\mathsf{dom}(R^{-1}) = \mathsf{rng}(R)$;   (d) $(R \cup S)^{-1} = R^{-1} \cup S^{-1}$;

(b) $\mathsf{rng}(R^{-1}) = \mathsf{dom}(R)$;   (e) $(R \cap S)^{-1} = R^{-1} \cap S^{-1}$;

(c) $(R^{-1})^{-1} = R$;   (f) $(R')^{-1} = (R^{-1})'$.

Composition of relations

Let $R \subseteq A \times B$ and $S \subseteq B \times C$. The **composition of R and S** is the binary relation $R \circ S \subseteq A \times C$ defined:

$$R \circ S = \{(a, c) \mid \exists b \in B(aRb \wedge bSc)\}.$$

In particular, when $R \subseteq A^2$, $R \circ R$ is defined and denoted R^2.

Note that we can always form the composition of two relations, even if their domains and ranges do not match, by suitably expanding these: if $R \subseteq A \times B_1$ and $S \subseteq B_2 \times C$, then $R \subseteq A \times (B_1 \cup B_2)$ and $S \subseteq (B_1 \cup B_2) \times C$.

Proposition 199 *The composition of binary relations is associative, that is,*

$$(R \circ S) \circ T = R \circ (S \circ T).$$

Note, however, that the composition of binary relations is *not commutative*.

Proposition 200 *For any binary relations R and S: $(R \circ S)^{-1} = S^{-1} \circ R^{-1}$.*

5.2.5   Special binary relations

Definition 201 *A binary relation $R \subseteq X^2$ is called:*

- **reflexive** *if it satisfies $\forall x (xRx)$;*
- **irreflexive** *if it satisfies $\forall x \neg (xRx)$, that is, if $X^2 - R$ is reflexive;*
- **serial** *if it satisfies $\forall x \exists y (xRy)$;*
- **functional** *if it satisfies $\forall x \exists! y (xRy)$, where $\exists! y$ means "there exists a unique y;"*
- **symmetric** *if it satisfies $\forall x \forall y (xRy \rightarrow yRx)$;*
- **asymmetric** *if it satisfies $\forall x \forall y (xRy \rightarrow \neg yRx)$;*
- **antisymmetric** *if it satisfies $\forall x \forall y (xRy \wedge yRx \rightarrow x = y)$;*
- **connected** *if it satisfies $\forall x \forall y (xRy \vee yRx \vee x = y)$;*
- **transitive** *if it satisfies $\forall x \forall y \forall z ((xRy \wedge yRz) \rightarrow xRz)$;*
- *an* **equivalence relation** *if it is reflexive, symmetric, and transitive;*
- **euclidean** *if it satisfies $\forall x \forall y \forall z ((xRy \wedge xRz) \rightarrow yRz)$;*
- *a* **pre-order** *(or* **quasi-order***) if it is reflexive and transitive;*
- *a* **partial order** *if it is reflexive, transitive, and antisymmetric, that is, an antisymmetric pre-order;*
- *a* **strict partial order** *if it is irreflexive and transitive;*
- *a* **linear order** *(or* **total order***) if it is a connected partial order; or*
- *a* **strict linear order** *(or* **strict total order***) if it is a connected strict partial order.*

Proposition 202 *For any set X and binary relation $R \subseteq X^2$:*

(a) R *is reflexive iff $E_X \subseteq R$;*
(b) R *is symmetric iff $R^{-1} \subseteq R$ iff $R^{-1} = R$;*
(c) R *is asymmetric iff $R^{-1} \cap R = \emptyset$;*
(d) R *is antisymmetric iff $R^{-1} \cap R \subseteq E_X$;*
(e) R *is connected iff $R \cup R^{-1} \cup E_X = X^2$;*
(f) R *is transitive iff $R^2 \subseteq R$.*

Functions can be regarded as special type of relations by means of their graphs: the **graph** of a function $f : A \rightarrow B$ can be defined as the binary relation $G_f \subseteq A \times B$ where $G_f = \{(a, f(a)) \mid a \in A\}$. A relation $R \subseteq A \times B$ is therefore functional iff it is the graph of a function from A to B (exercise).

Let $R \subseteq X \times X$ be a binary relation on a set X. Then

- the **reflexive closure** of R is the smallest by inclusion reflexive binary relation R^{ref} on X such that $R \subseteq R^{\text{ref}}$;
- the **symmetric closure** of R is the smallest by inclusion symmetric binary relation R^{sym} on X such that $R \subseteq R^{\text{sym}}$; and
- the **transitive closure** of R is the smallest by inclusion transitive binary relation R^{tran} on X such that $R \subseteq R^{\text{tran}}$.

As an exercise, show that each of these closures always exists.

Proposition 203 *Let $R \subseteq X \times X$. Then*

(a) $R^{\text{ref}} = \bigcap \{S \mid R \subseteq S \subseteq X^2, \text{ and } S \text{ is reflexive}\}$
(b) $R^{\text{sym}} = \bigcap \{S \mid R \subseteq S \subseteq X^2, \text{ and } S \text{ is symmetric}\}$
(c) $R^{\text{tran}} = \bigcap \{S \mid R \subseteq S \subseteq X^2, \text{ and } S \text{ is transitive}\}$.

5.2.5.1 Equivalence relations, quotient-sets, and partitions

Recall that a binary relation $R \subseteq X \times X$ on a set X is called an **equivalence relation** on X if it is reflexive, symmetric, and transitive. If $R \subseteq X \times X$ is an equivalence relation on X and $x \in X$, the subset $[x]_R = \{y \in X \mid xRy\}$ of X is called the **equivalence class** (or the **cluster**) of x generated by R.

Proposition 204 *For every equivalence relation R on a set X and $x, y \in X$ the following hold:*

(a) $x \in [x]_R$
(b) $x \in [y]_R$ *implies* $[x]_R = [y]_R$
(c) $x \notin [y]_R$ *implies* $[x]_R \cap [y]_R = \emptyset$
(d) *if* $[x]_R \neq [y]_R$ *then* $[x]_R \cap [y]_R = \emptyset$.

Let $R \subseteq X \times X$ be an equivalence relation. The set $X/_R = \{[x]_R \mid x \in X\}$, consisting of all equivalence classes of elements of X under R, is called the **quotient set of X by R**. The equivalence class $[x]_R$, also denoted $x/_R$, is the **quotient-element** of x by R. The mapping $\eta_R : X \to X_R$ defined by $\eta_R(x) = [x]_R$ is the **canonical mapping** of X onto $X/_R$.

A **partition** of a set X is any family $\mathcal{P}$ of non-empty and pairwise disjoint subsets of X, the union of which is X.

Proposition 205 *If $R \subseteq X^2$ is an equivalence relation then $X/_R$ is a partition of X.*

Proposition 206 *If $\mathcal{P}$ is a partition of X then the relation $\sim_{\mathcal{P}} \subseteq X \times X$ defined by $x \sim_{\mathcal{P}} y$ iff x and y belong to the same member of $\mathcal{P}$ is an equivalence relation on X.*

Equivalence relations and partitions of a set are therefore two faces of the same coin.

5.2.6 Ordered sets

5.2.6.1 Pre-orders and partial orders

Recall that a binary relation R on a set X is called a **pre-order** if it is reflexive and transitive. If R is also anti-symmetric, then it is called a **partial order**. If R is a partial order on X, then the pair (X, R) is called a **partially ordered set** or a **poset**.

An arbitrary poset is typically denoted $(X, \leq)$. Given a poset $(X, \leq)$ the following notations are standard: $x \geq y$ means $y \leq x$; $x < y$ means $x \leq y$ and $x \neq y$; and $x > y$ means $y < x$.

Proposition 207 *Let R be a pre-order on X. Then:*

(a) *The relation $\sim$ on X defined by*

$$x \sim y \ \textit{iff} \ xRy \ \textit{and} \ yRx$$

is an equivalence relation on X.

(b) *The relation $\tilde{R}$ on $X/_\sim$, defined by*

$$[x]_\sim \ \tilde{R} \ [y]_\sim \ \textit{iff} \ xRy.$$

*is a well-defined partial order on $X/_\sim$, called **the partial order induced by** R.*

5.2.6.2   Lower and upper bounds; minimal and maximal elements

Let $(X, \leq)$ be a poset and $Y \subseteq X$. An element $x \in X$ is:

- a **lower bound for** Y **in** X if $x \leq y$ for every $y \in Y$;

- the **greatest lower bound**, also called **infimum**, of Y in X if x is a lower bound for Y in X and $x' \leq x$ for every lower bound x' of Y in X;

- an **upper bound for** Y **in** X if $x \geq y$ for every $y \in Y$; or

- the **least upper bound**, also called **supremum**, of Y in X if x is an upper bound for Y in X and $x \leq x'$ for every upper bound x' of Y in X.

Proposition 208 *Let $(X, \leq)$ be a poset and $Y \subseteq X$.*

(a) *If Y has an infimum in X, then it is unique.*
(b) *If Y has a supremum in X, then it is unique.*

Let $(X, \leq)$ be a poset and $Y \subseteq X$. An element $x \in Y$ is called:

- **minimal** in Y if there is no element of Y strictly less than x, that is, for every $y \in Y$, if $y \leq x$ then $x = y$; or

- **maximal** in Y if there is no element of Y strictly greater than x, that is, for every $y \in Y$, if $y \geq x$ then $x = y$.

Note that the least (respectively, greatest) element of Y, if it exists, is the only minimal (respectively, maximal) element of Y.

5.2.6.3   Well-ordered sets

An **ascending** (respectively, **strictly ascending**) **chain** in a poset $(X, \leq)$ is any finite or infinite sequence $x_1, x_2, x_3, \ldots$ of elements of X such that $x_1 \leq x_2 \leq x_3 \leq \cdots$ (respectively, $x_1 < x_2 < \cdots$). A **descending** (respectively, **strictly descending**) **chain** in a poset

$(X, \leq)$ is any finite or infinite sequence $x_1, x_2, x_3, \ldots$ of elements of X such that $x_1 \geq x_2 \geq \cdots$ (respectively, $x_1 > x_2 > \cdots$).

A poset is called **well-founded** if it contains no infinite strictly descending chains. For instance, every finite poset is well-founded. On the other hand, the poset $(\mathcal{P}(X), \subseteq)$, where X is any infinite set, is not well-founded (exercise).

A well-founded linear order is called a **well-order**.

Proposition 209 *The **lexicographic order** in $\mathbb{N}^2$ defined by*

$$\langle x_1, y_1 \rangle \leq \langle x_2, y_2 \rangle \text{ iff } x_1 < x_2 \text{ or } (x_1 = x_2 \text{ and } y_1 \leq y_2)$$

is a well-order on $\mathbb{N}^2$.

Proposition 210 *A poset $(X, \leq)$ is well-founded if and only if every non-empty subset of X has a minimal element.*

In particular, a linear order $(X, \leq)$ is a well-order if and only if every non-empty subset of X has a least element.

We now generalize the principle of Mathematical Induction to induction on arbitrary well-founded sets; see also Section 1.4.4.

Theorem 211 (Induction principle for well-founded sets) *Let $(X, \leq)$ be a well-founded set and $P \subseteq X$ be such that for every $x \in X$, if all elements of X less than x belong to P, then x itself belongs to P. Then $P = X$.*

References for further reading

Most of the books listed in the previous section provide many concrete examples – including many of the exercises listed further below – of mathematical reasoning and proofs on sets, functions, and relations. See in particular: Sollow (1990), Garnier and Taylor (1996), Devlin (2004, 2012), Nederpelt and Kamareddine (2004), Velleman (2006), Makinson (2008), Day (2012), van Benthem *et al.* (2014), and Conradie and Goranko (2015).

See also Bornat (2005), Makinson (2008), and Conradie and Goranko (2015) for mathematical reasoning and proofs on some other topics arising in computer science (loop invariants, arrays, trees, etc.), as well as in number theory, combinatorics, probability, and graph theory. For a more involved treatment and discussion of axiomatic set theory and ZF, see Shoenfield (1967), Hamilton (1988), Mendelson (1997), and Enderton (2001).

Exercises

Do each of the following proofs either by formalizing the claims in $\mathcal{L}_{ZF}$ and using any deductive system for first-order logic, or using semi-formal mathematical reasoning. In the latter case, indicate the method of proof and explicate the logical structure of the argument. Distinguish the logical steps from the mathematical steps.

Exercises on sets and operations on sets

5.2.1   Using the axioms of ZF and logical reasoning, prove the following.
     (a) There is only one empty set, that is, every two empty sets are equal.
     (b) $\emptyset \subseteq x$ for any set x.
     (c) For any sets x, y there is only one pair set $\{x, y\}$ for them.
     (d) For any sets x, y there is only one union set $x \cup y$ for them.
     (e) For any sets x, y there exists exactly one intersection set $x \cap y$ for them.
     (f) For any sets x, y there exists exactly one Cartesian product $x \times y$ for them.
     (g) For any set x there is only one powerset $\mathcal{P}(x)$.
     (h) Prove that $x' \neq x$ for any set x, where $x' := x \cup \{x\}$. (Hint: use REG.)
     (i) Prove that no set may be an element of itself.

5.2.2   Prove that there is a unique least "infinite" set, that is, a set ω which satisfies the axiom INF when substituted for x and is included in all sets satisfying INF.

     Then prove also that ω is indeed infinite, in the sense that there is a bijection between ω and its proper subset $\omega - \{\emptyset\}$.

5.2.3   Recall the definition of an ordered pair as a set: $(a, b) := \{\{a\}, \{a, b\}\}$.
     Prove that it is formally correct, in a sense that for any objects (sets) a_1, b_1, a_2, b_2:

$$(a_1, a_2) = (b_1, b_2) \text{ iff } (a_1 = b_1 \wedge a_2 = b_2)$$

5.2.4   Formalize in $\mathcal{L}_{ZF}$ each of the properties of the set intersection listed in Proposition 186 and prove them.

5.2.5   Formalize in $\mathcal{L}_{ZF}$ each of the properties of the set union listed in Proposition 187 and prove them.

5.2.6   Formalize in $\mathcal{L}_{ZF}$ each of the properties of the set difference listed in Proposition 188 and prove them.

5.2.7   Formalize in $\mathcal{L}_{ZF}$ each of the properties of the interaction between union and intersection listed in Proposition 189 and prove them.

5.2.8   Formalize in $\mathcal{L}_{ZF}$ each of the properties of the powerset listed in Proposition 190 and prove them.

5.2.9   Formalize in $\mathcal{L}_{ZF}$ each of the properties of the Cartesian product listed in Proposition 191 and prove them.

Exercises on functions

For each of the following exercises consider a suitable first-order language extending $\mathcal{L}_{ZF}$ with the functional symbols mentioned in the exercise. Then formalize the respective property in that language and attempt to prove it using a deductive system of your choice. Provide semi-formal mathematical proofs, as required, by indicating the method of proof and explicating the logical structure of the argument. Distinguish the logical steps from the mathematical steps.

5.2.10 Prove Proposition 192.   **5.2.13** Prove Proposition 195.

5.2.11 Prove Proposition 193.   **5.2.14*** Prove Proposition 196.

5.2.12 Prove Proposition 194.

Exercises on binary relations

For each of the following exercises consider a suitable first-order language extending $\mathcal{L}_{ZF}$ with the relational symbols mentioned in the exercise. Then formalize the respective property in that language and attempt to prove it using a deductive system of your choice. Provide semi-formal mathematical proofs, as required, by indicating the method of proof and explicating the logical structure of the argument. Distinguish the logical from the mathematical steps.

5.2.15 Prove Proposition 197.   **5.2.18** Prove Proposition 200.

5.2.16 Prove Proposition 198.   **5.2.19** Prove Proposition 202.

5.2.17 Prove Proposition 199.

5.2.20 Show that a relation $R \subseteq A \times B$ is functional iff it is the graph of a function from A to B.

5.2.21 Show that each of the reflexive, symmetric, and transitive closures of any given binary relation exists.

5.2.22 Prove Proposition 203.

5.2.23* Prove each of the following for any binary relations $R, S, T \subseteq X^2$.
   (a) If R and S are transitive and $R \circ S = S \circ R = T$, then T is also transitive.
   (b) If R and S are transitive, R is symmetric, and $R \cup S = X^2$, then $R = X^2$ or $S = X^2$.
   (c) If R or S is reflexive and transitive and $R \circ S = E_X$, then $R = S = E_X$.

5.2.24 Prove Proposition 204.   **5.2.26** Prove Proposition 206.

5.2.25 Prove Proposition 205.   **5.2.27** Prove Proposition 207.

5.2.28 Prove that $R \subseteq X^2$ is an equivalence relation iff R is reflexive and euclidean.

5.2.29* Let E_1 and E_2 be equivalence relations on a set X. Then prove that:
   (a) $E_1 \cap E_2$ is an equivalence relation on X.
   (b*) The composition $E_1 \circ E_2$ is an equivalence relation if and only if the two relations commute, that is, if and only if $E_1 \circ E_2 = E_2 \circ E_1$.

5.2.30 Prove that if $(X, \leq)$ is a poset and $Y \subseteq X$, then $(Y, \leq|_Y)$ is also a poset.

Exercises on ordered sets

5.2.31   Consider the poset $(\mathcal{P}(X), \subseteq)$ where X is a non-empty set. Show that every subset $\mathcal{X} \subseteq \mathcal{P}(X)$ has both a supremum and an infimum in $(\mathcal{P}(X), \subseteq)$.

5.2.32   Let X be an infinite set. Prove that the poset $(\mathcal{P}(X), \subseteq)$ is not well-founded.

5.2.33   Prove Proposition 208.   **5.2.35**   Prove Proposition 210.

5.2.34   Prove Proposition 209   **5.2.36**   Prove Proposition 211.

Friedrich Wilhelm Karl Ernst Schröder (25.11.1841–16.06.1902) was a German mathematician mainly known for his work on algebraic logic, regarded as one of the creators of **mathematical logic** (a term he may have invented, according to Wikipedia).

Schröder studied mathematics at Heidelberg, Königsberg, and then Zürich. After teaching at school for a few years, he had a position at the Technische Hochschule Darmstadt in 1874. He took up a Chair in Mathematics at the Polytechnische Schule in Karlsruhe two years later, where he remained until the end of his life.

Schröder made important contributions to algebra, set theory, lattice theory, ordered sets, and ordinal numbers. Together with Cantor, he discovered the now called **Cantor–Bernstein–Schröder Theorem**, claiming that the cardinailities of any two sets are comparable. However, he is best known for his monumental three-volume *Vorlesungen über die Algebra der Logik* (*Lectures on the Algebra of Logic*), published during 1890 –1905 at his own expense. In this collection, he systematized and modernized the various systems of formal logic of his time, from the works of Boole, Morgan, MacColl, and especially of Peirce (from whom he took the idea of quantification). With this work he prepared to a great extent the groundwork for the emergence of mathematical logic as a new mathematical discipline in the late 19th–early 20th century. Schröder's contribution in popularizing Peirce's work and his influence for the emergence of mathematical logic is regarded as at least as strong as that of Frege and Peano.

Georg Cantor (19.02.1845–6.01.1918) was a German mathematician, famous for creating and developing the **theory of infinite sets**.

Cantor was born in in St Petersburg in Russia in 1845 and, after some initial resistance from his father, went on to study mathematics at the University of Zurich and then the University of Berlin. He was taught by some of the leading mathematicians of the time, including Kummer, Weierstrass, and Kronecker. Cantor's first paper on infinite sets was published shortly before he turned 30.

The notion of infinity has always been a tricky and elusive topic that very easily leads to paradoxes such as those of Zeno of Elea (c. 490–430 BC). Because of that, most mathematicians (including the genius Gauss) had previously preferred to work only with "potential" infinities (e.g., for every natural number there is a greater one) rather than actual "completed" infinities (e.g., the set of all natural numbers).

Cantor was the first to deal explicitly with actual infinite sets, perform operations on them, and compare them; although regarded as a matter of course in mathematics today, this was quite revolutionary then. It is therefore not surprising that Cantor's theory of infinite sets and transfinite numbers was originally regarded as counter-intuitive, even unbelievable, such that it encountered strong resistance from many authoritative mathematicians of the time such as Kronecker and Poincaré, and later Weyl and Brouwer. For instance, Poincaré referred to his ideas as a "grave disease infecting the discipline of mathematics," while Kronecker went even further and personally attacked included Cantor, describing him as a "scientific charlatan" and a "corrupter of youth."

Some strong philosophical and theological objections against Cantor's set theory were also raised against it, by some influential theologians and philosophers. For example, much later Wittgenstein still regarded it as "utter nonsense" and "laughable" and complained about its poisonous effect on mathematics. Russell's discovery in 1901 of the paradoxical set of all sets that are not elements of themselves certainly contributed to the strong suspicion and even plain rejection faced by Cantor's set theory for a long period. Possibly because of the harsh reaction of his contemporaries, Cantor had recurring bouts of depression from 1884 until the end of his life in a mental hospital.

The practical power of his theory was gradually recognized however, such that in the early 1890s he was elected President of the German Mathematical Society. He taught at the University of Halle in Germany from 1869 until his retirement in 1913. In 1904, the British Royal Society awarded Cantor its Sylvester Medal, the highest honor it confers for work in mathematics, while David Hilbert strongly defended his theory by famously proclaiming: "No one shall expel us from the Paradise that Cantor has created."

5.3 Mathematical Induction and Peano Arithmetic

This section is devoted to two related important topics which are specific to the arithmetic on natural numbers, where logical reasoning plays a crucial role.

The method of **Mathematical Induction** is a very important reasoning technique in mathematics which cannot be extracted from the purely logical rules of Natural Deduction for first-order logic. Even though it is not a rule of purely logical reasoning, Mathematical Induction is an indispensable reasoning tactic for proving universal statements about natural numbers, so I present and explain it here in some detail.

Probably the most popular formal mathematical theory is the first-order axiomatic theory for addition and multiplication in the set of natural numbers, known as **Peano Arithmetic** (PA) after the logician and mathematician Giuseppe Peano[2] who was the first to formalize and study it systematically. I present the axioms of PA, one of which is a partial formalization of the principle of Mathematical Induction in the first-order language of PA, list a number of basic arithmetical facts derivable in PA, and sketch semi-formal proofs of a few of them (leaving the rest as exercises on axiom-based mathematical reasoning).

5.3.1 Mathematical Induction

Mathematical Induction is a special case of the general Induction Principle for inductively defined sets formulated in Section 1.4; I recommend that the reader (re)visits that section to see the full generality and strength of the induction method. I present here its basic version with a few variations and equivalent statements.

5.3.1.1 Mathematical Induction: basic version

Recall that we assume the set of natural numbers to be $\mathbb{N} = \{0, 1, 2, \ldots \}$. Here is the most common version of the Principle of Mathematical Induction (PMI).

Theorem 212 (Principle of Mathematical Induction) *Suppose that some property P of natural numbers holds for 0, and whenever P holds for some natural number k then it also holds for $k + 1$. Then P holds for every natural number.*

We write $P(n)$ to say that the property P holds for the number n. Using the PMI to prove a statement of the type "$P(n)$ holds for every natural number n" therefore requires that the following two steps be proved:

1. **Base step**: $P(0)$ holds; and
2. **Induction step**: for any $k \in \mathbb{N}$, if $P(k)$ holds then $P(k + 1)$ also holds.

We illustrate the use of the PMI with the following example.

Example 213 *Prove that for every natural number n,*

$$0 + 1 + \ldots + n = \frac{n(n + 1)}{2}.$$

Proof. Let us denote the above statement $P(n)$. First, we have to verify the base step, that is, show that $P(0)$ is true:

$$0 = \frac{0(0 + 1)}{2}.$$

---

[2] Read more on Peano in the biographic box at the end of this section.

*To verify the induction step, we then assume that $P(k)$ is true for some $k \geq 0$. This assumption is called the **Inductive Hypothesis (IH)**. We now have to show that $P(k+1)$ is also true:*

$$0 + 1 + \ldots + k + (k+1) = (0 + 1 + \ldots + k) + (k+1)$$

$$= \frac{k(k+1)}{2} + (k+1)$$

$$= \frac{k(k+1) + 2(k+1)}{2}$$

$$= \frac{(k+1)((k+1)+1)}{2},$$

where we use the Inductive Hypothesis to justify the second equality.

By the Principle of Mathematical Induction, we therefore conclude that $P(n)$ holds for every natural number n. ∎

Sometimes, we only need to prove that a property holds for all natural numbers greater than some $m \geq 0$. PMI works just as well by starting with a base step $P(m)$ and applying the induction step only to natural numbers $k \geq m$. Indeed, we can reduce this case to the original by stating the property as an implication: $n \geq m \rightarrow P(n)$. As an exercise, show that this reduction works as intended.

Example 214 *Prove that $n^2 \leq 2^n$ for every natural number $n \geq 4$.*

Proof. Let $P(n)$ be the statement $n^2 \leq 2^n$.

Base step: Prove that $P(4)$ is true: $4^2 = 16 \leq 16 = 2^4$.

Induction step: Assume that $P(k)$ is true for some $k \geq 4$, that is, $k^2 \leq 2^k$. This is the Inductive Hypothesis (IH).

We now have to prove that $(k+1)^2 \leq 2^{(k+1)}$. We begin with:

$$(k+1)^2 = k^2 + 2k + 1 = k\left(k + 2 + \frac{1}{k}\right).$$

Our goal is to rewrite this equation so that we can use the IH.

Note that $\frac{1}{k} < 1$ since $k \geq 4$. We therefore have:

$$k\left(k + 2 + \frac{1}{k}\right) \leq k(k + 2 + 1) = k(k+3) \leq k(k+k) = 2k^2.$$

We can now use the IH to obtain $2k^2 \leq 2 \cdot 2^k = 2^{(k+1)}$.

We therefore have that $(k+1)^2 \leq 2^{(k+1)}$, as required.

By the PMI, we conclude that $n^2 \leq 2^n$ for all natural numbers $n \geq 4$. ∎

5.3.1.2 Equivalent versions of the principle of Mathematical Induction

Here I state two equivalent principles from which PMI can be deduced.

First, a fundamental property of the set of natural numbers, which we can take as an axiom, is the following **Principle of Well-Ordering (PWO)**:

Every non-empty set of natural numbers has a least element.

This is equivalent to the following **Principle of Descending Chains (PDC)**:

There is no infinite strictly descending sequence of natural numbers.

Indeed, assume the PWO, then every strictly descending sequence of natural numbers, considered as a set of its members, must have a least element; it therefore may not be infinite. Conversely, assume the PDC and suppose PWO does not hold. Then take any non-empty set X of natural numbers with no least element. Take any $n_0 \in X$. Since it is not least in X, there must be $n_1 \in X$ such that $n_1 < n_0$. Again, n_1 is not least in X, so there must be $n_2 \in X$ such that $n_2 < n_1$, etc. We construct an infinite strictly descending sequence of natural numbers, which contradicts the PDC.

Assuming any of the principles above, we can now prove the PMI.

Theorem 215 (Principle of Mathematical Induction) *Suppose that some property P of natural numbers holds for some natural number $n_0 \geq 0$, and whenever P holds for some natural number $k \geq n_0$ it also holds for $k + 1$. Then P holds for every natural number greater than or equal to n_0.*

Proof. The PMI can be derived from the PWO by assumption of the contrary, as follows. Suppose that there are natural numbers for which P does not hold. The collection of such numbers has a least element m. It must be greater than n_0 since P holds for n_0. Therefore $m - 1$ is a natural number less than m, hence P must hold for $m - 1$. But then P must also hold for m, since $m = (m - 1) + 1$. This is a contradiction, therefore our assumption was wrong. ■

Here is a seemingly stronger but in fact equivalent form of the Principle of Mathematical Induction.

Theorem 216 (Principle of Complete Mathematical Induction (PCMI)) *Suppose a property P of natural numbers holds for a natural number n_0, and also suppose that, for any natural number $k \geq n_0$, if P holds for all natural numbers greater than or equal to n_0 and less than k, then P holds for k. Then P holds for every natural number greater than or equal to n_0.*

Proof. Left as exercise (Exercise 2. in Section 5.3.3.1). ■

When using the PMCI we therefore need to prove the following.

1. $P(n_0)$ holds for some natural number n_0.
2. For any $k \geq n_0$, assuming that $P(k)$ holds for all natural numbers greater than or equal to n_0 and less than k, then $P(k)$ also holds.

Note that in the statement of Theorem 216, the base step, that is, the assumption that the property P holds for n_0, can be omitted (see Exercise 3.).

5.3.2 Peano Arithmetic

The language of Peano Arithmetic is the already well-known first-order language $\mathcal{L}_\mathcal{N}$ containing $=$, one constant symbol $\mathbf{0}$, one unary functional symbol s, and two binary functional symbols, $+$ and $\times$. Recall that *numerals* are the special ground terms of the type $\mathbf{n} := \underbrace{s(\cdots s(\mathbf{0}) \cdots)}_{n \text{ times}}$, respectively interpreted in the number n.

The axioms of Peano Arithmetic, where all free variables below are implicitly universally quantified, are:

(PA1) $s(x) \neq \mathbf{0}$

(PA2) $s(x) = sy \to x = y$

(PA3) $x + \mathbf{0} = x$

(PA4) $x + s(y) = s(x + y)$

(PA5) $x \times \mathbf{0} = \mathbf{0}$

(PA6) $x \times s(y) = x \times y + x$

(PA7) Induction Scheme:

$$\varphi(\mathbf{0}) \wedge \forall x(\varphi(x) \to \varphi(s(x))) \to \forall x \varphi(x)$$

for every formula $\varphi(x)$ of $\mathcal{L}_\mathcal{N}$.

This axiomatic system is denoted **PA**.

Note that the Induction Scheme formalizes the principle of Mathematical Induction, but restricted only to properties (respectively, sets) of natural numbers *definable in the language $\mathcal{L}_\mathcal{N}$*. There are only countably many such definable properties/sets, since there are only countably many formulae in the language $\mathcal{L}_\mathcal{N}$. On the other hand, by Cantor's Theorem there are uncountably many sets of natural numbers, so all but countably many of them are *not definable* in $\mathcal{L}_\mathcal{N}$ and the induction principle cannot be applied to them.

Still, **PA** is a very expressive and powerful axiomatic system. In fact, by the famous *Gödel's First Incompleteness Theorem*, the set of its theorems is not axiomatizable "effectively," that is, by an algorithmically recognizable (recursive) set of axioms. Furthermore, by *Gödel's Second Incompleteness Theorem*, even though **PA** can express (in a sense) its own consistency, it cannot prove it; this of course assumes it is consistent, otherwise *Ex Falso Quodlibet* applies. These results can be regarded as a testimony of both the expressive strength of **PA** and as an insurmountable limitation of axiomatic/deductive reasoning.

Here are some definable relations and formulae in $\mathcal{L}_\mathcal{N}$ that will be used further:

1. $x \leq y =_{\text{def}} \exists z(x + z = y)$
2. $x < y =_{\text{def}} x \leq y \wedge x \neq y$
3. $\exists y_{y \leq z} A(\bar{x}, y) =_{\text{def}} \exists y(y \leq z \wedge A(\bar{x}, y))$, for any formula $A(\bar{x}, y)$
4. $\forall y_{y \leq z} A(\bar{x}, y) =_{\text{def}} \forall y(y \leq z \to A(\bar{x}, y))$, for any formula $A(\bar{x}, y)$.

We can also define the **value** of any ground term t by recursion on the inductive definition of ground terms:

1. $v(0) = 0$

2. $v(s(t)) = v(t) + 1$

3. $v(t_1 + t_2) = v(t_1) + v(t_2)$, $v(t_1 \times t_2) = v(t_1) \times v(t_2)$.

In the exercises I will list many simple but important claims about the arithmetic of $\mathbb{N}$ which are derivable in **PA**.

References for further reading

Mathematical Induction is a very common topic, treated in just about any book on calculus or discrete mathematics. For more detailed and logically enhanced treatment of induction see Sollow (1990), Nederpelt and Kamareddine (2004), Velleman (2006), Makinson (2008), Conradie and Goranko (2015).

For in-depth treatment of Peano Arithmetic, see Shoenfield (1967), Mendelson (1997), van Oosten (1999), Enderton (2001), Boolos *et al.* (2007), and Smith (2013).

Exercises

Exercises on Mathematical Induction

5.3.1 Using the basic Principle of Mathematical Induction (PMI) and applying (and explicitly indicating) the reasoning tactics for the logical steps, prove that for every natural number n the following holds.

1.1 $n^2 + n$ is even, for all natural numbers n.

1.2 If $n \geq 4$ then $2^n < n!$.

1.3 If $n \geq 1$ then the powerset of the set $\{1, 2, 3, \ldots, n\}$ has 2^n elements.

1.4 $2^0 + 2^1 + 2^2 + \cdots + 2^n = 2^{n+1} - 1$.

1.5 $1 + 3 + 5 + \cdots + (2n - 1) = n^2$.

1.6 $1^2 + 2^2 + \cdots + n^2 = \frac{n(n+1)(2n+1)}{6}$.

1.7 $1^3 + 2^3 + \cdots + n^3 = \frac{n^2(n+1)^2}{4}$.

1.8 $1 \times 3 + 2 \times 4 + \cdots + n \times (n+2) = \frac{n(n+1)(2n+7)}{6}$.

1.9 $\frac{1}{1 \times 3} + \frac{1}{2 \times 4} + \cdots + \frac{1}{n \times (n+2)} = \frac{n(3n+5)}{4(n+1)(n+2)}$.

1.10 $\frac{1}{\sqrt{1}} + \frac{1}{\sqrt{2}} + \cdots + \frac{1}{\sqrt{n}} \leq 2\sqrt{n}$.

5.3.2 Prove the PMCI as stated in Theorem 216 using a variation of the proof of the PMI in Theorem 215.

5.3.3 Show that the base step in the statement of Theorem 216, that is, the assumption that the property P holds for n_0, can be omitted.

5.3.4[*] Show that the PMCI is equivalent to the PMI.

5.3.5* Prove that the PMCI is equivalent to the following form: "Let a property P of natural numbers be such that, for any natural number k, if it holds for all natural numbers less than k then it holds for k. Then P holds for all natural numbers."

5.3.6 Some natural numbers seem more special, like 0, 1, 100, 3^{3^3}, 123456789, etc. Let us call them *interesting*. Using the PWO, prove that *all* natural numbers are interesting.

Exercises on Peano Arithmetic

5.3.7 Derive the following formulae of $\mathcal{L}_\mathcal{N}$, all implicitly universally quantified, in the axiomatic system of Peano Arithmetic (PA). You may write the proofs either as formal derivations in (PA) or in a semi-formal mathematical style. Most of them use the Induction Scheme formalizing the method of Mathematical Induction in $\mathbb{N}$, which is the main specific reasoning strategy here. When performing the proofs, pay special attention to the separation of logical and arithmetical axioms used in the proof steps.

7.1 $x \neq \mathbf{0} \to \exists y(x = s(y))$
7.2 $\mathbf{0} + x = x$
7.3 $y + s(x) = s(y) + x$
7.4 $x + y = y + x$
7.5 $(x + y) + z = x + (y + z)$
7.6 $x \times y = y \times x$
7.7 $(x \times y) \times z = x \times (y \times z)$
7.8 $s(x) \times y = x \times y + y$
7.9 $x \times (y + z) = x \times y + x \times z$
7.10 $x + s(y) \neq x$
7.11 $x + y = x \to y = \mathbf{0}$
7.12 $x \neq \mathbf{0} \to x + y \neq \mathbf{0}$
7.13 $(x \leq y \land x \leq y) \to x = y$
7.14 $(x \leq y \land y \leq z) \to x \leq z$
7.15 $x \leq y \lor y \leq x$
7.16 $x \leq s(y) \leftrightarrow x \leq y \lor x = y$
7.17 $x \leq y \leftrightarrow (x < y \lor x = y)$
7.18 $x < y \lor x = y \lor y < x$

7.19 $x \leq x + y$
7.20 $y \neq \mathbf{0} \to x \leq x \times y$
7.21 $x + y = \mathbf{0} \to (x = \mathbf{0} \land y = \mathbf{0})$
7.22 $x \times y = \mathbf{0} \to (x = \mathbf{0} \lor y = \mathbf{0})$
7.23 $x + z = y + z \to x = y$
7.24 $(z \neq \mathbf{0} \land x \times z = y \times z) \to$
      $\quad x = y$
7.25 $x \leq \mathbf{n} \leftrightarrow (\phi(\mathbf{0}) \land \ldots \land \phi(\mathbf{n}))$
7.26 $\exists x_{x \leq \mathbf{n}} \phi(x) \leftrightarrow (\phi(\mathbf{0}) \lor \ldots \lor$
      $\quad \phi(\mathbf{n}))$
7.27 $\forall x_{x \leq \mathbf{n}} \phi(x) \leftrightarrow (\phi(\mathbf{0}) \land \ldots \land$
      $\quad \phi(\mathbf{n}))$
7.28 $\exists x_{x < \mathbf{n}} \phi(x) \leftrightarrow (\phi(\mathbf{0}) \lor \ldots \lor$
      $\quad \phi(\mathbf{n-1}))$
7.29 $\forall x_{x < \mathbf{n}} \phi(x) \leftrightarrow (\phi(\mathbf{0}) \land \ldots \land$
      $\quad \phi(\mathbf{n-1}))$
7.30 $\exists x \phi(x) \to \exists x(\phi(x) \land$
      $\quad \forall y_{y < x} \neg \phi(y))$.

5.3.8 Recall that for any ground term t, $v(t)$ denotes its value in $\mathbb{N}$. Prove that the following hold.
   (a) $v(\mathbf{n}) = n$.
   (b) If $v(t) = n$ then $PA \vdash t = \mathbf{n}$ and if $v(t) \neq n$ then $PA \vdash t \neq \mathbf{n}$.
   (c) If $v(t_1) = n$ and $v(t_2) = m$ then:
        if $n \leq m$ then $PA \vdash t_1 \leq t_2$ and if $n \not\leq m$ then $PA \vdash \neg t_1 \leq t_2$.
   (d) Consequently, if $v(t) = n$ then $PA \vdash \exists x_{x \leq t} \phi(x) \leftrightarrow (\phi(\mathbf{0}) \lor \ldots \lor \phi(\mathbf{n}))$.

 **Giuseppe Peano** (27.8.1858–20.4.1932) was an Italian mathematician, one of the founders of mathematical logic, best known for the axiom system, now called **Peano Arithmetic**, that he proposed.

Peano was born and grew up in a farm in north Italy. He graduated from the University of Turin in 1880 and then took up a position and worked there for most of his career as Professor in Mathematics.

Peano was strongly influenced by the logical system developed by Frege and, along with Russell, did a lot to popularize and develop further his ideas. In 1889 Peano published his famous axioms, which defined the natural numbers in terms of sets. Peano's axioms, where "number" means "natural number," are as follows.

1. Zero is a number.

2. The successor of any number is another number.

3. There are no two numbers with the same successor.

4. Zero is not the successor of a number.

5. (Axiom of Induction) If a set S of numbers contains zero and also the successor of every number in S, then every number is in S.

Much later in 1930 Kurt Gödel proved his *First Incompleteness Theorem*, showing not only that Peano's system of axioms is incomplete but that it cannot be extended to a complete system in any "reasonable" way, so that the axioms are effectively (algorithmically) recognizable.

Another mathematical result for which Peano is known is his construction of a **space-filling curve** – a continuous curve filling the entire unit square, an early example of a *fractal* – as a counterexample to the claim that a continuous curve can be enclosed in an arbitrarily small region.

Since 1892 Peano embarked on an extremely ambitious project called *Formulario Mathematico*, intended as an encyclopedia of all mathematical formulae and theorems expressed in a symbolic language. Many symbols that Peano used in the *Formulario* are still in use today. His monumental work, completed in 1908, had a strong influence due to its rigorous exposition and modern style, but Peano's use of it for teaching was not liked by his students and colleagues. In 1903 Peano started developing an international language called *Latino sine flexione*, later also called *Interlingua*, using Latin vocabulary but with simplified grammar to make it easier to learn. He published the final edition of *Formulario Mathematico* in Latino sine flexione, which was an added reason for its lack of popularity.

Eventually, Peano received wide recognition for his crucial contributions to modernizing mathematics and making mathematical notation and reasoning more rigorous, based on mathematical logic.

5.4   Applications: automated reasoning and logic programming

This section discusses briefly two of the most popular areas of logic-based applications to computer science and artificial intelligence:

- automated reasoning and automated theorem proving; and
- logic programming and Prolog.

5.4.1   Automated reasoning and automated theorem proving

Automated reasoning is about the theory and development of mechanizable and computer implementable methods and systems for logical deduction. The origin of this very important area of applications dates back to the Middle Ages and is probably rooted in Ramon Llull's *Ars Magna* in the late 13th century. Automated reasoning in the modern sense has been under active development since the mid 20th century. It has become ever more intense and successful with the explosive development of computer hardware and software and has found numerous important applications not only in logic and mathematics, but also in computer science, engineering, and in artificial intelligence, where it was initially regarded as the most fundamental approach.

5.4.1.1   Automated and interactive theorem proving

A system of automated reasoning is based on a formal logical language for description of the axioms, facts, assumptions, conjectures, and inference rules by means of formally constructed expressions in that language plus an algorithmically implemented system for deduction and formal reasoning. When such a system is used for proving (or disproving) explicitly formulated conjectures, we talk about *automated theorem proving*. The proofs produced by systems for automated reasoning usually describe how and why the conjecture follows from the axioms and hypotheses in a manner that can be understood and agreed upon by everyone, and can be verified by means of a computer. The proof output may not only be a convincing argument that the conjecture is a logical consequence of the axioms and hypotheses, but it often also describes a process that may be implemented to solve some open problems in the field.

Systems for automated theorem proving are powerful computer programs, potentially capable of solving very difficult problems of high computational complexity and practically impossible to solve manually. Furthermore, the underlying notion of logical consequence – as in first-order logic – is often algorithmically undecidable, which means that no completely automatic procedure can always guarantee successful derivation of every logically valid consequence. For these reasons, the applications of systems for automated theorem proving often need to be guided by an expert in the domain of application, who has the intuition and insight to formulate intermediate conjectures (lemmas) and goals to optimize and direct the proof search in order to be completed successfully in a reasonable amount of time and memory space. This approach is known as **interactive theorem proving**.

A number of very successful systems for automated or interactive reasoning, such as *ACL2, Agda, Coq, HOL, Isabelle, LEGO, Mizar, NuPRL, PVS, TPS*, and *Twelf* have been designed and used to develop practically useful **computer-aided proof assistants**, while many of them can also be run in a fully automatic mode. Some of these systems, such as Isabelle, can be used as generic platforms for implementing other specific deductive systems for object logics, the axioms and rules of which can be formulated within Isabelle's metalogical language. The most commonly used formal logical language for automated reasoning is first-order logic, but there are also several highly successful systems for fully automated or interactive reasoning for *non-classical* or *higher-order* logics, based on various deductive approaches such as *lambda calculus, functional programming, term rewriting, resolution,* and *tableaux*.

Automated and interactive theorem-proving systems have been very useful in many areas of mathematics. In particular, they have been instrumental for solving several open problems in algebra, graph theory, logic, geometry, combinatorics, and calculus. Some notable examples include the proofs of: the *Four Color Theorem,* stating that every map in a plane can be colored using at most four colors in such a way that regions sharing a common boundary do not share the same color, in 1976; *the non-existence of a finite projective plane of order 10*, in 1989; *Robbins conjecture* that every "Robbins algebra" (that is, algebra satisfying certain algebraic identities) is a Boolean algebra, in 1996; and *Kepler's Conjecture* (stating that the optimally dense sphere packing in a box is for cubic or hexagonal arrangements), in 1998.

Most of the practically efficient systems for automated reasoning in first-order logic have their inference engines based on various refinements and improvements of resolution. The currently most popular implemented systems for resolution-based automated theorem proving include *E-SETHEO, Prover9* (successor of *Otter*, one of the first implemented systems for automated reasoning), *SNARK, SPASS, Vampire*. Some of these are well supported and easy both to install and use online; some of the exercises ask you to make use of them.

5.4.2   *Logic programming and Prolog*

Logic programming is a declarative programming style where the idea is to describe, using a formal logic-based language, *what* the program is supposed to compute rather than explicitly providing the instructions of *how* this should be executed, typical of the imperative style of programming languages. A program written in a logic programming language is therefore just a set of sentences in a formally specified logical language, expressing facts, assumptions, and conjectures or queries about the problem domain.

The concept of logic programming can be traced back to the late 1960s–early 1970s, just after the invention of the method of resolution with unification by J.A. Robinson, when an intensive debate was underway on the topic of declarative versus procedural knowledge representation and processing in the area of artificial intelligence. Currently, the most popular languages in logic programming include **Prolog** (which is short for for "*Pro*gramming in *log*ic"), **Datalog**, and **Answer Set Programming** (ASP). I only briefly discuss Prolog here. It is one of the first and most popular such languages originally developed in the early 1970s by Alain Colmerauer and Philippe Roussel in Marseille, France. Prolog has been extensively used since then for automated theorem proving, expert systems, and other systems for intelligent control.

A typical program construction in a logic programming language is a **clause-based rule** in the form:

$$A_1 \wedge \ldots \wedge A_n \implies H$$

or, written in alternative notation:

$$H \leftarrow A_1, \ldots, A_n$$

or in Prolog:

$$H : - A_1, \ldots, A_n$$

where $n \geq 0$. H is called the **head** of the rule and $A_1, \ldots, A_n$ form the **body** of the rule. The formulae $A_1, \ldots, A_n$ in the body are called **goals**, for reasons that will become clear. The rule above has the declarative meaning of the logical implication:

$$\text{if } A_1 \text{ and } \ldots \text{ and } A_n, \text{ then } H$$

and can be formally identified with the clause $\{\neg A_1, \ldots, \neg A_n, H\}$. All variables occurring in a rule are assumed implicitly universally quantified over.

The body of the rule can be empty, when $n = 0$. Such a rule is written:

$$\implies H$$

or in Prolog:

$$H : -$$

or just as H. It is called a **fact**, as it simply states the truth of H.

Further, when the head H of a rule is $\perp$ (the falsum), the rule is usually written:

$$A_1 \wedge \ldots \wedge A_n \implies$$

or in Prolog:

$$: - A_1, \ldots, A_n$$

and is said to have an *empty head*. Such rule is called a **query**, explained further in the following.

In most logic programming languages, such as the standard (pure) Prolog, the formulae in the head and body of a rule can only be atomic formulae. Such rules represent **definite clauses**, also called **Horn clauses**. An extension of Prolog, suitable for *non-monotonic reasoning*, also allows negations of atomic formulae in the body of a rule. A **logic program** is just a finite set of Horn clauses in first-order language (without equality). A logic program represents, in the same way as Resolution, a refutation-based deductive procedure (the meaning of which is explained below). Derivations in a logic program are essentially defined as for the system of first-order Resolution by successively applying Resolution with unification to derive new clauses (rules) that are added to the program. The purpose is to eventually derive the empty clause $: -$ (sometimes also denoted $\square$) which has the same meaning as in Resolution: a contradiction.

Due to the special shape of the clauses in a logic program, the application of Resolution here is more intuitive: it consists of unifying the head of one rule with a goal in the body of another, and then replacing that goal with the body of the latter rule after applying the unifying substitution. To be precise, it works as follows. Given the rules

$$H : -A_1, A_2, \ldots, A_n$$

and

$$C : -B_1, \ldots, B_m$$

such that σ is a most general unifier (MGU) of A_1 and C (the fact that A_1 is the first goal in the body of the rule is not essential), Prolog resolves the rule above to produce the new rule

$$\sigma(H : -B_1, \ldots, B_m, A_2, \ldots, A_n).$$

In particular, if applied to rules $: -A_1$ and $C : -$ it produces the empty clause $: -$.

Let me explain briefly how a logic program works.

First, a query $: -A_1, \ldots, A_n$ represents the negated conjunction of the goals $A_1, \ldots, A_n$ and asks for **answers** or **solutions**: instantiations of the occurring variables that satisfy the conjunction of the goals. The idea of a logic program P is that, given a query G, the program P attempts to answer that query by deriving the empty clause from its goals, possibly after suitable substitutions. The declarative meaning (semantics) of a logic program P is given as follows. The program takes as an input a query G (or its negation represented by the empty-headed rule with body G) and keeps applying Resolution steps on its rules, as above. If it eventually derives the empty clause, the output of the program is a **solution of G in P**, formally being the composite substitution σ of (usually ground) terms for the variables occurring in G, applied in the derivation of the empty clause. The purpose of that substitution is to suitably instantiate the variables occurring in the query, in order to unify its goals with facts and heads of other rules in the program, eventually computing an answer to the query.

In the case when no variables occur in the query, the program P must simply decide whether the query follows logically from its clauses by trying to derive the empty clause from it (i.e., from the rule representing its negation). In this case, the only possible solution is the identity substitution when the empty clause is derivable, thus indicating an answer "Yes." Otherwise, the answer (if the program terminates at all) is "No," which in fact means "I don't know" (see more on that further). In the general case, the solution is computed by taking the composition of all unifying substitutions applied in the derivation of the empty clause, and then restricting it to the variables occurring in the query G. In other words, since the rule represents the negated query G, all variables in the original query G are implicitly *existentially* quantified, so a solution to the query consists of explicit *witnesses* for the existence of objects named by the terms in the query goals, satisfying these goals.

Let us look at an example of a logic program written in Prolog. In what follows, ancestor and parent are binary predicates and Elisabeth, Charles,

Harry are constants/names (interpreted in the familiar domain of humans). The program P consists of the following four facts and two rules:

(F1) `parent(Elisabeth, Charles).`
(F2) `parent(Charles, William).`
(F3) `parent(Diana, William).`
(F4) `parent(William, George).`
(R1) `ancestor(X,Y) :- parent(X,Y).`
(R2) `ancestor(X,Y) :- parent(X,Z),ancestor(Z,Y).`

1. First, consider the following query:

   (Q0): `:- parent(Charles, William).`

   A derivation in the program $P^{Q0} = P + (Q0)$ consist of a single step, resolving the query (Q0) with the fact (F2) and producing the empty clause, thereby returning the identity substitution, that is, the answer "Yes."

2. Next, consider the following query:

   (Q1): `:- ancestor(Charles, William).`

   Here is a derivation in the program $P^{Q1} = P + (Q1)$: first, the substitution [`Charles/X, William/Y`] unifies the head of (R1) with the body of (Q1), thus resolving these two rules to produce the new query

   (Q1′): `:- parent(Charles, William).`

   This, again, is immediately derivable, returning again the answer "Yes."

3. On the other hand, running P on the query

   (Q2): `:- parent(Elisabeth, William).`

   cannot produce the empty clause, simply because none of the rules in P can be resolved with (Q2) since none of their heads can be unified with the only goal in the body of Q2. P will therefore return an answer "No," which actually means "*I don't know whether* `parent(Elisabeth, William)` *holds or not.*"
   This feature of Prolog, known as **Negation-by-failure**, is discussed further below.

4. Now consider the query:

   (Q3): `:- ancestor(Elisabeth, William).`

   A derivation in the program $P^{Q3} = P + (Q3)$ is as follows.
   (a) The substitution [`Elisabeth/X, William/Y`] unifies the head of (R2) with the body of (Q3), thus resolving these two rules to produce the new query

   (Q31): `:- parent(Elisabeth,Z), ancestor(Z,William).`

   (b) The substitution [`Charles/Z`] unifies the fact (F1) with the first goal in the body of (Q31), thus resolving these rules to produce the new query

   (Q32): `:- ancestor(Charles, William).`

(c) The substitution [Charles/X, William/Y] unifies the head of (R1) with the body of (Q32), thus resolving these rules to produce the new query

(Q33): :- parent(Charles, William).

(d) The fact (F2) is identical to the body of (Q33), thus resolving these rules to produce the empty query/clause

(Q34): :- .

The derivation of the empty clause in P^{Q3} means that the program P has answered the original query (Q3) with "Yes."

5. Let us now run P on the following query:

(Q4): :- ancestor(Elisabeth, Y).

(a) The substitution [Elisabeth/X] unifies the head of (R1) with the body of (Q4), thus resolving these two rules to produce the new query

(Q41): :- parent(Elisabeth, Y).

(b) The substitution [Charles/Y] unifies the fact (F1) with the body of (Q41), thus resolving these rules to produce the empty query

(Q42): :- .

This completes the derivation, yielding a solution [Charles/Y].
    If we are interested in only one solution, we can stop here. However, if we want to look for other solutions we can re-run the program by instructing it to look for *another* solution. This can be done by using a special feature of Prolog that cuts the execution branch producing the previous solution and forces Prolog to look for a new solution.

(c) In this case, re-running the program will cause Prolog to look for another way to unify a rule with the body of (Q4). Such an alternative is provided by rule (R2) and the substitution [Elisabeth/X, Wiliam/Y] will, as before, unify the head of (R2) with the body of (Q4). This causes a further execution repeating the previous one, which will also succeed, producing another solution: [Wiliam/Y].

(d) Repeating the re-running after disallowing the previous two solutions will likewise produce yet another solution: [George/Y]. It is not difficult to see that no more solutions can be obtained.
    As expected, the program computes consecutively three answers – Charles, William, George – to the query (Q4), which essentially asks "*To whom is Elisabeth an ancestor?*"

How exactly does Prolog run? The derivation engine of Prolog is based on the earlier mentioned SLD-resolution strategy, enhanced with a special mechanism for controlled backtracking, which allows for a search of multiple solutions as above. As for input resolution, SLD-resolution is not refutation-complete for first-order logic but it is complete for the Horn logic of Prolog programs. However, because of the depth-first search strategy inbuilt to the SLD-resolution, pure Prolog does not always terminate even when a derivation exists; it is therefore practically incomplete. To remedy that problem, Prolog employs some non-logical features such as `cut` and `fail`, implementing "Negation-by-Failure" in a more aggressive way. This feature introduces an obvious problem: a Prolog program implicitly assumes that only what can be derived from its facts and rules is true; everything else is declared to be false. Negation-by-failure therefore yields a non-standard semantics of the negation: $not A$ succeeds iff deriving A fails. This semantics is *non-monotonic*: adding more facts or rules to the program may enable more derivations and therefore falsify some negated statements which were previously declared true.

Another problem with the semantics of Prolog programs arises from the control of the execution of such programs. The execution, and sometimes the outcome, of a Prolog program can be sensitive to the order in which the rules are listed within the program or how the literals are ordered within a rule; the programmer must take this into account when ordering these. Furthermore, the predicates `cut` and `fail` enable a mechanism for preventing the deductive engine from revisiting certain paths in the search space (by using `cut`) or for pruning them altogether (by using `fail`), which alters the purely logical semantics of a Prolog program. However, skillful use of these features can produce very elegant and efficient Prolog programs. Despite these strange effects, Prolog has remained one of the top choices of logic programming language for over 40 years.

References for further reading
There is abundant literature on automated reasoning and theorem proving, including classical texts and references on the topic (some freely available on the internet) such as: Robinson and Voronkov (2001), Ben-Ari (2012), Fitting (1996), Chang and Lee (1997), Nerode and Shore (1993), and Gallier (1986). In addition, some specific useful references include:

- https://en.wikipedia.org/wiki/Automated_reasoning;
- Geoff Sutcliffe's Overview of Automated Theorem Proving http://www.cs.miami.edu/tptp/OverviewOfATP.html; and
- The Thousand Problems for Theorem Provers (TPTP) library www.tptp.org.

For more on the history, theory, and art of logic programming and Prolog, see Nilsson and Małuszyński (1995), Spivey (1995), Ebbinghaus *et al.* (1996), Fitting (1996), and Ben-Ari (2012).

Exercises

5.4.1   Use the online version WebSPASS of the automated theorem prover SPASS http://www.spass-prover.org, or any other implementation of an automated theorem prover of your choice, to formalize and solve exercises 7, 8, 9, and 14 from Section 4.5.8, as well as the formalized exercises from Section 5.2.7.

5.4.2 For many more exercises on automated theorem proving, using SPASS or any of the other implemented tools mentioned in the text, visit the Thousand Problems for Theorem Provers (TPTP) library www.tptp.org.

5.4.3 Consider the following Prolog program:

(F1) `parent(Elisabeth, Charles).`
(F2) `parent(Charles, William).`
(F3) `parent(Diana, William).`
(F3) `parent(Diana, Harry).`
(F4) `parent(William, George).`
(R1) `ancestor(X,Y) :- parent(X,Y).`
(R2) `ancestor(X,Y) :- parent(X,Z),ancestor(Z,Y).`

Run the program on the following input queries to produce all solutions (if any):

(a) (Q1) `:- ancestor(Elisabeth, Harry).`
(b) (Q2) `:- ancestor(Diana, Y).`
(c) (Q3) `:- ancestor(X, Harry).`
(d) (Q4) `:- ancestor(X, George).`

Alan Mathison Turing (23.6.1912–7.6.1954) was a British mathematician, logician, computer scientist, and crypto-analyst, regarded as one of the founders of both theoretical computer science, for his creation of the universal model of computations, and artificial intelligence, for proposing the idea of using a special test (**Turing's test**) for deciding if an artificial agent has intelligence. Turing was instrumental in breaking the secret codes of the German Enigma coding machine used during World War II.

In 1935 Turing studied Gödel's incompleteness results and Hilbert's **Entscheidungsproblem**, which queries the existence of an algorithm that takes as input a sentence of a first-order logic and answers if it is true in every structure. The following year he wrote his groundbreaking paper *On Computable Numbers, with an Application to the Entscheidungsproblem*, where he introduced an abstract universal computing device, now called a **Turing machine**, and argued that such a machine can perform any conceivable mathematical computation which can be represented as an algorithm. Remarkably, Turing machines described the concept of modern programmable computers before they were actually invented. Moreover, Turing described in his 1936 paper the construction of a **universal Turing machine** which can simulate the work of any Turing machine. He then showed that the **halting problem** for Turing machines is algorithmically undecidable in a sense that there cannot exist an algorithm (represented by a Turing machine) that can take the code of any Turing machine and an input for it and determine as an output whether that Turing machine will eventually halt for that input. Turing then concluded that Hilbert's *Entscheidungsproblem* has no solution, thus reproving this fundamental result which had recently been published by Church.

Turing completed a doctorate in mathematical logic in Princeton University under Church's supervision in 1938. Church and Turing showed that the Church's lambda calculus and the Turing machine were equivalent as models of computation, and argued that they capture precisely the intuitive notion of algorithm; this claim became known as the **Church–Turing thesis**.

In 1938, Turing was invited to work for the British crypto-analytic department to help break the secret codes of the German encrypting machines Enigma. Turing was instrumental in developing methods and constructing special devices, called *Bombe*, for decrypting messages sent by Enigma coders, saving many lives during the course of the World War II. At the end of the war Turing was invited by the National Physical Laboratory in London to design a real, physical computer and he submitted a report proposing the Automatic Computing Engine (ACE) in 1946.

Turing was a homosexual and was arrested and convicted for violation of the British anti-homosexuality law in 1952. To avoid going to prison, he agreed to chemical castration. Turing died in 1954 of cyanide poisoning, widely believed to be suicide following his prosecution and treatment. In 2009 the British Prime Minister offered an official apology from the British Government for Turing's prosecution, and in 2013 he was granted a royal pardon by the Queen.

In 1966 the Association for Computing Machinery established the *Turing Award* as the most prestigious award for scientific contributions in computer science, regarded as the equivalent to the Nobel Prize.

6

Answers and Solutions to Selected Exercises

Section 1.1

1.1.1 (a) Yes. (b) Yes. (c) No. (d) No. (e) Yes. (f) No. (Imperative sentence.) (g) No. ("her" is not defined in this sentence, so truth value cannot be assigned to it.) (h) Yes. (i) No. (Question.) (i) Yes. (k) Yes. (True.) (l) No. (Either truth value can be assigned.)

1.1.2 (a) `true`. (c) `true`. (e) `false`. (g) `true`.

1.1.3 (a) Antecedent: "John talks." Consequent: "Everyone else listens."

   (c) Antecedent: "John talks." Consequent: "Everyone else listens."

   (e) Antecedent: "The cube of the integer n is positive."
   Consequent: "An integer n is positive."

   (g) Antecedent: "A function is differentiable."
   Consequent: "The function is continuous."

   (i) Antecedent: "A function is continuous."
   Consequent: "The function is differentiable."

1.1.4 (a) Neither. (c) Sufficient. (e) Necessary and sufficient.

   (g) Necessary. (i) Necessary and sufficient.

1.1.5 Denote:

"The Earth rotates around itself" by A (`true`),
"The Moon rotates around the Earth" by B (`true`),
"The Sun rotates around the Moon" by C (`false`),
"The Sun rotates around the Earth" by D (`false`),
"The Earth rotates around the Moon" by E (`false`).

Then we have:
(a) $A \wedge (B \to C)$: `false`. (c) $(\neg D \wedge \neg E) \to \neg B$: `false`.
(e) $A \leftrightarrow (\neg B \vee \neg E)$: `true`.

1.1.6 (a) By the truth definition of the implication, A must be true.

   (c) $\neg B$ is true so B is false. Then, for $A \vee B$ to be true, A must be true.

Logic as a Tool: A Guide to Formal Logical Reasoning, First Edition. Valentin Goranko.
© 2016 John Wiley & Sons, Ltd. Published 2016 by John Wiley & Sons, Ltd.

(e) Since $\neg C \land B$ is true then B is true and $\neg C$ is true, hence C is false. Since $\neg(A \lor C) \to C$ is true and C is false, then $\neg(A \lor C)$ is false, that is, $A \lor C$ is true. Now since C is false, A must be true.

1.1.7   (b) Suppose R is false. Since $P \lor R$ is true, then P is true. For $(P \lor Q) \to R$ to be true, $P \lor Q$ must be false. Hence P and Q is false. Contradiction. Hence R is true.

(d) Suppose Q is true. Since $\neg P \to \neg Q$ must be true, then $\neg P$ must be false and thus P is true. For $P \land \neg R$ to be false, $\neg R$ must be false. But this is a contradiction with $\neg R$ true. Hence Q is false.

(e) Suppose P is true. Since $P \to Q$ is true then $\neg Q$ is false. Hence $S \land \neg Q$ is false. But $R \lor (S \land \neg Q)$ is true so R must be true. This contradicts $\neg R$ is true, which was given, hence P must be false.

(i) Since $P \to R$ is false then P is true and R is false. But with $P \to Q$ true, this means Q is true. Since $Q \to (R \lor S)$ is true then $R \lor S$ must be true. Since R is false we have that S is true.

1.1.8   (g) Tautology:

p	q	$(p$	$\lor$	$\neg q)$	$\to$	$\neg$	$(q$	$\land$	$\neg p)$
F	F	F	T	T	T	T	F	F	T
F	T	F	F	F	T	F	T	T	T
T	F	T	T	T	T	T	F	F	F
T	T	T	T	F	T	T	T	F	F

(main connective $\to$, $\uparrow$)

(i) Neither a tautology nor a contradictory formula:

p	q	r	$\neg$	$(\neg p$	$\leftrightarrow$	$q)$	$\land$	$(r$	$\lor$	$\neg q)$
F	F	F	T	T	F	F	T	F	T	T
F	F	T	T	T	F	F	T	T	T	T
F	T	F	F	T	T	T	F	F	F	F
F	T	T	F	T	T	T	F	T	T	F
T	F	F	F	F	T	F	F	F	T	T
T	F	T	F	F	T	F	F	T	T	T
T	T	F	T	F	F	T	F	F	F	F
T	T	T	T	F	F	T	T	T	T	F

(main connective $\land$, $\uparrow$)

(j) Contradictory formula:

p	q	r	$\neg$	$((p$	$\land$	$\neg q)$	$\to$	$r)$	$\leftrightarrow$	$(\neg$	$(q$	$\lor$	$r)$	$\to$	$\neg p)$
F	F	F	F	F	F	T	T	F	F	T	F	F	F	T	T
F	F	T	F	F	F	T	T	T	F	F	F	T	T	T	T
F	T	F	F	F	F	F	T	F	F	F	T	T	F	T	T
F	T	T	F	F	F	F	T	T	F	F	T	T	T	T	T
T	F	F	T	T	T	T	F	F	F	T	F	F	F	F	F
T	F	T	F	T	T	T	T	T	F	F	F	T	T	T	F
T	T	F	F	T	F	F	T	F	F	F	T	T	F	T	F
T	T	T	F	T	F	F	T	T	F	F	F	F	T	T	F

(main connective $\leftrightarrow$, $\uparrow$)

1.1.9   (d) If $((p \to q) \lor (p \to \neg q)) \to \neg p$ is not a tautology, then:

$$((p \to q) \lor (p \to \neg q)): \text{T and } \neg p: \text{F}$$

$$[p \to q: \text{T or } p \to \neg q: \text{T}] \text{ and } p: \text{T}$$

$$[p \to q: \text{T and } p: \text{T}] \text{ or } [p \to \neg q: \text{T and } p: \text{T}]$$

$$[q: \text{T and } p: \text{T}] \text{ or } [q: \text{F and } p: \text{T}]$$

The formula is not a tautology. Either of the assignments $p : \text{T}$, $q : \text{T}$ or $p : \text{T}$, $q : \text{F}$ will render the formula false.

(j) In order to falsify the formula $((p \lor q) \to r) \to ((p \to r) \land (q \to r))$, we must have $(p \lor q) \to r : \text{T}$ and $(p \to r) \land (q \to r) : \text{F}$. For $(p \lor q) \to r$ to be true, there are two possible cases:

Case 1: $p \lor q : \text{F}$. Then $p : \text{F}$ and $q : \text{F}$. But then $p \to r : \text{T}$ and $q \to r : \text{T}$ so that $(p \to r) \land (q \to r) : \text{T}$, a contradiction with $(p \to r) \land (q \to r) : \text{F}$.

Case 2: $r : \text{T}$. Then $p \to r : \text{T}$ and $q \to r : \text{T}$ so that $(p \to r) \land (q \to r) : \text{T}$, a contradiction with $(p \to r) \land (q \to r) : \text{F}$.

All attempts to falsify the formula end in contradictions, so the formula is a tautology.

(l) If $((p \to r) \lor (q \to r)) \to ((p \lor q) \to r)$ is not a tautology then:

$$((p \to r) \lor (q \to r)) : \text{T and } (p \lor q) \to r: \text{F}$$

$$[p \to r: \text{T or } q \to r: \text{T}] \text{ and } [p \lor q: \text{T and } r: \text{F}]$$

$$[p \to r: \text{T and } p \lor q: \text{T and } r: \text{F}] \text{ or } [q \to r: \text{T and } p \lor q: \text{T and } r: \text{F}]$$

$$[p: \text{F and } r: \text{F and } p \lor q: \text{T}] \text{ or } [q: \text{F and } r: \text{F and } p \lor q: \text{T}]$$

$$[p: \text{F and } q: \text{T and } r: \text{F}] \text{ or } [p: \text{T and } q: \text{F and } r: \text{F}]$$

The formula is not a tautology. For example, the assignment $p : \text{F}$, $q : \text{T}$, $r : \text{F}$ will render it false.

1.1.10   (a) P cannot be a liar, otherwise he would be telling the truth. So, P is a truth-teller and therefore Q must be a liar.

(b) If the answer was "yes," the inhabitant he asked is a truth-teller and the other inhabitant a liar. If the answer was "no," then either both are truth-tellers or the one who answered was a liar, so in this case the stranger would not be able to determine who is a liar and who is not. The answer must therefore have been "yes".

(c) If A is a liar, then his claim must have been false, therefore B must be a liar while A is not. So, A cannot be a liar. Therefore, B cannot be a liar either. Therefore both A and B are truth-tellers.

(d) Note that no local can say that he is a liar, as this would result in the Liar's paradox; Y's statement must therefore be false, making him a liar. But this means that Z's statement is true, so he is a truth-teller. Since X's answer invariably must be "No, I'm a truth-teller," it is not possible to determine whether X is a truth-teller or liar.

(e) If D is a truth-teller, then E must be a liar. C's answer is indeed "just one," so C must also be a liar; otherwise, there would be two of them and C's answer would have been false. There is therefore only one truth-teller; C must therefore have said the truth – a contradiction. D must therefore have been a liar, so E must have been a truth-teller and the stranger should go with E since E has truthfully said that he was not a man-eater.

(f) The stranger could ask either inhabitant, "What would your answer be if I asked you whether your road leads to the spaceship port?" Note that a liar and a truth-teller would answer such a question in the same way. If the answer is "yes" the stranger should take that road, otherwise, the stranger should take the other road.

Section 1.2

1.2.1 (1) $\Rightarrow$ (3): Assume $A_1, A_2, \ldots, A_n \models B$. Then, for all assignments of truth values to the variables occurring in $A_1, A_2, \ldots, A_n, B$, if $A_1, A_2, \ldots, A_n$ are true, then B is also true. Hence, for all such assignments, $A_1 \wedge A_2 \wedge \ldots \wedge A_n \rightarrow B$ is true, so $A_1 \wedge A_2 \wedge \ldots \wedge A_n \rightarrow B$ is a tautology.

(3) $\Rightarrow$ (2): It is sufficient to show that $\models A \rightarrow B$ implies $A \models B$. Indeed, suppose $\models A \rightarrow B$. Then any assignment of truth values to the variables occurring in A, B that make A true, also make – by the truth definition of the implication – B true. Therefore, $A \models B$.

1.2.2 (a) $q \rightarrow p$ is not a consequence of $p \rightarrow q$ since the third row contains premises with truth values T but with a truth value F for the conclusion:

p	q	$p \rightarrow q$	$q \rightarrow p$
T	T	T T T	T T T
T	F	T F F	F T T
F	T	F T T	T F F
F	F	F T F	F T F

Hence, the rule is not sound.

(c) $\neg q \rightarrow \neg p$ is a consequence of $p \rightarrow q$ since wherever the premises have truth values T, the conclusion also has a truth value T:

p	q	$p \rightarrow q$	$\neg q \rightarrow \neg p$
T	T	T T T	F T F
T	F	T F F	T F F
F	T	F T T	F T T
F	F	F T F	T T T

Hence, the rule is sound.

1.2.3 (a) The inference rule is sound since wherever the premises have truth values T, the conclusion also has a truth value T:

p	q	r	$p \rightarrow q$	$\neg q \vee r$	$\neg r$	$\neg p$
F	F	F	F T F	T T F	T	T
F	F	T	F T F	T T T	F	T
F	T	F	F T T	F F F	T	T
F	T	T	F T T	F T T	F	T
T	F	F	T F F	T T F	T	F
T	F	T	T F F	T T T	F	F
T	T	F	T T T	F F F	T	F
T	T	T	T T T	F T T	F	F
			↑	↑	↑	

(c) The inference rule is *not* sound since the third and seventh rows contain premises with truth values T but with a truth value F for the conclusion:

p	q	r	$p \rightarrow q$	$p \vee \neg r$	$\neg r$	$\neg q \vee r$
F	F	F	F T F	F T T	T	T T F
F	F	T	F T F	F F F	F	T T T
F	T	F	F T T	F T T	T	F F F
F	T	T	F T T	F F F	F	F T T
T	F	F	T F F	T T T	T	T T F
T	F	T	T F F	T T F	F	T T T
T	T	F	T T T	T T T	T	F F F
T	T	T	T T T	T T F	F	F T T
			↑	↑	↑	↑

1.2.4 (b) Denote the propositions "Y is greater than -1" by p and "Y is greater than -2" by q. Then the inference rule for the argument can be written as:

$$\frac{p \rightarrow q, \neg q}{\neg p}.$$

This inference rule is sound:

p	q	$p \rightarrow q$	$\neg q$	$\neg p$
F	F	F T F	T	T
F	T	F T T	F	T
T	F	T F F	T	F
T	T	T T T	F	F
		↑	↑	

(d) Denote "Victor is good at logic" by p, "Victor is clever" by q, and "Victor is rich" by r. Then the inference rule for the argument can be written as:

$$\frac{p \rightarrow q, q \rightarrow r}{p \rightarrow r}.$$

The truth table:

p	q	r	$p \to q$	$q \to r$	$(p \to r)$
F	F	F	F T F	F T F	F T F
F	F	T	F T F	F T T	F T T
F	T	F	F T T	T F F	F T F
F	T	T	F T T	T T T	F T T
T	F	F	T F F	F T F	T F F
T	F	T	T F F	F T T	T T T
T	T	F	T T T	T F F	T F F
T	T	T	T T T	T T T	T T T
			↑	↑	↑

This inference rule is sound.

1.2.5 (a) **Converse:** If a is greater than -2, then a is greater than -1.
Inverse: If a is not greater than -1, then a is not greater than -2.
Contrapositive: If a is not greater than -2, then a not is greater than -1.

(c) **Converse:** The square of x is positive only if x is positive.
Inverse: x is not positive only if its square is not positive.
Contrapositive: The square of x is not positive only if x is not positive.

(e) **Converse:** For the function f to be differentiable it is sufficient that it is continuous.
Inverse: For the function f to be discontinuous it is sufficient that it is not differentiable.
Contrapositive: For the function f to be not differentiable it is sufficient that it is discontinuous.

(g) **Converse:** For the integer n to be indivisible by 10 it is necessary that it is prime.
Inverse: For the integer n to be not prime it is necessary that it is divisible by 10.
Contrapositive: For the integer n to be divisible by 10 it is necessary that it is not prime.

Section 1.3

1.3.2 (a)

p	q	$\neg$	$(p \leftrightarrow q)$	$(p \wedge \neg q)$	$\vee$	$(q \wedge \neg p)$
F	F	F	F T F	F F T	F	F F T
F	T	T	F F T	F F F	T	T T T
T	F	T	T F F	T T T	T	F F F
T	T	F	T T T	T F F	F	T F F
		↑			↑	

(f)

p	q	r	p	$\to$	$(q \to r)$	$(p \land q)$	$\to$	r
F	F	F	F	T	F T F	F F F	T	F
F	F	T	F	T	F T T	F F F	T	T
F	T	F	F	T	T F F	F F T	T	F
F	T	T	F	T	T T T	F F T	T	T
T	F	F	T	T	F T F	T F F	T	F
T	F	T	T	T	F T T	T F F	T	T
T	T	F	T	F	T F F	T T T	F	F
T	T	T	T	T	T T T	T T T	T	T
				↑			↑	

(h)

p	q	r	$(p \to r)$	$\land$	$(q \to r)$	$(p \lor q)$	$\to$	r
F	F	F	F T F	T	F T F	F F F	T	F
F	F	T	F T T	T	F T T	F F F	T	T
F	T	F	F T F	F	T F F	F T T	F	F
F	T	T	F T T	T	T T T	F T T	T	T
T	F	F	T F F	F	F T F	T T F	F	F
T	F	T	T T T	T	F T T	T T F	T	T
T	T	F	T F F	F	T F F	T T T	F	F
T	T	T	T T T	T	T T T	T T T	T	T
				↑			↑	

1.3.3   (h)

p	q	r	$(p \to r)$	$\lor$	$(q \to r)$	$(p \lor q)$	$\to$	r
F	F	F	F T F	T	F T F	F F F	T	F
F	F	T	F T T	T	F T T	F F F	T	T
F	T	F	F T F	T	T F F	F T T	F	F
F	T	T	F T F	T	T F F	F T T	T	T
T	F	F	T F F	T	F T F	T T F	F	F
T	F	T	T T T	T	F T T	T T F	T	T
T	T	F	T F F	F	T F F	T T T	F	F
T	T	T	T T T	T	T T T	T T T	T	T
				↑			↑	

The respective (indicated) columns of truth values are not the same, hence the formulae are *not* equivalent.

(k)

p	q	r	p	$\leftrightarrow$	$(q \leftrightarrow r)$	q	$\leftrightarrow$	$(p \leftrightarrow r)$
F	F	F	F	F	F T F	F	F	F T F
F	F	T	F	T	F F T	F	T	F F T
F	T	F	F	T	T F F	T	T	F T F
F	T	T	F	F	T T T	T	F	F F T
T	F	F	T	T	F T F	F	T	T F F
T	F	T	T	F	F F T	F	F	T T T
T	T	F	T	F	T F F	T	F	T F F
T	T	T	T	T	T T T	T	T	T T T
				↑			↑	

The respective (indicated) columns of truth values are the same, hence the formulae are equivalent.

1.3.4 (a)

$$\neg((p \vee \neg q) \wedge \neg p)$$
$$\equiv \neg(p \vee \neg q) \vee \neg\neg p$$
$$\equiv (\neg p \wedge \neg\neg q) \vee p$$
$$\equiv (\neg p \wedge q) \vee p$$

(f)

$$\neg(p \to (\neg q \leftrightarrow r))$$
$$\equiv p \wedge \neg(\neg q \leftrightarrow r)$$
$$\equiv p \wedge ((\neg\neg q \wedge r) \vee (\neg q \wedge \neg r))$$
$$\equiv p \wedge ((q \wedge r) \vee (\neg q \wedge \neg r))$$

Section 1.4

No solutions are provided for these exercises, but the interested reader can find many of these in the references provided at the end of the section.

Section 2.2

2.2.2 (c) 1. $\vdash_{\mathbf{H}} (p \wedge q) \to q$ by Axiom $(\wedge 1)$
2. $\vdash_{\mathbf{H}} (p \wedge q) \to p$ by Axiom $(\wedge 2)$
3. $\vdash_{\mathbf{H}} ((p \wedge q) \to q) \to (((p \wedge q) \to p) \to ((p \wedge q) \to (q \wedge p)))$ by Axiom $(\wedge 3)$
4. $\vdash_{\mathbf{H}} ((p \wedge q) \to p) \to ((p \wedge q) \to (q \wedge p))$ by 1, 3, and MP
5. $\vdash_{\mathbf{H}} (p \wedge q) \to (q \wedge p)$ by 2, 4, and MP

(e) 1. $\vdash_{\mathbf{H}} p \to ((p \to p) \to p)$ by Axiom $(\to 1)$
2. $\vdash_{\mathbf{H}} (p \to ((p \to p) \to p)) \to ((p \to (p \to p)) \to (p \to p))$ by Axiom $(\to 2)$
3. $\vdash_{\mathbf{H}} ((p \to (p \to p)) \to (p \to p)))$ by 1, 2, and MP
4. $\vdash_{\mathbf{H}} p \to (p \to p)$ by Axiom $(\to 1)$
5. $\vdash_{\mathbf{H}} p \to p$ by 3, 4, and MP

2.2.5 Let $\Gamma, A \vdash_{\mathbf{H}} B$ and $\Gamma, B \vdash_{\mathbf{H}} C$. Then, $\Gamma \vdash_{\mathbf{H}} B \to C$ by the Deduction Theorem (DT). Therefore, $\Gamma, A \vdash_{\mathbf{H}} B \to C$, hence $\Gamma, A \vdash_{\mathbf{H}} C$ by using Modus Ponens.

2.2.6 Let $A \vdash_{\mathbf{H}} B$. Then $B \to C, A \vdash_{\mathbf{H}} B$. Besides, $B \to C, A \vdash_{\mathbf{H}} B \to C$. Therefore $B \to C, A \vdash_{\mathbf{H}} C$ by MP, hence $B \to C \vdash_{\mathbf{H}} A \to C$ by DT.

2.2.9 (e)
1. $P \to (Q \to R) \vdash_{\mathbf{H}} P \to (Q \to R)$
2. $P \to (Q \to R), P \vdash_{\mathbf{H}} (Q \to R)$ by 1 and DT
3. $P \to (Q \to R), P, Q \vdash_{\mathbf{H}} R$ by 2 and DT
4. $P \to (Q \to R), Q \vdash_{\mathbf{H}} (P \to R)$ by 3 and DT
   $P \to (Q \to R) \vdash_{\mathbf{H}} Q \to (P \to R)$ by 4 and DT.

(g)
1. $P \to (Q \to R) \vdash_{\mathbf{H}} P \to (Q \to R)$
2. $P \to (Q \to R), P \vdash_{\mathbf{H}} (Q \to R)$ by 1 and DT
3. $P \to (Q \to R), P \wedge Q \vdash_{\mathbf{H}} P$ by Axiom ($\wedge$1)
4. $P \to (Q \to R), P \wedge Q \vdash_{\mathbf{H}} (Q \to R)$ by 2, 3, and Exercise 5.
5. $P \to (Q \to R), P \wedge Q, Q \vdash_{\mathbf{H}} R$ by 4 and DT
6. $P \to (Q \to R), P \wedge Q \vdash_{\mathbf{H}} Q$ by Axiom ($\wedge$2)
7. $P \to (Q \to R), P \wedge Q \vdash_{\mathbf{H}} R$ by 5, 6, and Exercise 5.
8. $P \to (Q \to R) \vdash_{\mathbf{H}} (P \wedge Q) \to R$ by 7 and DT.

(i)
1. $\vdash_{\mathbf{H}} (\neg P \to \neg Q) \to ((\neg P \to Q) \to P)$ by Axiom ($\to$ 3)
2. $\neg P \to \neg Q \vdash_{\mathbf{H}} (\neg P \to \neg Q) \to ((\neg P \to Q) \to P)$ by 1
3. $\neg P \to \neg Q \vdash_{\mathbf{H}} \neg P \to \neg Q$
4. $\neg P \to \neg Q \vdash_{\mathbf{H}} ((\neg P \to Q) \to P)$ by 2, 3, and MP.
5. $Q \vdash_{\mathbf{H}} \neg P \to Q$ by Axiom ($\to$ 1)
6. $(\neg P \to Q) \to P \vdash_{\mathbf{H}} Q \to P$ by 5 and Exercise 6
7. $\neg P \to \neg Q \vdash_{\mathbf{H}} Q \to P$ by 4, 6, and Exercise 5.

(m)
1. $\vdash_{\mathbf{H}} (\neg P \to \neg\neg P) \to ((\neg \to \neg P) \to P)$ by Axiom ($\to$ 3)
2. $\neg P \to \neg\neg P \vdash_{\mathbf{H}} (\neg P \to \neg P) \to P$ by 1 and DT
3. $\vdash_{\mathbf{H}} \neg\neg P \to (\neg P \to \neg\neg P)$ by Axiom ($\to$ 1)
4. $\neg\neg P \vdash_{\mathbf{H}} \neg P \to \neg\neg P$ by 3 and DT
5. $\neg\neg P \vdash_{\mathbf{H}} (\neg P \to \neg P) \to P$ by 2, 4, and Exercise 5.
6. $\neg P \vdash_{\mathbf{H}} \neg P$
7. $\vdash_{\mathbf{H}} \neg P \to \neg P$ by 6 and DT
8. $\neg\neg P \vdash_{\mathbf{H}} P$ by 5, 7, and MP.

(o) Sketch:
1. $P \vdash_{\mathbf{H}} P \vee \neg P$ by Axiom ($\vee$1)
2. $\neg(P \vee \neg P) \vdash_{\mathbf{H}} \neg P$ by 1 and Exercise 9k.
3. $\neg P \vdash_{\mathbf{H}} P \vee \neg P$ by Axiom ($\vee$1)
4. $\neg(P \vee \neg P) \vdash_{\mathbf{H}} \neg\neg P$ by 3 and Exercise 9k.
5. $\vdash_{\mathbf{H}} \neg(P \vee \neg P) \to \neg\neg P$ by 4 and DP
6. $\vdash_{\mathbf{H}} (\neg(P \vee \neg P) \to \neg\neg P) \to ((\neg(P \vee \neg P) \to \neg P) \to (P \vee \neg P))$.   by Axiom ($\to$ 3)
7. $\vdash_{\mathbf{H}} (\neg(P \vee \neg P) \to \neg P) \to (P \vee \neg P)$ by 5, 6, and MP
8. $\vdash_{\mathbf{H}} \neg(P \vee \neg P) \to \neg P$ by 2 and DP
9. $\vdash_{\mathbf{H}} P \vee \neg P$ by 7, 8, and MP.

Section 2.3

2.3.1   (a) The formula is not a tautology:

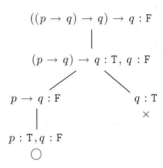

$$((p \to q) \to q) \to q : F$$

$$(p \to q) \to q : T, \ q : F$$

$$p \to q : F \qquad\qquad q : T$$
$$\times$$

$$p : T, q : F$$
$$\bigcirc$$

(c) Tautology:

$$((p \to q) \land (p \to \neg q)) \to \neg p : F$$

$$(p \to q) \land (p \to \neg q) : T, \ \neg p : F$$

$$p : T$$

$$p \to q : T, p \to \neg q : T$$

$$p : F \qquad\qquad q : T$$
$$\times$$

$$p : F \qquad\qquad \neg q : T$$
$$\times$$

$$q : F$$
$$\times$$

(e) Tautology:

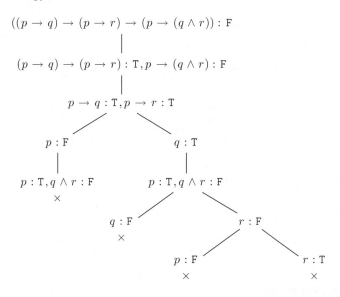

$$((p \rightarrow q) \rightarrow (p \rightarrow r) \rightarrow (p \rightarrow (q \wedge r)) : F$$

$$(p \rightarrow q) \rightarrow (p \rightarrow r) : T, p \rightarrow (q \wedge r) : F$$

$$p \rightarrow q : T, p \rightarrow r : T$$

$p : F \qquad\qquad q : T$

$p : T, q \wedge r : F \qquad p : T, q \wedge r : F$
$\times$

$q : F \qquad\qquad r : F$
$\times$

$p : F \qquad\qquad r : T$
$\times \qquad\qquad \times$

(g) Not a tautology:

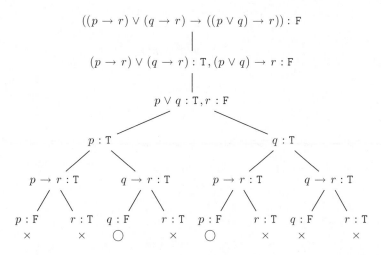

$$((p \rightarrow r) \vee (q \rightarrow r) \rightarrow ((p \vee q) \rightarrow r)) : F$$

$$(p \rightarrow r) \vee (q \rightarrow r) : T, (p \vee q) \rightarrow r : F$$

$$p \vee q : T, r : F$$

$p : T \qquad\qquad q : T$

$p \rightarrow r : T \qquad q \rightarrow r : T \qquad p \rightarrow r : T \qquad q \rightarrow r : T$

$p : F \quad r : T \quad q : F \quad r : T \quad p : F \quad r : T \quad q : F \quad r : T$
$\times \qquad \times \qquad \bigcirc \qquad \times \qquad \bigcirc \qquad \times \qquad \times \qquad \times$

(i) Tautology:

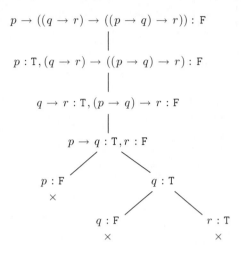

(k) Not a tautology.

2.3.2 (b) The logical consequence is valid.

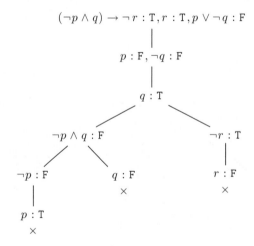

(d) The logical consequence is not valid.

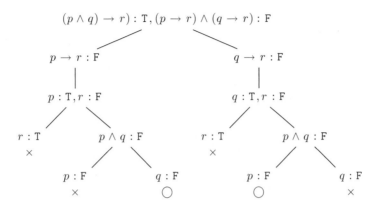

(g) The logical consequence is valid.

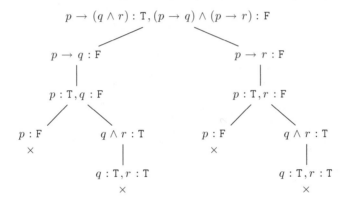

2.3.3 For each of these problems we construct the tableau with all premises assigned the required truth values and the conclusion assigned the opposite truth value. If the tableau closes, then such truth assignment is impossible; the stated consequence therefore holds. If the tableau does not close, then the stated consequence does not necessarily hold.

(a) The tableau closes, so the consequence holds:

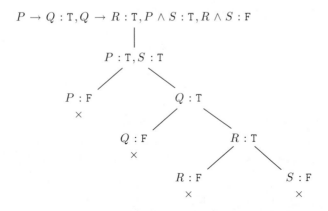

(c) The tableau closes, so the consequence holds:

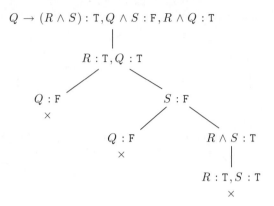

(e) The tableau closes, so the consequence holds:

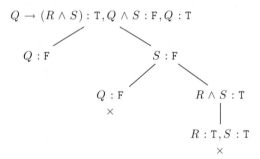

2.3.4   (a) The inference rule is sound:

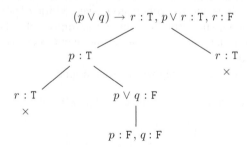

(c) The inference rule is not sound:

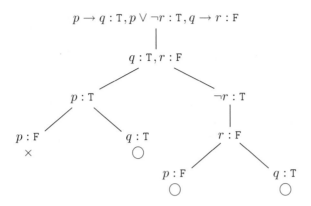

(e) The inference rule is not sound:

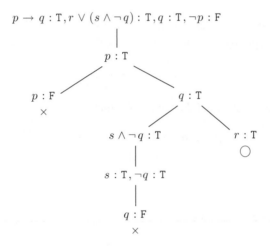

2.3.5 We first formalize each of the propositional arguments by identifying the atomic propositions in it and replacing them with propositional variables. We then check the soundness of the resulting inference rule.

(b) Denote "n is divisible by 2" by p, "n is divisible by 3" by q, and "n is divisible by 6" by r. Then the inference rule on which the argument is based is:

$$\frac{(p \wedge q) \to r, r \to p, \neg q}{\neg r}.$$

The rule is not sound, hence the argument is not correct.

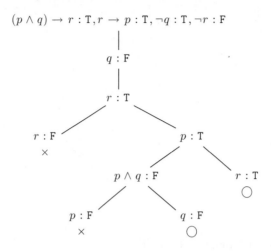

(d) Denote "Socrates is wise" by p, "Socrates is happy" by q, and "Socrates is a philosopher" by r. Then the inference rule on which the argument is based is:

$$\frac{p \vee \neg q, \neg p \rightarrow \neg r}{r \rightarrow \neg q}.$$

The rule is not sound, hence the argument is not correct.

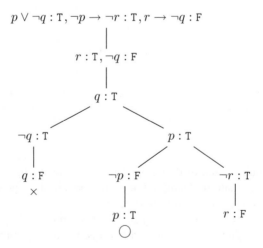

(f) Denote

" Victor studies hard" by p,
" Victor is clever" by q,
" Victor will go to a university" by r, and
" Victor will get a good job" by s.

Then the inference rule on which the argument is based is:

$$\frac{(p \land q) \rightarrow r, \neg p \rightarrow \neg s}{q \rightarrow (r \lor \neg s)} .$$

The rule is sound, hence the argument is correct.

$(p \land q) \rightarrow r : \mathsf{T}, \neg p \rightarrow \neg s : \mathsf{T}, q \rightarrow (r \lor \neg s) : \mathsf{F}$

$|$

$q : \mathsf{T}, r \lor \neg s : \mathsf{F}$

$|$

$r : \mathsf{F}, \neg s : \mathsf{F}$

$|$

$s : \mathsf{T}$

$\neg p : \mathsf{F}$ $\neg s : \mathsf{T}$

$|$ $|$

$p : \mathsf{T}$ $s : \mathsf{F}$
                $\times$

$p \land q : \mathsf{F}$ $r : \mathsf{T}$
                 $\times$

$p : \mathsf{F}$ $q : \mathsf{F}$

$\times$ $\times$

(h) Denote

" Johnnie learns logic" by p,
" Johnnie skips lectures" by q, and
" Johnnie passes the exam" by r.

Then the inference rule on which the argument is based is:

$$\frac{q \to \neg p,\, p \wedge \neg q \to r}{q \vee \neg r \to \neg p}.$$

The rule is sound, hence the argument is correct.

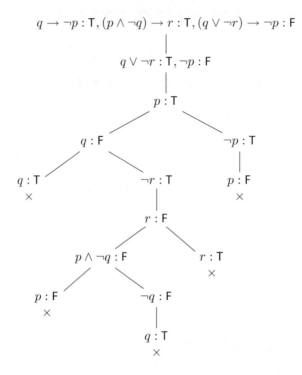

$q \to \neg p : \mathsf{T},\, (p \wedge \neg q) \to r : \mathsf{T},\, (q \vee \neg r) \to \neg p : \mathsf{F}$

$q \vee \neg r : \mathsf{T},\, \neg p : \mathsf{F}$

$p : \mathsf{T}$

$q : \mathsf{F}$ $\neg p : \mathsf{T}$

$q : \mathsf{T}$ $\neg r : \mathsf{T}$ $p : \mathsf{F}$
$\times$ $\times$

$r : \mathsf{F}$

$p \wedge \neg q : \mathsf{F}$ $r : \mathsf{T}$
                                    $\times$

$p : \mathsf{F}$ $\neg q : \mathsf{F}$
$\times$

$q : \mathsf{T}$
$\times$

(j) Denote

"Bonnie comes/will come to the party" by p,
"Clyde comes/will come to the party" by q, and
"Alice comes/will come to the party" by r.

Then the argument is based on the following inference rule:

$$\frac{\neg p \to (\neg q \vee r),\, q \to \neg r}{q \to p}.$$

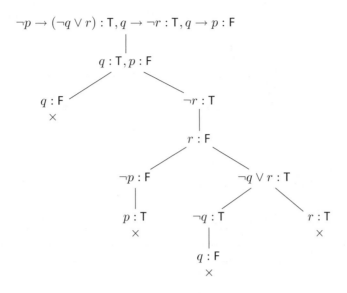

The rule is sound, hence the argument is correct.

Section 2.4

2.4.2

Introduction rule   **Elimination rules**

$$(\leftrightarrow I) \quad \dfrac{\begin{array}{cc}[A] & [B]\\ \vdots & \vdots \\ B & A\end{array}}{A \leftrightarrow B} \qquad\qquad (\leftrightarrow E) \quad \dfrac{A, \quad A \leftrightarrow B}{B}, \quad \dfrac{B, \quad A \leftrightarrow B}{A}$$

2.4.3 (c)

$$\dfrac{\dfrac{[\neg\neg p]^2 \quad \dfrac{[\neg p]^1 \quad \neg p}{\bot}_1}{p}}{\neg\neg p \to p}_2$$

(d)

$$\cfrac{[\neg(p \vee \neg p)]^3 \quad \cfrac{[p]^1}{p \vee \neg p}}{\cfrac{\bot}{\neg p}\,1} \qquad \cfrac{[\neg(p \vee \neg p)]^3 \quad \cfrac{[\neg p]^2}{p \vee \neg p}}{\cfrac{\bot}{\neg\neg p}\,2}$$
$$\cfrac{\bot}{p \vee \neg p}\,3$$

(f) $\vdash_{\mathbf{ND}} ((p \to q) \wedge (p \to \neg q)) \to \neg p$:

$$\cfrac{[p]^1 \quad \cfrac{[(p \to q) \wedge (p \to \neg q)]^2}{p \to q}}{q} \qquad \cfrac{[p]^1 \quad \cfrac{[(p \to q) \wedge (p \to \neg q)]^2}{p \to \neg q}}{\neg q}$$
$$\cfrac{\bot}{\neg p}\,1$$
$$\cfrac{}{((p \to q) \wedge (p \to \neg q)) \to \neg p}\,2$$

2.4.4 (a) Suppose $A \vdash_{\mathbf{ND}} B$.

$$\cfrac{\neg B, \quad \begin{matrix}[A]^1 \\ \vdots \\ B\end{matrix}}{\cfrac{\bot}{\neg A}\,1}$$

Hence $\neg B \vdash_{\mathbf{ND}} \neg A$.
(b) Suppose $\neg B \vdash_{\mathbf{ND}} \neg A$.

$$\cfrac{A, \quad \begin{matrix}[\neg B]^1 \\ \vdots \\ \neg A\end{matrix}}{\cfrac{\bot}{B}\,1}$$

Hence $A \vdash_{\mathbf{ND}} B$.

2.4.5 (a) $A \to C, B \to C \quad \vdash_{\textbf{ND}} \quad (A \lor B) \to C$:

$$\cfrac{[A \lor B]^2 \quad \cfrac{\cfrac{[A]^1, \ A \to C}{C} \quad \cfrac{[B]^1, \ B \to C}{C}}{C} \, 1}{\cfrac{C}{(A \lor B) \to C} \, 2}$$

(c) $(A \lor B) \to C \quad \vdash_{\textbf{ND}} \quad (A \to C) \land (B \to C)$:

$$\cfrac{\cfrac{\cfrac{\cfrac{[A]^1}{A \lor B} \quad (A \lor B) \to C}{C}}{A \to C} \, 1 \quad \cfrac{\cfrac{\cfrac{[B]^2}{A \lor B} \quad (A \lor B) \to C}{C}}{B \to C} \, 2}{(A \to C) \land (B \to C)}$$

(e) $A \to \neg A \quad \vdash_{\textbf{ND}} \quad \neg A$:

$$\cfrac{\cfrac{\cfrac{\cfrac{[\neg\neg A]^2, \ [\neg A]^1}{\bot}}{A} \, 1 \quad A \to \neg A}{\neg A} \quad [\neg\neg A]^2}{\cfrac{\bot}{\neg A} \, 2}$$

NB: there is a simpler derivation. Find it!

(g) $\neg A \to B, \neg A \to \neg B \quad \vdash_{\textbf{ND}} \quad A$:

$$\cfrac{\cfrac{[\neg A]^1, \ \neg A \to B}{B} \quad \cfrac{[\neg A]^1, \ \neg A \to \neg B}{\neg B}}{\cfrac{\bot}{A} \, 1}$$

(i) $(\neg A \wedge B) \to \neg C, C \vdash_{\mathbf{ND}} A \vee \neg B$:

$$
\cfrac{
\cfrac{
\cfrac{
\cfrac{[\neg(A \vee \neg B)]^2}{\vdots} \leftarrow \text{use 4(a)} \quad\quad
\cfrac{\cfrac{[\neg(A \vee \neg B)]^2}{\vdots \;\leftarrow \text{use 4(a)}}{\neg\neg B} \quad [\neg B]^1}{\cfrac{\bot}{B}\,1}
}{\neg A \wedge B} \quad\quad (\neg A \wedge B) \to \neg C
}{\quad C \quad\quad\quad\quad \neg C \quad}
}{\cfrac{\bot}{A \vee \neg B}\,2}
$$

(k) $\neg(A \vee B) \quad \vdash_{\mathbf{ND}} \quad \neg A \wedge \neg B$:

$$
\cfrac{
\cfrac{\neg(A \vee B)}{\vdots} \quad \longleftarrow \text{using 2(a)} \longrightarrow \quad \cfrac{\neg(A \vee B)}{\vdots}
}{\neg A \wedge \neg B}
$$

Left branch ends in $\neg A$, right branch ends in $\neg B$.

(m) $\neg A \wedge \neg B \vdash_{\mathbf{ND}} \neg(A \vee B)$:

$$
\cfrac{
[A \vee B]^2 \quad\quad
\cfrac{[A]^1 \quad \cfrac{\neg A \wedge \neg B}{\neg A}}{\bot} \quad\quad
\cfrac{[B]^1 \quad \cfrac{\neg A \wedge \neg B}{\neg B}}{\bot}
}{\cfrac{\bot}{\neg(A \vee B)}\,2}\;1
$$

(o) $A \wedge \neg B \vdash_{\mathbf{ND}} \neg(A \to B)$:

$$
\cfrac{
\cfrac{\cfrac{A \wedge \neg B}{A} \quad [A \to B]^1}{B} \quad\quad \cfrac{A \wedge \neg B}{\neg B}
}{\cfrac{\bot}{\neg(A \to B)}\,1}
$$

2.4.6 (a)

$$
\cfrac{
\cfrac{
[\neg r]^2 \qquad
\cfrac{
\cfrac{[\neg p]^1,\ [\neg p \to r]^3}{r}
}{
\cfrac{\cfrac{\bot}{p}\ 1 \qquad p \to \neg q}{\neg q}
}
}{\neg r \vee p \qquad \neg q}
\qquad
\cfrac{[p]^2 \quad p \to \neg q}{\neg q}
}{
\cfrac{\neg q}{(\neg p \to r) \to \neg q}\ 3
}\ 2
$$

2.4.7 We formalize each of the propositional arguments by identifying the atomic propositions in them and replacing them with propositional variables. For a selection of them, we then will prove the soundness of the resulting inference rule by deriving it in **ND**.

(b) Denote
  "Nina will go to a party" by p, and
  "Nina will go to office" by q.

Then the argument becomes:

$$
\frac{p \vee \neg q,\ \neg p \vee \neg q}{\neg q}.
$$

The rule is derivable in **ND** and therefore sound, so the argument is correct.

$$
\cfrac{
\neg q \vee p \qquad
\cfrac{
[\neg q]^2 \qquad
\cfrac{
\neg q \vee \neg p \cfrac{[\neg q]^1}{\neg q} \qquad
\cfrac{\cfrac{[\neg p]^1,\ [p]^2}{\bot}}{\neg q}
}{\neg q}\ 1
}{\neg q}
}{\neg q}\ 2
$$

(d) Denote
  "Socrates is happy" by p,
  "Socrates is stupid" by q, and
  "Socrates is a philosopher" by r.

Then the inference rule on which the argument is based is:

$$
\frac{p \vee \neg q,\ p \to \neg r}{r \to \neg q}.
$$

The rule is derivable in **ND** and therefore sound, so the argument is correct.

$$
\cfrac{
\cfrac{
p \vee \neg q \qquad
\cfrac{
\cfrac{\cfrac{[p]^1,\ p \to \neg r}{\neg r, \qquad r]^2}}{\bot}
}{\neg q} \qquad [\neg q]^1
}{\neg q}\ 1
}{
\cfrac{\neg q}{r \to \neg q}\ 2
}
$$

(f) Denote

"Olivia is sleeping" by p,
"Olivia is eating" by q, and
"Olivia is smiling" by r.

The underlying inference rule:

$$\frac{\neg r \to \neg p, \neg q \lor r}{(\neg q \to p) \to r}.$$

Here is a derivation in **ND**:

$$\frac{\neg q \lor p \quad \dfrac{\dfrac{[\neg q]^3, [\neg q \to p]^4}{p} \quad \dfrac{[\neg r]^1, \neg r \to \neg p}{\neg p}}{\dfrac{\bot}{r}\,1} \quad [p]^3 \dfrac{[\neg r]^2, \neg r \to \neg p}{\neg p} \\ \dfrac{\bot}{r}\,2}{\dfrac{r}{(\neg q \to p) \to r}\,4}\,3$$

(h) Denote

"Bill smokes regularly" by p,
"Bill is a heavy smoker" by q,
"Bill will quit smoking" by r, and
"Bill will get lung cancer" by s.

The underlying inference rule:

$$\frac{p \to q, \neg(p \to r), q \to (r \lor s)}{s}.$$

The rule is valid, hence the inference is correct.

(j) Denote

"Kristina is a good student" by p,
"Kristina is clever" by q, and
"Kristina is lazy" by r.

Then the argument can be written as:

$$\frac{\neg p \to (\neg q \lor r), \neg q \lor \neg r}{q \to p}.$$

Derivation in **ND**:

$$\frac{q]^2, \dfrac{\dfrac{[\neg p]^1, \neg p \to (\neg q \lor r)}{\neg q \lor r} \quad \neg q \lor \neg r}{\neg q}(*)}{\dfrac{\dfrac{\bot}{p}\,1}{q \to p}\,2}$$

(*) Here we use the derivation in Exercise 7b.

Section 2.5

2.5.3 Hint: given a truth table of a formula, consider the conjunction of elementary disjunctions obtained as follows. For every row where the formula is false take the disjunction of all variables assigned value F in that row and all negations of variables assigned value T in that row.

2.5.4 (a) First method:

$$\neg(p \leftrightarrow q)$$
$$\equiv \neg((p \to q) \wedge (q \to p))$$
$$\equiv \neg((\neg p \vee q) \wedge (\neg q \vee p))$$
$$\equiv (p \wedge \neg q) \vee (q \wedge \neg p) \textbf{ (DNF)}$$
$$\equiv ((p \wedge \neg q) \vee q) \wedge ((p \wedge \neg q) \vee \neg p)$$
$$\equiv (p \vee q) \wedge \top \wedge \top \wedge (\neg q \vee \neg p)$$
$$\equiv (p \vee q) \wedge (\neg q \vee \neg p) \textbf{ (CNF)}$$

Second method:

p	q	$\neg$	$(p \leftrightarrow q)$		
F	F	F	F	T	F
F	T	T	F	F	T
T	F	T	T	F	F
T	T	F	T	T	T

DNF: $(\neg p \wedge q) \vee (p \wedge \neg q)$
CNF: $(\neg p \vee \neg q) \wedge (p \vee q)$

(c) First method:

$$(p \leftrightarrow \neg q) \leftrightarrow r$$
$$\equiv ((p \leftrightarrow \neg q) \to r) \wedge (r \to (p \leftrightarrow \neg q))$$
$$\equiv (\neg(p \leftrightarrow \neg q) \vee r) \wedge (\neg r \vee (p \leftrightarrow \neg q))$$
$$\equiv (\neg((p \to \neg q) \wedge (\neg q \to p)) \vee r) \wedge (\neg r \vee ((p \to \neg q) \wedge (\neg q \to p)))$$
$$\equiv (\neg((\neg p \vee \neg q) \wedge (\neg\neg q \vee p)) \vee r) \wedge (\neg r \vee ((\neg p \vee \neg q) \wedge (\neg\neg q \vee p)))$$
$$\equiv ((p \wedge q) \vee (\neg q \wedge \neg p) \vee r) \wedge (\neg r \vee (\neg p \wedge q) \vee (\neg p \wedge p) \vee (\neg q \wedge q) \vee (\neg q \wedge p))$$
$$\equiv (p \wedge q \wedge \neg r) \vee (p \wedge q \wedge \neg p \wedge q) \vee (p \wedge q \wedge \neg q \wedge p) \vee (\neg q \wedge \neg p \wedge \neg r) \vee$$
$$(\neg q \wedge \neg p \wedge \neg p \wedge q) \vee (\neg q \wedge \neg p \wedge \neg q \wedge p) \vee (r \wedge \neg r) \vee (r \wedge \neg p \wedge q) \vee (r \wedge \neg q \wedge p)$$
$$\equiv (p \wedge q \wedge \neg r) \vee (\neg q \wedge \neg p \wedge \neg r) \vee (r \wedge \neg p \wedge q) \vee (r \wedge \neg q \wedge p) \textbf{ (DNF)}$$

$$\equiv (q \wedge ((p \wedge \neg r) \vee (r \wedge \neg p))) \vee (\neg q \wedge ((\neg p \wedge \neg r) \vee (r \wedge p)))$$
$$\equiv (q \wedge (p \vee r) \wedge (p \vee \neg p) \wedge (\neg r \vee r) \wedge (\neg r \vee \neg p)) \vee$$
$$(\neg q \wedge (\neg p \vee r) \wedge (\neg p \vee p) \wedge (\neg r \vee r) \wedge (\neg r \vee p))$$
$$\equiv (q \wedge (p \vee r) \wedge (\neg r \vee \neg p)) \vee (\neg q \wedge (\neg p \vee r) \wedge (\neg r \vee p))$$
$$\equiv (q \vee \neg q) \wedge (q \vee \neg p \vee r) \wedge (q \vee \neg r \vee p) \wedge (p \vee r \vee \neg q) \wedge$$
$$(p \vee r \vee \neg p \vee r) \wedge (p \vee r \vee \neg r \vee p) \wedge (\neg r \vee \neg p \vee \neg q) \wedge$$
$$(\neg r \vee \neg p \vee \neg p \vee r) \wedge (\neg r \vee \neg p \vee \neg r \vee p)$$
$$\equiv (q \vee \neg p \vee r) \wedge (q \vee \neg r \vee p) \wedge (p \vee r \vee \neg q) \wedge (\neg r \vee \neg p \vee \neg q) \textbf{ (CNF)}$$

Second method:

p	q	r	$(p \leftrightarrow \neg q)$	$\leftrightarrow$	r
F	F	F	F F T	T	F
F	F	T	F F T	F	T
F	T	F	F T F	F	F
F	T	T	F T F	T	T
T	F	F	T T T	F	F
T	F	T	T T T	T	T
T	T	F	T F F	T	F
T	T	T	T F F	F	T

DNF: $(\neg p \wedge \neg q \wedge \neg r) \vee (\neg p \wedge q \wedge r) \vee (p \wedge \neg q \wedge r) \vee (p \wedge q \wedge \neg r)$
CNF: $(p \vee q \vee \neg r) \wedge (p \vee \neg q \vee r) \wedge (\neg p \vee q \vee r) \wedge (\neg p \vee \neg q \vee \neg r)$

(e) First method:

$$(\neg p \wedge (\neg q \leftrightarrow p)) \rightarrow ((q \wedge \neg p) \vee p)$$
$$\equiv \neg(\neg p \wedge (\neg q \leftrightarrow p)) \vee (q \wedge \neg p) \vee p$$
$$\equiv p \vee \neg((\neg q \rightarrow p) \wedge (p \rightarrow \neg q)) \vee (q \wedge \neg p) \vee p$$
$$\equiv p \vee \neg((q \vee p) \wedge (\neg p \vee \neg q)) \vee (q \wedge \neg p)$$
$$\equiv p \vee (\neg q \wedge \neg p) \vee (p \wedge q) \vee (q \wedge \neg p) \textbf{ (DNF)}$$
$$\equiv ((p \vee \neg q) \wedge (p \vee \neg p)) \vee ((p \vee q) \wedge (p \vee \neg p) \wedge (q \vee q) \wedge (q \vee \neg p))$$
$$\equiv (p \vee \neg q) \vee ((p \vee q) \wedge q \wedge (q \vee \neg p))$$
$$\equiv (p \vee \neg q \vee p \vee q) \wedge (p \vee \neg q \vee q) \wedge (p \vee \neg q \vee q \vee \neg p)$$
$$\equiv \neg q \vee q \textbf{ (CNF)}$$

Second method:

p	q	$(\neg p$	$\wedge$	$(\neg q \leftrightarrow p))$	$\rightarrow$	$((q \wedge \neg p) \vee p)$
F	F	T	F	T F F	T	F F T F F
F	T	T	T	F T F	T	T T T T F
T	F	F	F	T T T	T	F F F T T
T	T	F	T	F F T	T	T F F T T

DNF: $(\neg p \wedge \neg q) \vee (\neg p \wedge q) \vee (p \wedge \neg q) \vee (p \wedge q)$
CNF: $p \vee \neg p$

2.5.5 (a) First, we transform $\neg(((p \rightarrow q) \rightarrow q) \rightarrow q)$ into a CNF:

$$\neg(((p \rightarrow q) \rightarrow q) \rightarrow q)$$
$$\equiv ((p \rightarrow q) \rightarrow q) \wedge \neg q$$
$$\equiv (\neg(p \rightarrow q) \vee q) \wedge \neg q$$
$$\equiv ((p \wedge \neg q) \vee q) \wedge \neg q$$
$$\equiv (p \vee q) \wedge (\neg q \vee q) \wedge \neg q$$
$$\equiv (p \vee q) \wedge \neg q.$$

Now, transform the result to clausal form:

$$C_1 = \{p, q\} \text{ and } C_2 = \{\neg q\}.$$

Applying Resolution, we get $C_3 = Res(C_1, C_2) = \{p\}$.
No more clauses are derivable. Therefore, the empty clause cannot be derived, so the formula is *not* a tautology.

(c) Transform $\neg(((p \to q) \land (p \to \neg q)) \to \neg p)$ into a CNF:

$$\neg(((p \to q) \land (p \to \neg q)) \to \neg p)$$
$$\equiv ((p \to q) \land (p \to \neg q)) \land p$$
$$\equiv (\neg p \lor q) \land (\neg p \lor \neg q) \land p.$$

Now, transform the result to clausal form:

$$C_1 = \{\neg p, q\}, C_2 = \{\neg p, \neg q\}, \text{and } C_3 = \{p\}.$$

Finally, applying Resolution successively, we get

$$C_4 = Res(C_1, C_2) = \{\neg p\}$$
$$C_5 = Res(C_3, C_4) = \{\}.$$

The empty clause has been derived, therefore the formula is a tautology.

(e) Transform $\neg(((p \to q) \land (p \to r)) \to (p \to (q \land r)))$ into a CNF:

$$\neg(((p \to q) \land (p \to r)) \to (p \to (q \land r)))$$
$$\equiv ((p \to q) \land (p \to r)) \land \neg(p \to (q \land r))$$
$$\equiv (\neg p \lor q) \land (\neg p \lor r) \land p \land (\neg q \lor \neg r).$$

Now, transform the result to clausal form:

$$C_1 = \{\neg p, q\}, C_2 = \{\neg p, r\}, C_3 = \{p\}, \text{ and } C_4 = \{\neg q, \neg r\}.$$

Finally, applying Resolution successively, we get

$$C_5 = Res(C_2, C_4) = \{\neg p, \neg q\}$$
$$C_6 = Res(C_1, C_5) = \{\neg p\}$$
$$C_7 = Res(C_3, C_6) = \{\}.$$

The empty clause has been derived, therefore the formula is a tautology.

(g) First, transform $\neg(((p \to r) \lor (q \to r)) \to ((p \lor q) \to r))$ into a CNF:

$$\neg(((p \to r) \lor (q \to r)) \to ((p \lor q) \to r))$$
$$\equiv ((p \to r) \lor (q \to r)) \land \neg((p \lor q) \to r)$$
$$\equiv (\neg p \lor r \lor \neg q \lor r) \land (p \lor q) \land \neg r$$
$$\equiv (\neg p \lor r \lor \neg q) \land \neg r \land (p \lor q).$$

Now, transform the result to clausal form:

$$C_1 = \{\neg p, \neg q, r\}, C_2 = \{p, q\}, C_3 = \{\neg r\}.$$

Now, applying Resolution successively:
$C_4 = Res(C_1, C_2)$ (on the pair $\neg p, p$) $= \{\neg q, q, r\}$.
$C_5 = Res(C_1, C_2)$ (on the pair $\neg q, q$) $= \{\neg p, p, r\}$.
$C_6 = Res(C_1, C_3) = \{\neg p, \neg q\}$.
$C_7 = Res(C_2, C_6)$ (on the pair $\neg p, p$) $= \{\neg q, q\}$.
$C_8 = Res(C_2, C_6)$ (on the pair $\neg q, q$) $= \{\neg p, p\}$.
$C_9 = Res(C_4, C_2) = Res(C_5, C_2) = \{p, q, r\}$.

At this stage, no new clauses can be obtained by applying Resolution again and the empty clause has not been derived; the formula is therefore not a tautology.

(i) First, transform $\neg(p \to ((q \to r) \to ((p \to q) \to r)))$ into a CNF:

$$\neg(p \to ((q \to r) \to ((p \to q) \to r)))$$
$$\equiv p \wedge \neg((q \to r) \to ((p \to q) \to r))$$
$$\equiv p \wedge (q \to r) \wedge \neg((p \to q) \to r)$$
$$\equiv p \wedge (\neg q \vee r) \wedge (p \to q) \wedge \neg r$$
$$\equiv p \wedge (\neg q \vee r) \wedge (\neg p \vee q) \wedge \neg r.$$

Now, transform the result to clausal form:

$$C_1 = \{p\}, C_2 = \{\neg q, r\}, C_3 = \{\neg p, q\}, \text{ and } C_4 = \{\neg r\}.$$

Applying Resolution successively, we get

$$C_5 = Res(C_1, C_3) = \{q\}$$
$$C_6 = Res(C_2, C_4) = \{\neg q\}$$
$$C_7 = Res(C_5, C_6) = \{\}.$$

The empty clause has been derived, so the formula is a tautology.

2.5.6 (a) Transform the formulae to clausal form:

$$C_1 = \{p, q\}, C_2 = \{p, \neg q\}, \text{ and } C_3 = \{\neg p\}.$$

Applying Resolution successively, we get

$$C_4 = Res(C_1, C_2) = \{p\}$$
$$C_5 = Res(C_3, C_4) = \{\}.$$

The empty clause has been derived, so the consequence holds.

(c) Transform the formulae to clausal form:

$$C_1 = \{\neg p, r\}, C_2 = \{\neg q, r\}, C_3 = \{p, q\}, \text{and } C_4 = \{\neg r\}.$$

Applying Resolution successively, we get

$$C_5 = Res(C_2, C_3) = \{p, r\}$$
$$C_6 = Res(C_1, C_5) = \{r\}$$
$$C_7 = Res(C_4, C_6) = \{\}.$$

The empty clause has been derived, so the consequence holds.

(e) Transform the formulae to clausal form:

$$C_1 = \{\neg p, q\}, C_2 = \{p, \neg r\}, C_3 = \{\neg r\}, \text{and } C_4 = \{q\}.$$

Finally, applying Resolution, we get

$$C_5 = Res(C_1, C_2) = \{q, \neg r\}.$$

No more clauses are derivable. The empty clause cannot be derived, so the consequence does not hold.

(g) Transform the formulae to clausal form:

$$C_1 = \{\neg p, q\}, C_2 = \{r, s\}, C_3 = \{\neg q, r\}, C_4 = \{\neg r\}, \text{and } C_5 = \{p\}.$$

Applying Resolution successively, we get

$$C_6 = Res(C_1, C_5) = \{q\}$$
$$C_7 = Res(C_3, C_4) = \{\neg q\}$$
$$C_8 = Res(C_6, C_7) = \{\}.$$

The empty clause has been derived, so the consequence holds.

2.5.7 (b) Denote
"Nina will wear a red dress at the dinner" by p,
"Nina will wear high heels at the dinner" by q, and
"Nina will wear a silk scarf at the dinner" by r.
The argument is then formalized as

$$\frac{p \vee \neg r, p \rightarrow q, \neg p}{q \wedge \neg r}.$$

We now transform the set $\{p \vee \neg r, p \rightarrow q, \neg p, \neg(q \wedge \neg r)\}$ to clausal form:

$$C_1 = \{\neg p, q\}, C_2 = \{p, \neg r\}, C_3 = \{\neg p\}, C_4 = \{\neg q, r\}.$$

Now, applying Resolution:
$$C_5 = Res(C_2, C_3) = \{\neg r\}$$
$$C_6 = Res(C_4, C_5) = \{\neg q\}$$

$C_7 = Res(C_1, C_6) = \{\neg p\}$
$C_8 = Res(C_1, C_2) = \{q, \neg r\}$
$C_9 = Res(C_4, C_8)$(resolving on the pair$\{r, \neg r\}) = \{q, \neg q\}$
$C_{10} = Res(C_4, C_8)$(resolving on the pair$\{q, \neg q\}) = \{r, \neg r\}$.

At this stage, no new clauses can be obtained by applying Resolution and the empty clause has not been derived; the argument is therefore not logically correct.

(d) Denote

"The property prices increase" by p,
"The interest rates go down" by q, and
"The economy is doing well" by r.

The underlying inference rule:

$$\frac{q \to p, q \vee \neg r, \neg p}{\neg r}.$$

This inference rule is sound, so the argument is correct.

(f) Denote

p: Alice comes/will come to the party.
q: Bonnie comes/will come to the party.
r: Clyde comes/will come to the party.

The respective inference rule is:

$$\frac{q \to \neg p, \quad r \to (p \wedge \neg q)}{(q \vee \neg r) \to \neg p}.$$

Transformation of the premises and the negated conclusion to clausal form:

$q \to \neg p \equiv \neg q \vee \neg p$
$C_1 := \{\neg p, \neg q\}$

$r \to (p \wedge \neg q) \equiv \neg r \vee (p \wedge \neg q) \equiv (\neg r \vee p) \wedge (\neg r \vee \neg q)$
$C_2 := \{p, \neg r\}; C_3 := \{\neg q, \neg r\}$

$\neg((q \vee \neg r) \to \neg p) \equiv (q \vee \neg r) \wedge p$
$C_4 := \{q, \neg r\}; C_5 := \{p\}$.

Applying Resolution successively:

$C_6 := Res(C_1, C_5) = \{\neg q\}$
$C_7 := Res(C_4, C_6) = \{\neg r\}$.

No new clauses can be derived anymore. The empty clause is therefore not derivable and the inference rule is not logically valid.

(h) Notation:

W: Hans does his job well.
C: Hans is clever.
P: Hans will be promoted.
F: Hans will be fired.

The argument in symbolic form is:

$$\frac{W \wedge C \to P, \quad \neg W \to F}{C \to P \vee F}.$$

Transforming the premises and the negated conclusion to clausal form:

$W \wedge C \rightarrow P \equiv \neg W \vee \neg C \neg F. C_1 = \{\neg W, \neg C, F\}$

$\neg W \rightarrow F \equiv W \vee F\ C_2 = \{W, F\};$

$\neg(C \rightarrow P \vee F) \equiv C \wedge \neg P \wedge \neg F\ C_3 := \{C\}, C_4 := \{\neg P\}, C_5 := \{\neg F\}.$

Applying Resolution successively:

$C_6 = Res(C_2, C_5) = \{W\}$

$C_7 = Res(C_1, C_6) = \{\neg C, P\}$

$C_8 = Res(C_3, C_7) = \{P\}$

$C_9 = (C_4, C_8) = \{\}.$

The empty clause has been derived; the inference rule is therefore sound and the argument is logically correct.

Section 3.1

3.1.1 We can view any graph as the structure $\mathcal{G} = (V, R)$ in the language with one binary relational symbol $\{E\}$. The domain of $\mathcal{G}$ is the set of vertices V, and E is interpreted as the edge relation R.

We can view any group as the structure for the language $\mathcal{L}_G$ with non-logical symbols $\{\cdot, ', e\}$, where $\cdot$ is a binary functional symbol (multiplication), $'$ is a unary functional symbol (the inverse operation), and e is a constant symbol. The domain of $\mathcal{G}$ is the set of group elements and the interpretations of $\cdot, ', e$ satisfy the defining axioms of a group.

We can likewise view any ring as a structure for the language $\mathcal{L}_R$ with non-logical symbols $\{\cdot, +, 0\}$, where $\cdot$ and $+$ are both binary functional symbols (multiplication and addition), and 0 is a constant symbol, such that their interpretations in the set of ring elements satisfy the defining axioms of a ring.

A vector space over a field of scalars S can be formalized as a first-order structure for the language $\mathcal{L}_V$ with non-logical symbols $\{+, \mathbf{0}\} \cup \{c_s \mid s \in S\}$, where $+$ is a binary functional symbol (vector addition), $\mathbf{0}$ is a constant symbol, and each c_s, where $s \in S$ is a unary functional symbol (multiplication by the scalar s). The domain of a vector space is a set of vectors V and the interpretations of $+, \{c_s \mid s \in S\}, \mathbf{0}$ in V satisfy the defining axioms of a vector space.

3.1.3 (a) Yes; $x \times y$.

(b) No: $\times$ is a binary functional symbol.

(c) No: parentheses and comma are missing.

(d) No: one right parenthesis is missing.

(e) Yes; $((x \times (\mathbf{3} + y)) + ((\mathbf{3} + y) \times x))$.

(f) No, mismatch of arity: $\times$ is a binary functional symbol, while s is a unary one.

(g) No: s is a name of a unary functional symbol, not a constant or variable.

(h) Yes; $s((\mathbf{2} \times x) \times (\mathbf{3} + y))$.

(i) Yes; $(s(\mathbf{2} \times x) \times s(s(\mathbf{3}) + y))$.

(j) No, mismatch of arity: s is a unary functional symbol.

3.1.4 (a) Term: $t = \times(x, y)$

      Subterms: $\{t, x, y\}$

Parsing tree:

(e)

Term:

$t = +(\times(x, +(\mathbf{3}, y)), \times(+(\mathbf{3}, y), x))$

Subterms:

$\{x,$

$y,$

$\mathbf{0},$

$\mathbf{1},$

$\mathbf{2},$

$\mathbf{3},$

$+(\mathbf{3},y),$

$\times(x, +(\mathbf{3},y)),$

$\times(+(\mathbf{3},y), x),$

$+(\times(x, +(\mathbf{3},y)), \times(+(\mathbf{3},y), x))\}.$

Parsing tree:

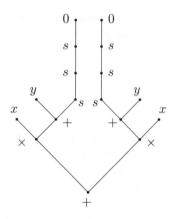

(h)

Term: $s(\times(\times(\mathbf{2},x), +(\mathbf{3},y)))$

Subterms:

$\{x,$

$y,$

$\mathbf{0},$

$\mathbf{1},$

$\mathbf{2},$

$\mathbf{3},$

$\times(\mathbf{2},x),$

$+(\mathbf{3},y),$

$\times(\times(\mathbf{2},x),$

$+(\mathbf{3},y)),$

$s(\times(\times(\mathbf{2},x), +(\mathbf{3},y)))\}.$

Parsing tree:

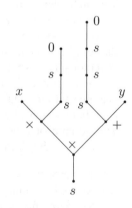

3.1.6 (b) No, a quantifier must be followed immediately by a variable x.

      (d) No, $=$ is an infix predicate symbol, so it may not be preceded by $\neg$.

      (f) Yes.

      (h) No, x is a variable, not a formula.

      (j) Yes.

      (l) Yes (though formally $x < \mathbf{0}$ should be in parentheses).

      (n) No, there is no connective between $y < y$ and $\neg \forall x(x < x)$.

3.1.7 (a)   Parsing tree:

Formula: $A = \neg\forall x(x = \mathbf{0})$

The set of subformulae:

$\{A,$

$\forall x(x = \mathbf{0}),$

$x = \mathbf{0}\}$

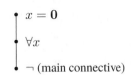

(f)   The parsing tree:

Formula: $A = \neg\forall x(\neg x = \mathbf{0})$.

The main connective is $\neg$.

The set of subformulae:

$\{x = \mathbf{0},$

$\neg(x = \mathbf{0}),$

$\forall x(\neg x = \mathbf{0}),$

$\neg\forall x(\neg x = \mathbf{0})\}$

(j)   Parsing tree:

Formula:

$A = \forall x((x < \mathbf{0}) \to \forall x(x > \mathbf{0}))$

Subformulae:

$\{A,$

$(x < \mathbf{0}) \to \forall x(x > \mathbf{0}),$

$\forall x(x > \mathbf{0}),$

$x < \mathbf{0},$

$x > \mathbf{0}\}$

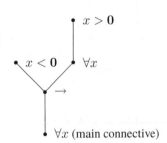

(m)   Parsing tree:

Formula:

$A = \forall y(x < x \to \forall x(x < x))$

Subformulae:

$\{A,$

$x < x \to \forall x(x < x),$

$\forall x(x < x), x < x\}$

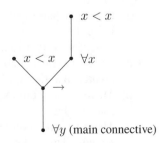

3.1.10 (a) $A[\neg(f(x) = y)/p, \exists x(x > y)/q)] =$
$(\neg\neg(f(x) = y) \rightarrow (\exists x(x > y) \vee \neg(\neg(f(x) = y) \wedge \exists x(x > y))))$.
(b) $A[\neg(x = y)/p, \neg(x = y)/q)] =$
$(\neg\neg(x = y) \rightarrow (\neg(x = y) \vee \neg(\neg(x = y) \wedge \neg(x = y))))$.

3.1.11 No, because $(x = y)$ is substituted for the first occurrence of p and $(y = x)$ is substituted for the second occurrence of p. Note that $(x = y)$ and $(y = x)$ are *different* formulae.

Section 3.2

3.2.1 (a) $10 = 12$: false.
(b) $4 > 4$: false.
(c) The formula is $x^3 + 1 = (x + 1)(x^2 - x + 1)$.
This is an algebraic identity, so it is true for every value of x.

3.2.4 Some answers and hints:
(a) $z > y \rightarrow x > y$: false
(c) $\exists y(z > y \rightarrow x > y)$: true, take $y = z$
(e) $\forall y(z > y \rightarrow x > y)$: false, take $y = x$
(g) $\exists x\forall z(z > y \rightarrow x > y)$: true, take $x = y + 1$
(i) $\exists z\forall x(z > y \rightarrow x > y)$: true, take $z = y$
(k) $\exists y\forall z(z > y \rightarrow x > y)$. yes, take $y = 0$
(m) $\forall x\exists z(z > y \rightarrow x > y)$: true, take $z = y$
(o) $\forall y\exists z(z > y \rightarrow x > y)$: true, take $z = y$
(q) $\forall z\exists y(z > y \rightarrow x > y)$: true, take $y = z$
(s) $\exists y\forall x\forall z(z > y \rightarrow x > y)$: false
(u) $\forall y\exists x\forall z(z > y \rightarrow x > y)$: true
(w) $\forall z\exists x\forall y(z > y \rightarrow x > y)$: true
(y) $\forall y\forall z\exists x(z > y \rightarrow x > y)$: true, take $x = y + 1$.

3.2.5 (a) Not every real number is different from 0. True.
(c) False, take $x = 0$.
(e) There exists a real number which, if that number equals its square, then it is negative. True: take any x such that $x \neq x^2$. Then the antecedent is false so the implication is true.
(g) True.
(i) For every pair of real numbers x and y, one of them is less than the other. False: take $x = y$.
(k) True, for every x take for example $y = 1$.
(m) For every real x there is a real y such that if x is greater than y then it is also greater than the square of y. True: given x take $y = x$.
(p) There is a real number x which can be added to any real number y to obtain x again. False.
(q) There is a real number x which can be added to any real number y to obtain y again. True: take $x = 0$.

(r) There is a real number x such that, for every real number y, x is greater than y or $-x$ is greater than y. False.

(t) There is a real x such that every real y greater than x will have a square also greater than x. True: take for example $x = 1$.

(w) For any real number x, there is some real number y so that xy will equal yz for all real numbers z. True: given x, take $y = 0$.

(y) Between any two distinct real numbers there lies another real number. True: density of $\mathbf{R}$.

3.2.6 (b) No natural number is greater than 6 or not greater than 5.
            False: for example, 0 falsifies the claim.

(d) Every natural number that is not greater than 4 is not greater than 3.
     False: take $x = 4$.

(f) Every natural number that is not 1 or 31 is not a divisor of 31.
     True: 31 is a prime number.

(h) For every natural number x there is a greater number which is less than or equal to any natural number greater than x.
     True: every natural number has an immediate successor.

(k) For every natural number x there is a natural number z that is greater than x and less than every natural number greater than x.
     False: take any x. Then whatever z such that $x < z$ is chosen, take $y = z$.

(l) If for every natural number x, the property P holds for x whenever it holds for all natural numbers less than x, then P holds for every natural number.
     True: this is the principle of (complete) mathematical induction.

(n) Every non-empty set of natural numbers (the interpretation of the predicate P) has a least element.
     True. Suppose that for some interpretation of P in some set $X \subseteq \mathbb{N}$ it is the case that X has no least element. It can then be proven by induction that no natural number belongs to X, and therefore X must be empty.
     In fact, the property above is equivalent to the principle of mathematical induction.

3.2.7 (a) "Every father knows all of his children." False.

(c) "There is no woman who does not know any of her children."
     False: there are women who have no children. They (vacuously) do not know any of their children.

(e) "Every man is the father of every child of his." True.

Section 3.3

3.3.1 (b) $\forall x(\neg\mathsf{I}(x) \to x^2 > 0)$

(d) $\forall x(\mathsf{I}(x) \to \exists y(\mathsf{I}(y) \land (x = y + y \lor x = y + y + 1)))$

(f) $\forall x(\mathsf{I}(x) \to \exists y(\mathsf{I}(y) \land x < y))$

(h) $\neg\forall x\exists y(\mathsf{I}(y) \wedge x > y)$

(j) $\neg\exists x\forall y(\mathsf{I}(y) \rightarrow x > y)$

(l) $\exists x\forall y(xy = y)$

3.3.2 (a) $\forall x\mathsf{T}(x) \vee \forall x\mathsf{L}(x)$

(c) $T(\mathrm{John}) \rightarrow \forall x\mathsf{L}(x)$

(e) $\forall x(\mathsf{T}(x) \rightarrow \forall y(\neg y = x \rightarrow \mathsf{L}(y)))$

3.3.3 (a) $\neg\forall x\exists y\mathsf{T}_2(x, y)$

(c) $\forall x(\exists y\mathsf{T}_2(x, y) \rightarrow \mathsf{L}_2(\mathsf{John}, x))$

(e) $\forall x(\neg\mathsf{L}_2(x, \mathsf{John}) \rightarrow \mathsf{T}_2(\mathsf{John}, x))$

(g) $\forall x\forall y(\neg\mathsf{L}_2(y, x) \rightarrow \neg\mathsf{L}_2(x, y))$

(i) $\forall x\exists y\mathsf{T}_2(x, y) \rightarrow \neg\exists x\exists y\mathsf{L}_2(x, y)$

3.3.4 (a) $\exists x(\mathsf{M}(x) \wedge \forall y(\mathsf{W}(y) \rightarrow \mathsf{L}(x, y)))$

(c) $\forall x(\mathsf{M}(x) \rightarrow \exists y(\mathsf{W}(y) \wedge \mathsf{L}(x, y) \wedge \neg\mathsf{L}(y, x)))$

(e) $\exists x(\mathsf{M}(x) \wedge \forall y(\mathsf{W}(y) \rightarrow (\mathsf{L}(x, y) \rightarrow \neg\mathsf{L}(y, x))))$
$\equiv \exists x(\mathsf{M}(x) \wedge \forall y((\mathsf{W}(y) \wedge \mathsf{L}(x, y)) \rightarrow \neg\mathsf{L}(y, x)))$.

(g) $\forall x(\mathsf{W}(x) \rightarrow \forall y(\mathsf{C}(y, x) \rightarrow \mathsf{L}(x, y)))$

(i) Two possible readings:
"Everyone loves any child": $\forall x\forall y(\exists z\mathsf{C}(y, z) \rightarrow \mathsf{L}(x, y))$
or
"Everyone loves some child": $\forall x\exists y(\exists z\mathsf{C}(y, z) \wedge \mathsf{L}(x, y))$

(k) $\forall x(\exists y\mathsf{C}(x, y) \rightarrow \exists z\mathsf{L}(z, x))$

(m) $\forall x(\exists y\mathsf{C}(x, y) \rightarrow (L(x, \mathsf{m}(x)) \vee L(x, \mathsf{f}(x))))$

(o) $\neg\exists x(\exists yC(x, y) \wedge \exists z(\exists u(z = m(u) \wedge \neg L(z, u)) \wedge L(x, z)))$

3.3.5 (a) $\forall x(\exists yC(y, x) \rightarrow \forall z(\exists u(m(u) = z) \rightarrow R(z, x)))$

(c) $\exists x\forall y(C(x, y) \rightarrow \neg K(x, y))$

(e) $\exists x\forall y(C(x, y) \rightarrow \neg R(x, y))$ $\quad$ or $\quad$ $\exists x(\exists yC(x, y) \wedge \forall y(C(x, y) \rightarrow \neg R(x, y)))$

(g) $\neg\exists x(W(x) \wedge \exists y(M(y) \wedge L(x, y) \wedge \exists z(C(z, x) \wedge \neg K(y, z))))$

(i) $\neg\exists x(M(x) \wedge \exists y(W(y) \wedge R(x, y) \wedge \exists z(C(z, y) \wedge \neg K(y, z))))$

(k) $\forall x(\exists yC(x, y) \rightarrow \forall z(\exists u(m(u) = z) \rightarrow L(z, x)))$

(l) Either $\neg\exists x(K(x, \mathsf{Eve}) \wedge \neg x = \mathsf{Adam})$,
which is equivalent to $\forall x(K(x, \mathsf{Eve}) \rightarrow x = \mathsf{Adam})$,
or $K(\mathsf{Adam}, \mathsf{Eve}) \wedge \forall x(K(x, \mathsf{Eve}) \rightarrow x = \mathsf{Adam})$

(m) $\forall x(\forall y((K(x, y) \wedge W(y)) \rightarrow L(x, y)) \rightarrow x = \mathsf{Adam})$

(o) $L(\mathsf{m}(\mathsf{John}), \mathsf{John}) \wedge \forall x(L(x, \mathsf{John}) \rightarrow x = \mathsf{m}(\mathsf{John}))$

(p) $L(\mathsf{m}(\mathsf{John}), \mathsf{John}) \wedge \exists x(L(x, \mathsf{John}) \wedge \neg x = \mathsf{m}(\mathsf{John}))$

3.3.6 (a) $\forall x((Q(x) \wedge f(x) = 0) \rightarrow 0 < x)$

(c) $\neg\exists x(x < 0 \wedge \forall y(f(y) = 0 \rightarrow y < x))$

(e) This sentence is possibly ambiguous. There are two non-equivalent translations:
$\exists x(x < 0 \wedge f(x) = 0 \wedge \neg\forall y((f(y) = 0 \wedge \neg Q(y)) \rightarrow x < y))$ and
$\exists x(x < 0 \wedge f(x) = 0 \wedge \forall y((f(y) = 0 \wedge \neg Q(y)) \rightarrow \neg x < y))$.

3.3.7   (a) $x|y := \neg x = 0 \wedge \exists z(x \times z = y)$
   (b) $P(x) := 1 < x \wedge \forall y \forall z(y \times z = x \rightarrow (y = 1 \wedge z = x) \vee (y = x \wedge z = 1))$
   (d) $\forall x(2 < x \wedge \exists y(2 \times y = x) \rightarrow \exists z \exists v(P(z) \wedge P(v) \wedge z + v = x))$
   (f) $\forall x \exists y(P(y) \wedge \neg(x < y \times y) \wedge \exists u(u \times y = x))$
   (h) $z|x \wedge z|y \wedge \forall u((u|x \wedge u|y) \rightarrow u \le z)$

3.3.10   (a) The only free occurrence of a variable is the first occurrence of y.
   The scopes of the respective occurrences of quantifiers:
   $\exists x \forall z(Q(z, y) \vee \neg \forall y(Q(y, z) \rightarrow P(x)))$,
   $\forall z(Q(z, y) \vee \neg \forall y(Q(y, z) \rightarrow P(x)))$,
   $\forall y(Q(y, z) \rightarrow P(x))$.
   (d) The free occurrence of variables are: the first occurrence of y, the last occurrence of z, and the last occurrence of x.
   The scopes of the respective occurrences of quantifiers:
   $\exists x(\forall z Q(z, y) \vee \neg \forall y Q(y, z))$, $\forall z Q(z, y)$,
   $\forall y Q(y, z)$.
   (g) The free occurrence of variables are: all occurrences of y.
   The scopes of the respective occurrences of quantifiers:
   $\exists x(\forall x Q(x, y) \vee \neg \forall z(Q(y, z) \rightarrow P(x)))$,
   $\forall x Q(x, y)$,
   $\forall z(Q(y, z) \rightarrow P(x))$.

3.3.11   (a) $\exists x \forall z(Q(z, y) \vee \neg \forall w(Q(w, z) \rightarrow P(x)))$
   (c) $\exists x(\forall z Q(z, y) \vee \neg \forall w(Q(y, w) \rightarrow P(x)))$
   (e) $\exists u \forall z(Q(z, y) \vee \neg \forall w Q(u, z)) \rightarrow P(x)$
   (g) $\exists x(\forall z Q(z, y) \vee \neg(\forall w Q(y, w) \rightarrow P(x)))$
   (i) $\exists u(\forall y(Q(u, y) \vee \neg \forall v Q(v, z))) \rightarrow P(x)$

3.3.13   (a) No.
   (c) Yes; $\exists x(\forall z P(f(y)) \vee \neg \forall y(Q(y, z) \rightarrow P(x)))$.
   (f) Yes; $\forall y(\neg(\forall x \exists z(\neg P(z) \wedge \exists y Q(z, x))) \wedge (\neg \forall x Q(x, y) \vee P(f(x))))$.
   (g) No.
   (h) Yes; $(\forall y \exists z \neg P(z) \wedge \forall x Q(g(f(z), y), x)) \rightarrow (\neg \exists y Q(x, y) \vee P(g(f(z), y)))$.
   (i) No.

Section 3.4

3.4.1   (a) $\forall x(P(x) \vee \neg P(x))$: **Yes**.
   (c) $\forall x P(x) \vee \neg \forall x P(x)$: **Yes**.
   (e) $\exists x P(x) \rightarrow \forall x P(x)$: **No**.
   (g) $\exists x(P(x) \rightarrow \forall y P(y))$: **Yes**.
   Here is a short intuitive argument: either every element of the domain satisfies P or not. In the latter case, the implication $P(x) \rightarrow \forall y P(y)$ can be made true for a suitable x which falsifies the antecedent. In the former case, the implication $P(x) \rightarrow \forall y P(y)$ is true for any x because the consequent is true. Hence, the validity.

Now, to prove this validity by formal semantic argument, take any structure S for the first-order language introduced above and any variable assignment v in S. (We know that the variable assignment is irrelevant here because the formula is a sentence, but we need it in order to process the quantifiers.)
We have to show that:
$S, v \models \exists x (P(x) \to \forall y P(y))$.
Consider two cases:
(i) There is a variable assignment v' obtained from v by redefining it on x such that $P(x)$ is false. Then:
(1) $S, v' \nvDash P(x)$,
but by the definition of $\to$ it follows that:
(2) $S, v' \models P(x) \to \forall y P(y)$.
According to the truth definition of $\exists$, it follows from (2) that
(3) $S, v \models \exists x (P(x) \to \forall y P(y))$.

(ii) For every variable assignment v:
(7) $v, S, v \models P(x)$.
Then, clearly also:
(8) $S, v \models P(y)$ for every variable assignment v.
According to the truth definition of $\forall$, it follows from (8) that
(9) $S, v \models \forall y P(y)$,
but by the definition of $\to$ it follows from (7) and (9) that:
(10) $S, v \models P(x) \to \forall y P(y)$.
From (10) it follows from the truth definition of $\exists$ that
(11) $S, v \models \exists x (P(x) \to \forall y P(y))$.
Thus, in either case $S, v \models \exists x (P(x) \to \forall y P(y))$, which concludes the proof.

(i) **Yes.**

(k) **Yes.**

(m) **No.**

(o) **Yes.**

(q) **No.** Consider for example the set $\mathbb{N}$ where $P(x, y)$ is interpreted as $|x - y| = 1$.

(r) **No.** Consider for example the set $\mathbb{N}$ where $P(x, y)$ is interpreted as $x < y$.

3.4.6 (a) Suppose $\forall x A(x) \nvDash \neg \exists x \neg A(x)$. Then there is a structure S and an assignment v such that $S, v \models \forall x A(x)$ but $S, v \nvDash \neg \exists x \neg A(x)$. $S, v \models \forall x A(x)$ implies $S, v[x := s] \models A(x)$ for all $s \in S$. From $S, v \nvDash \neg \exists x \neg A(x)$, we have $S, v \models \exists x \neg A(x)$, which means that there is an $s' \in S$ such that $S, v[x := s'] \models \neg A(x)$. Hence, $S, v[x := s'] \nvDash A(x)$, contradicting the fact that $S, v[x := s] \models A(x)$ for all $s \in S$.

(b) Likewise.

(c) Suppose $\exists x A(x) \nvDash \neg \forall x \neg A(x)$. Then there is a structure S and an assignment v such that $S, v \models \exists x A(x)$ but $S, v \nvDash \neg \forall x \neg A(x)$. From $S, v \models \exists x A(x)$ it follows that there is $s' \in S$ such that $S, v[x := s'] \models A(x)$. Besides, $S, v \nvDash \neg \forall x \neg A(x)$ implies that $S, v \models \forall x \neg A(x)$. Hence, $S, v[x := s] \models \neg A(x)$ for all $s \in S$, and so $S, v[x := s] \nvDash A(x)$ for all $s \in S$, contradicting the fact that $S, v[x := s'] \models A(x)$.

(d) Likewise.

(e) Suppose $\exists x \exists y A(x, y) \nVDash \exists y \exists x A(x, y)$. Then there is a structure $\mathcal{S}$ and an assignment v such that $\mathcal{S}, v \models \exists x \exists y A(x, y)$ but $\mathcal{S}, v \nVDash \exists y \exists x A(x, y)$. $\mathcal{S}, v \models \exists x \exists y A(x, y)$ implies that there are $s', s'' \in \mathcal{S}$ such that $\mathcal{S}, v[x := s'][y := s''] \models A(x, y)$. From $\mathcal{S}, v \nVDash \exists y \exists x A(x, y)$, we have $\mathcal{S}, v[x := s][y := t] \nVDash A(x, y)$ for all $s, t \in \mathcal{S}$. This means that $\mathcal{S}, v[x := s'][y := s''] \nVDash A(x, y)$, which is a contradiction.

3.4.7   Hint: note the equivalences in Exercise 10 and also use the fact that if x does not occur free in B, then $B \equiv \forall x B \equiv \exists x B$. Here we prove the case 7f.

(f) $\exists x A(x) \rightarrow B \models \forall x (A(x) \rightarrow B)$

To prove this logical consequence, consider any structure $\mathcal{S}$ and a variable assignment v, such that

(1) $\mathcal{S}, v \models \exists x A(x) \rightarrow B$.

Now, suppose that

(2) $\mathcal{S}, v \nVDash \forall x (A(x) \rightarrow B)$.

By the truth definition of $\forall$, there is a variable assignment v' which possibly only differs from v in the value for x, such that

(2) $\mathcal{S}, v' \nVDash A(x) \rightarrow B$.

Then, by the truth definition of $\rightarrow$ it follows that

(3) $\mathcal{S}, v' \models A(x)$, and

(4) $\mathcal{S}, v' \nVDash B$.

Since x does not occur free in B, it then follows that

(4) $\mathcal{S}, v \nVDash B$.

On the other hand, by (3) and the truth definition of $\exists$, it follows that

(5) $\mathcal{S}, v \models \exists x A(x)$.

From (4) and (5) it follows that

$\mathcal{S}, v \nVDash \exists x A(x) \rightarrow B$,

which contradicts (1).

Therefore, the assumption (2) must be wrong, hence

$\mathcal{S}, v \models \forall x (A(x) \rightarrow B)$.

This proves the logical consequence $\exists x A(x) \rightarrow B \models \forall x (A(x) \rightarrow B)$.

3.4.8   (a) **Yes.** For the sake of a contradiction, suppose $\forall x A(x), \forall x B(x) \nVDash \forall x (A(x) \wedge B(x))$. Then there exist a structure $\mathcal{S}$ and an assignment v in $\mathcal{S}$ such that $\mathcal{S}, v \models \forall x A(x)$ and $\mathcal{S}, v \models \forall x B(x)$ but $\mathcal{S}, v \nVDash \forall x (A(x) \wedge B(x))$. We therefore have that $\mathcal{S}, v[x := s] \models A(x)$ and $\mathcal{S}, v[x := s] \models B(x)$ for all $s \in \mathcal{S}$, while $\mathcal{S}, v[x := c] \nVDash A(x) \wedge B(x)$ for some $c \in \mathcal{S}$. Hence, $\mathcal{S}, v[x := c] \nVDash A(x)$ or $\mathcal{S}, v[x := c] \nVDash B(x)$ for some $c \in \mathcal{S}$, which contradicts the fact that $A(x)$ and $B(x)$ are true for all s in $\mathcal{S}$.

(c) **Yes.** For the sake of a contradiction, suppose $\forall x A(x) \vee \forall x B(x) \nVDash \forall x (A(x) \vee B(x))$. Then there exist a structure $\mathcal{S}$ and an assignment v in $\mathcal{S}$ such that $\mathcal{S}, v \models \forall x A(x) \vee \forall x B(x)$ but $\mathcal{S}, v \nVDash \forall x (A(x) \vee B(x))$. We therefore have that $\mathcal{S}, v[x := s] \models A(x)$ or $\mathcal{S}, v[x := s] \models B(x)$ for all $s \in \mathcal{S}$, while $\mathcal{S}, v[x := c] \nVDash A(x)$ and $\mathcal{S}, v[x := c] \nVDash B(x)$ for some $c \in \mathcal{S}$, contradicting the fact that $A(x)$ or $B(x)$ is true for all s in $\mathcal{S}$.

(e) **No.** Consider for example the structure $\mathcal{H}$ with A and B taken respectively as the predicates M and W.

(g) **Yes**. For the sake of a contradiction, suppose $\exists x(A(x) \wedge B(x))$ $\not\models \exists x A(x) \wedge \exists x B(x)$. Then there exist a structure $\mathcal{S}$ and an assignment v in $\mathcal{S}$ such that $\mathcal{S}, v \models \exists x(A(x) \wedge B(x))$, while $\mathcal{S}, v \not\models \exists x A(x) \wedge \exists x B(x)$. Hence, $\mathcal{S}, v \not\models \exists x A(x)$ or $\mathcal{S}, v \not\models \exists x B(x)$. This means $\mathcal{S}, v[x := s] \not\models A(x)$ for all s or $\mathcal{S}, v[x := s] \not\models B(x)$ for all s, while $\mathcal{S}, v[x := c] \models A(x)$ and $\mathcal{S}, v[x := c] \models B(x)$ for some $c \in \mathcal{S}$, which is a contradiction.

(i) **Yes**. For the sake of a contradiction, suppose $\exists x(A(x) \vee B(x))$ $\not\models \exists x A(x) \vee \exists x B(x)$. Then there exist a structure $\mathcal{S}$ and an assignment v in $\mathcal{S}$ such that $\mathcal{S}, v \models \exists x(A(x) \vee B(x))$, while $\mathcal{S}, v \not\models \exists x A(x) \vee \exists x B(x)$. Hence, $\mathcal{S}, v \not\models \exists x A(x)$ and $\mathcal{S}, v \not\models \exists x B(x)$. This means that there is some $c \in \mathcal{S}$ such that $\mathcal{S}, v[x := c] \models A(x)$ or $\mathcal{S}, v[x := c] \models B(x)$, while $\mathcal{S}, v[x := s] \not\models A(x)$ and $\mathcal{S}, v[x := s] \not\models B(x)$ for all $s \in \mathcal{S}$, which is a contradiction.

(k) **No**. We will show that $\exists x(A(x) \to B(x)) \not\models \exists x A(x) \to \exists x B(x)$. It is sufficient to find any counter-model, that is, a structure $\mathcal{S}$ and a variable assignment v such that:
$\mathcal{S}, v \models \exists x(A(x) \to B(x))$ and $\mathcal{S}, v \not\models \exists x A(x) \to \exists x B(x)$.
Take for example the structure $\mathcal{S}$ with domain the set $\mathbf{Z}$ of all integers, with an interpretation of the unary predicate A to be $\{0\}$ and interpretation of the unary predicate B to be $\emptyset$. (We know that the variable assignment is irrelevant here because all formulae involved in the logical consequence are sentences, but we will nevertheless need it in order to process the truth definitions of the quantifiers.)
Now let v_1 be a variable assignment obtained from v by redefining it on x as follows: $v'(x) := 1$. Then:
(1) $\mathcal{S}, v_1 \not\models A(x)$.
(2) $\mathcal{S}, v_1 \not\models B(x)$.
According to the truth definition of $\to$ it follows from (1) and (2) that:
(3) $\mathcal{S}, v_1 \models A(x) \to B(x)$.
According to the truth definition of $\exists$ it therefore follows from (3) that:
(4) $\mathcal{S}, v \models \exists x(A(x) \to B(x))$.
Now let v_2 be a variable assignment obtained from v by redefining it on x as follows: $v_2(x) := 0$. Then:
(5) $\mathcal{S}, v_2 \models A(x)$.
From (5) it follows according to the truth definition of $\exists$ that:
(6) $\mathcal{S}, v \models \exists x A(x)$.
It follows directly from the definition of B and the truth definition of $\exists$ that:
(7) $\mathcal{S}, v \not\models \exists x B(x)$.
According to the truth definition of $\to$ it follows that:
(8) $\mathcal{S}, v \not\models \exists x A(x) \to \exists x B(x)$.
$\mathcal{S}$ is therefore a counter-model, falsifying the logical consequence $\exists x(A(x) \to B(x)) \models \exists x A(x) \to \exists x B(x)$.

3.4.12 (a) We show that each side logically implies the other.
First, to show $\forall x A(x) \models \neg \exists x \neg A(x)$, suppose the contrary. Then there is a structure $\mathcal{S}$ and an assignment v such that $\mathcal{S}, v \models \forall x A(x)$ but $\mathcal{S}, v \not\models$

$\neg \exists x \neg A(x)$. $S, v \models \forall x A(x)$ implies $S, v[x := s] \models A(x)$ for all $s \in S$. From $S, v \not\models \neg\exists x\neg A(x)$, we have $S, v \models \exists x\neg A(x)$, which means that there is an $s' \in S$ such that $S, v[x := s'] \models \neg A(x)$. Hence, $S, v[x := s'] \not\models A(x)$, contradicting the fact that $S, v[x := s] \models A(x)$ for all $s \in S$.
Likewise, $\neg\exists x\neg A(x) \models \forall x A(x)$.

(c) Again, we show that each side logically implies the other.
Suppose $\exists x A(x) \not\models \neg\forall x\neg A(x)$. Then there is a structure S and an assignment v such that $S, v \models \exists x A(x)$ but $S, v \not\models \neg\forall x\neg A(x)$. From $S, v \models \exists x A(x)$ it follows that there is $s' \in S$ such that $S, v[x := s'] \models A(x)$. Besides, $S, v \not\models \neg\forall x\neg A(x)$ implies that $S, v \models \forall x\neg A(x)$. Hence, $S, v[x := s] \models \neg A(x)$ for all $s \in S$, and so $S, v[x := s] \not\models A(x)$ for all $s \in S$, contradicting the fact that $S, v[x := s'] \models A(x)$.
Likewise, $\neg\forall x\neg A(x) \models \exists x A(x)$.

(h) Again, we show that each side logically implies the other.
Suppose first that $\exists x\exists y A(x, y) \not\models \exists y\exists x A(x, y)$. Then there is a structure S and an assignment v such that $S, v \models \exists x\exists y A(x, y)$ but $S, v \not\models \exists y\exists x A(x, y)$. $S, v \models \exists x\exists y A(x, y)$ implies that there are $s', s'' \in S$ such that $S, v[x := s'][y := s''] \models A(x, y)$. From $S, v \not\models \exists y\exists x A(x, y)$, we have $S, v[x := s][y := t] \not\models A(x, y)$ for all $s, t \in S$. This means that $S, v[x := s'][y := s''] \not\models A(x, y)$, which is a contradiction.
Likewise, $\exists y\exists x A(x, y) \models \exists x\exists y A(x, y)$.

3.4.13 All equivalences follow from the fact that if x does not occur free in Q, then $S, v \models Q$ iff $S, v' \models Q$ for every x-variant v' of v iff $S, v \models \forall x Q$.

3.4.14 (a) No. Consider for example the structure $\mathcal{H}$ with P and Q interpreted as the predicates M and W, respectively.

(c) Yes.

(e) No. Consider for example the structure $\mathcal{N}$ with Q interpreted as the predicate $<$.

(g) No. Consider for example the structure $\mathcal{Z}$ with $Q(x, y)$ interpreted as "$x \leq y$."

(i) Yes. The two formulae are essentially contrapositives to each other.

3.4.15 (b) $\neg\forall x((x = x^2 \land x > 1) \to x^2 < 1)$

$\equiv \neg\forall x(\neg(x = x^2 \land x > 1) \lor x^2 < 1)$

$\equiv \exists x(x = x^2 \land x > 1 \land \neg x^2 < 1)$.

(d) $\neg\forall x(x = 0 \lor \exists y\neg(xy = x))$

$\equiv \exists x(\neg x = 0 \land \neg\exists y\neg(xy = x))$

$\equiv \exists x(\neg x = 0 \land \forall y(xy = x))$.

(f) $\neg\exists x\exists y(x > y \lor -x > y)$

$\equiv \forall x\forall y(\neg x > y \land \neg(-x > y))$.

(h) $\neg(\forall x(P(x) \to Q(x)) \to (\forall x P(x) \to \forall x Q(x)))$

$\equiv \forall x(P(x) \to Q(x)) \land \neg(\neg\forall x P(x) \lor \forall x Q(x))$

$\equiv \forall x(P(x) \to Q(x)) \land (\forall x P(x) \land \exists x \neg Q(x)).$

3.4.16 Using predicates $L(x)$ for "x is a lawyer" and $G(x)$ for "x is greedy," we can formalize A and B as follows:

$$A = \neg\forall x(L(x) \to G(x)),$$

$$B = \exists x L(x) \land \neg\forall x G(x).$$

First, we show that $A \models B$. Suppose $\neg\forall x(L(x) \to G(x)) \not\models \exists x L(x) \land \neg\forall x G(x)$. Then there is a structure $\mathcal{S}$ and an assignment v on $\mathcal{S}$ such that $\mathcal{S}, v \models \neg\forall x(L(x) \to G(x))$, while $\mathcal{S}, v \not\models \exists x L(x) \land \neg\forall x G(x)$. But then $\mathcal{S}, v \not\models \forall x(L(x) \to G(x))$, while $\mathcal{S}, v \not\models \exists x L(x)$ or $\mathcal{S}, v \models \forall x G(x)$. $\mathcal{S}, v \not\models \forall x(L(x) \to G(x))$ implies there is some c in $\mathcal{S}$ such that $\mathcal{S}, v[x := c] \not\models L(x) \to G(x)$. This means that $\mathcal{S}, v[x := c] \models L(x)$ but $\mathcal{S}, v[x := c] \not\models G(x)$. Hence, $\mathcal{S}, v \models \exists x L(x)$ and $\mathcal{S}, v \not\models \forall x G(x)$, which is a contradiction.

On the other hand, $B \not\models A$. To see this, consider for instance the structure $\mathcal{N}$ with $L(x)$ taken to mean "x is divisible by 6" and $G(x)$ to mean "x is divisible by 3."

3.4.17 Using predicates $H(x)$ for "x is happy" and $D(x)$ for "x is drunk," we can formalize A and B as follows: $A = \neg\exists x(H(x) \to D(x))$ and $B = \exists x H(x) \land \exists x \neg D(x)$.
Claims: $A \models B$ and $B \not\models A$.

3.4.18 (a) Using predicates $P(x)$ for "x is a philosopher," $H(x)$ for "x is human," and $M(x)$ for "x is mortal," we can formalize the argument in first-order logic as follows:

$$\frac{\forall x(H(x) \to M(x)), \forall x(P(x) \to H(x))}{\forall x(P(x) \to M(x))}.$$

The argument is correct. To see this, suppose $\forall x(P(x) \to H(x)), \forall x(H(x) \to M(x)) \not\models \forall x(P(x) \to M(x))$. Then there is a structure $\mathcal{S}$ and an assignment v on $\mathcal{S}$ such that $\mathcal{S}, v \models \forall x(P(x) \to H(x))$ and $\mathcal{S}, v \models \forall x(H(x) \to M(x))$ but $\mathcal{S}, v \not\models \forall x(P(x) \to M(x))$. Now, $\mathcal{S}, v \not\models \forall x(P(x) \to M(x))$ implies that $\mathcal{S}, v[x := c] \not\models P(x) \to M(x)$ for some c in $\mathcal{S}$, and so $\mathcal{S}, v[x := c] \models P(x)$ while $\mathcal{S}, v[x := c] \not\models M(x)$. On the other hand, since $\mathcal{S}, v \models \forall x(P(x) \to H(x))$, $\mathcal{S}, v[x := c] \models P(x) \to H(x)$, which means $\mathcal{S}, v[x := c] \not\models P(x)$ or $\mathcal{S}, v[x := c] \models H(x)$. If $\mathcal{S}, v[x := c] \not\models P(x)$, we have a contradiction with $\mathcal{S}, v[x := c] \models P(x)$. If $\mathcal{S}, v[x := c] \models H(x)$, note that $\mathcal{S}, v \models \forall x(H(x) \to M(x))$ implies $\mathcal{S}, v[x := c] \not\models H(x)$ or $\mathcal{S}, v[x := c] \models M(x)$, again a contradiction.

(c) Using predicates $I(x)$ for "x is an integer," $Q(x)$ for "x is a rational," and $N(x)$ for "x is negative," we can formalize the argument as follows:

$$\frac{\exists x(N(x) \wedge Q(x)), \forall x(I(x) \rightarrow Q(x))}{\exists x(I(x) \wedge N(x))}.$$

This argument is not logically valid. To see this, consider for instance the structure of integers $\mathcal{Z}$ and take $I(x)$ to mean "x is positive," $Q(x)$ to mean "x is an integer" (i.e., always true in $\mathcal{Z}$), and $N(x)$ to mean "x is negative."

(e) Not valid.

(g) Using predicates $P(x)$ for "x is a penguin," $B(x)$ for "x is a bird," and $W(x)$ for "x is white," we can formalize the argument as follows:

$$\frac{\forall x(P(x) \rightarrow B(x)), \exists x(P(x) \wedge W(x))}{\exists x(B(x) \wedge W(x) \wedge P(x))}.$$

The argument is correct. To see this, suppose $\forall x(P(x) \rightarrow B(x)), \exists x(P(x) \wedge W(x)) \nvDash \exists x(B(x) \wedge W(x) \wedge P(x))$. Then there is some structure $\mathcal{S}$ and an assignment on $\mathcal{S}$ such that $\mathcal{S}, v \models \forall x(P(x) \rightarrow B(x))$ and $\mathcal{S}, v \models \exists x(P(x) \wedge W(x))$ but $\mathcal{S}, v \nvDash \exists x(B(x) \wedge W(x) \wedge P(x))$. Now, $\mathcal{S}, v \models \exists x(P(x) \wedge W(x))$ implies that $\mathcal{S}, v[x := c] \models P(x) \wedge W(x)$ for some c in $\mathcal{S}$. Hence, $\mathcal{S}, v[x := c] \models P(x)$ and $\mathcal{S}, v[x := c] \models W(x)$. But since $\mathcal{S}, v \models \forall x(P(x) \rightarrow B(x))$, $\mathcal{S}, v[x := c] \nvDash P(x)$ or $\mathcal{S}, v[x := c] \models B(x)$. Furthermore, $\mathcal{S}, v \nvDash \exists x(B(x) \wedge W(x) \wedge P(x))$ implies that $\mathcal{S}, v[x := c] \nvDash B(x)$ or $\mathcal{S}, v[x := c] \nvDash W(x)$ or $\mathcal{S}, v[x := c] \nvDash P(x)$. It is not difficult to see that in each of these cases we get a contradiction.

(j) Using predicates $P(x)$ for "x is a politician," $S(x)$ for "x is successful," and $Q(x)$ for "x is poor," we can formalize the argument as follows:

$$\frac{\neg\exists x(P(x) \wedge S(x) \wedge Q(x)), \exists x(Q(x) \rightarrow P(x))}{\neg\exists x(S(x) \wedge Q(x))}.$$

The argument is not logically correct. For a counter-model consider for instance the structure $\mathcal{N}$ and take $P(x)$ to mean "x is negative," $S(x)$ to mean "x is positive," and $Q(x)$ to mean "x is odd."
(There are poor and successful philosophers, for instance.)

Section 3.5

3.5.2 (a) The major term is "mortal," the minor term is "philosopher," and the middle term is "human." The first premise is the minor one and the second premise is the major one. This syllogism is of type AAA-1.

(c) This syllogism is of type IAI-3.

(e) This syllogism is of type EIO-2.

3.5.4   The nine syllogistic forms whose validity is conditional on the assumption of existential import are:

Figure 1   **Figure 2**   **Figure 3**   **Figure 4**

(a) AAI-1 Barbari   (a) EAO-2 Cesaro   (a) AAI-3 Darapti   (a) AAI-4 Bramantip
(b) EAO-1 Celaront   (b) AEO-2 Camestros   (b) EAO-3 Felapton   (b) EAO-4 Fesapo
                                                                   (c) AEO-4 Camenop

Section 4.1

For this section I only provide a small selection of solutions. More can be found in some of the references listed at the end of the section.

4.1.4   In the derivations below we will skip some purely propositional derivations. For exercises on these, see Section 2.2.

(a) 1. $\vdash_{\mathbf{H}} \forall x P(x) \rightarrow P(y)$ by (Ax$\forall$2)
2. $\vdash_{\mathbf{H}} \forall y(\forall x P(x) \rightarrow P(y))$ by 1 and Generalization
3. $\vdash_{\mathbf{H}} \forall y(\forall x P(x) \rightarrow P(y)) \rightarrow (\forall y \forall x P(x) \rightarrow \forall y P(y))$ by (Ax$\forall$1)
4. $\vdash_{\mathbf{H}} \forall y \forall x P(x) \rightarrow \forall y P(y)$ by 2, 3, and Modus Ponens
5. $\vdash_{\mathbf{H}} \forall x P(x) \rightarrow \forall y \forall x P(x)$ by (Ax$\forall$3)
6. $\vdash_{\mathbf{H}} \forall x P(x) \rightarrow \forall y P(y)$ by 4, 5, Deduction Theorem, and Modus Ponens.

(c) 1. $A(x) \vdash_{\mathbf{H}} \neg\neg A(x)$ by propositional derivation
                                                                                          (see Section 2.2.3, Exercise 9n)

2. $\forall x A(x) \vdash_{\mathbf{H}} \forall x \neg\neg A(x)$ by 1 and Generalization (Exercise 2)
3. $\forall x \neg\neg A(x) \vdash_{\mathbf{H}} \neg\neg \forall x \neg\neg A(x)$ by propositional derivation
                                                                                          (Section 2.2.3, Exercise 9n)

4. $\forall x A(x) \vdash_{\mathbf{H}} \neg\neg \forall x \neg\neg A(x)$ by 2, 3, and transitivity
                                                                                          (Section 2.2.3, Exercise 5)

that is, $\forall x A(x) \vdash_{\mathbf{H}} \neg\exists x \neg A(x)$.

(e) 1. $\forall y A(x, y) \vdash_{\mathbf{H}} A(x, y)$ by (Ax$\forall$2)
2. $\forall x \forall y A(x, y) \vdash_{\mathbf{H}} \forall x A(x, y)$ by 1 and Generalization
                                                                                          (Section 4.1.4, Exercise 2)

3. $\forall x \forall y A(x, y) \vdash_{\mathbf{H}} \forall y \forall x A(x, y)$ by 2 and (Ax$\forall$3).

4.1.5   (h) 1. $\forall x P(x) \wedge \neg Q \vdash_{\mathbf{H}} \forall x P(x)$ by Axiom ($\wedge$1)
2. $\forall x P(x) \vdash_{\mathbf{H}} P(x)$ by (Ax$\forall$2)
3. $\forall x P(x) \wedge \neg Q \vdash_{\mathbf{H}} P(x)$ by 1, 2 , and transitivity
4. $\forall x P(x) \wedge \neg Q \vdash_{\mathbf{H}} \neg Q$ by Axiom ($\wedge$2)
5. $\forall x P(x) \wedge \neg Q \vdash_{\mathbf{H}} P(x) \wedge \neg Q$ by 3, 4, Axiom ($\wedge$2) and
                                                                                          Propositional Derivation
6. $\forall x P(x) \wedge \neg Q \vdash_{\mathbf{H}} \forall x(P(x) \wedge \neg Q)$ by 5 and (Ax$\forall$3)
7. $\neg\forall x(P(x) \wedge \neg Q) \vdash_{\mathbf{H}} \neg(\forall x P(x) \wedge \neg Q)$ by 6 and contraposition
                                                                                          (Section 2.2.3, Exercise 9k)

8. $\neg(\forall x P(x) \wedge \neg Q) \vdash_{\mathbf{H}} \neg\forall x P(x) \vee Q$ by propositional derivation
                                                                                          (Section 2.2.3, Exercises 9u and 9m)

9. $\neg\forall x P(x) \vee Q, \forall x P(x) \vdash_{\mathbf{H}} Q$ by propositional derivation
(Section 2.2.3, Exercise 9q and Modus Ponens)
10. $\neg\forall x (P(x) \wedge \neg Q), \forall x P(x) \vdash_{\mathbf{H}} Q$ by 7, 8, 9, and transitivity
11. $P(x) \wedge \neg Q \vdash_{\mathbf{H}} \neg(P(x) \rightarrow Q)$ by propositional derivation
(using Section 2.2.3, Exercises 9r and 9t)
12. $\forall x (P(x) \wedge \neg Q) \vdash_{\mathbf{H}} \forall x \neg(P(x) \rightarrow Q)$ by 11 and Generalization
13. $\neg\forall x \neg(P(x) \rightarrow Q) \vdash_{\mathbf{H}} \neg\forall x (P(x) \wedge \neg Q)$ by 12 and contraposition
14. $\neg\forall x \neg(P(x) \rightarrow Q), \forall x P(x) \vdash_{\mathbf{H}} Q.$ by 13, 10, and transitivity
that is, $\exists x (P(x) \rightarrow Q), \forall x P(x) \vdash_{\mathbf{H}} Q.$

4.1.6 (p) 1. $\vdash_{\mathbf{H}} \neg(P(x) \rightarrow \forall y P(y)) \rightarrow (P(x) \wedge \neg\forall y P(y))$ propositional derivation
2. $\vdash_{\mathbf{H}} \forall x (\neg(P(x) \rightarrow \forall y P(y)) \rightarrow (P(x) \wedge \neg\forall y P(y)))$ by 1 and
Generalization
3. $\vdash_{\mathbf{H}} \forall x \neg(P(x) \rightarrow \forall y P(y)) \rightarrow \forall x (P(x) \wedge \neg\forall y P(y))$ by 2, (Ax$\forall$1), and
Modus Ponens
4. $\vdash_{\mathbf{H}} (P(x) \wedge \neg\forall y P(y)) \rightarrow \neg\forall y P(y)$ instance of a propositional axiom
5. $\vdash_{\mathbf{H}} \forall x ((P(x) \wedge \neg\forall y P(y)) \rightarrow \neg\forall y P(y))$ by 4 and Generalization
6. $\vdash_{\mathbf{H}} \forall x (P(x) \wedge \neg\forall y P(y)) \rightarrow \forall x \neg\forall y P(y)$ by 5, (Ax$\forall$1), and
Modus Ponens
7. $\vdash_{\mathbf{H}} \forall x (P(x) \wedge \neg\forall y P(y)) \rightarrow \forall x P(x)$ analogous to steps 4–6
8. $\vdash_{\mathbf{H}} \forall x \neg\forall y P(y) \rightarrow \neg\forall y P(y)$ by (Ax$\forall$2)
9. $\vdash_{\mathbf{H}} \forall x (P(x) \wedge \neg\forall y P(y)) \rightarrow \neg\forall y P(y)$ by 6, 8, and transitivity
10. $\vdash_{\mathbf{H}} \forall x P(x) \rightarrow \forall y P(y)$ by Exercise (1) above.
11. $\vdash_{\mathbf{H}} \forall x (P(x) \wedge \neg\forall y P(y)) \rightarrow \forall y P(y)$ by 7, 10, and transitivity
12. $\vdash_{\mathbf{H}} \neg\forall x (P(x) \wedge \neg\forall y P(y))$ by 9, 11, and propositional derivation using
Section 2.2.3, Exercise 9q
13. $\vdash_{\mathbf{H}} \neg\forall x (P(x) \wedge \neg\forall y P(y)) \rightarrow \neg\forall x \neg(P(x) \rightarrow \forall y P(y))$ by
contraposition of 3
14. $\vdash_{\mathbf{H}} \neg\forall x \neg(P(x) \rightarrow \forall y P(y))$ by 12, 13, and Modus Ponens
that is, $\vdash_{\mathbf{H}} \exists x (P(x) \rightarrow \forall y P(y))$. Therefore, $\models \exists x (P(x) \rightarrow \forall y P(y))$.

4.1.9 (b) 1. $x = f(x) \vdash_{\mathbf{H}} x = f(x)$
2. $\vdash_{\mathbf{H}} \forall x \forall y (x = y \rightarrow f(x) = f(y))$ by (Ax$_f$)
3. $\vdash_{\mathbf{H}} \forall y (x = y \rightarrow f(x) = f(y))$ by 2 and (Ax$\forall$2)
4. $\vdash_{\mathbf{H}} x = f(x) \rightarrow f(x) = f(f(x))$ by 3 and (Ax$\forall$2)
5. $x = f(x) \vdash_{\mathbf{H}} f(x) = f(f(x))$ by 4 and Deduction Theorem
6. $x = f(x) \vdash_{\mathbf{H}} x = f(f(x))$ by 1,4, (Ax$_=$3) and
Propositional Derivation
7. $x = f(x) \vdash_{\mathbf{H}} f(f(x)) = x$ by 6 and (Ax$_=$2)
8. $f(f(x)) = x \vdash_{\mathbf{H}} P(f(f(x))) = P(x)$ by 7 and (Ax$_r$)
9. $x = f(x) \vdash_{\mathbf{H}} P(f(f(x))) = P(x)$ by 7, 8, and (Ax$_r$)
10. $\vdash_{\mathbf{H}} x = f(x) \rightarrow P(f(f(x))) = P(x)$ by 9 and Deduction Theorem
11. $\vdash_{\mathbf{H}} \forall x (x = f(x) \rightarrow (P(f(f(x))) \rightarrow P(x)))$ by 10 and Generalization.

Section 4.2

4.2.1 (a)

$$\forall x P(x) \rightarrow \forall y P(y) : \text{F}$$

$$|$$

$$\forall x P(x) : \text{T}^1, \forall y P(y) : \text{F}$$

$$|$$

$$P(c) : \text{F}$$

$$|$$

$$P(c) : \text{T}$$

$$\times$$

The tableau closes, hence $\models \forall x P(x) \rightarrow \forall y P(y)$.

(d)

$$\neg \exists x \neg A(x) : \text{T}, \forall x A(x) : \text{F}$$

$$|$$

$$A(c) : \text{F}$$

$$|$$

$$\exists x \neg A(x) : \text{F}$$

$$|$$

$$\neg A(c) : \text{F}$$

$$|$$

$$A(c) : \text{T}$$

$$\times$$

The tableau closes, hence $\neg \exists x \neg A(x) \models \forall x A(x)$.

(g)

$$\exists x \exists y Q(x, y) : \text{T}, \exists y \exists x Q(x, y) : \text{F}$$

$$|$$

$$\exists y Q(c_1, y) : \text{T}$$

$$|$$

$$Q(c_1, c_2) : \text{T}$$

$$|$$

$$\exists x Q(x, c_2) : \text{F}$$

$$|$$

$$Q(c_1, c_2) : \text{F}$$

$$\times$$

The tableau closes, hence $\exists x \exists y Q(x, y) \models \exists y \exists x Q(x, y)$.

4.2.2   (a)

$$\forall x(P(x) \vee Q) : \text{T}, \forall x P(x) \vee Q : \text{F}$$

$$\forall x P(x) : \text{F}, Q : \text{F}$$

$$P(c) : \text{F}$$

$$P(c) \vee Q : \text{T}$$

$$P(c) : \text{T} \qquad\qquad Q : \text{T}$$
$$\times \qquad\qquad\qquad \times$$

The tableau closes, hence $\forall x(P(x) \vee Q) \models \forall x P(x) \vee Q$.

(d)

$$\exists x P(x) \wedge Q : \text{T}, \exists x(P(x) \wedge Q) : \text{F}$$

$$\exists x P(x) : \text{T}, Q : \text{T}$$

$$P(c) : \text{T}$$

$$P(c) \wedge Q : \text{F}$$

$$P(c) : \text{F} \qquad\qquad Q : \text{F}$$
$$\times \qquad\qquad\qquad \times$$

The tableau closes, hence $\exists x P(x) \wedge Q \models \exists x(P(x) \wedge Q)$.

(f)

$$\exists x P(x) \to Q : \text{T}, \forall x(P(x) \to Q) : \text{F}$$

$$P(c) \to Q : \text{F}$$

$$P(c) : \text{T}, Q : \text{F}$$

$$\exists x P(x) : \text{F} \qquad\qquad Q : \text{T}$$
$$\qquad\qquad\qquad\qquad\qquad \times$$

$$P(c) : \text{F}$$
$$\times$$

The tableau closes, hence $\exists x P(x) \to Q \models \forall x(P(x) \to Q)$.

(h)

$$\exists x(Q \to P(x)) : T, Q : T, \exists x P(x) : F$$

$$|$$

$$Q \to P(c) : T$$

$$|$$

$$P(c) : F$$

$$P(c) : T \qquad\qquad Q : F$$
$$\times \qquad\qquad\qquad \times$$

The tableau closes, hence $\exists x(Q \to P(x)), Q \models \exists x P(x)$.

4.2.3 (a)

$$\exists x(P(x) \to \forall y P(y)) : F$$

$$|$$

$$P(d) \to \forall y P(y) : F$$

$$|$$

$$P(d) : T, \forall y P(y) : F$$

$$|$$

$$P(c) : F$$

$$|$$

$$P(c) \to \forall y P(y) : F$$

$$|$$

$$P(c) : T, \forall y P(y) : F$$
$$\times$$

The tableau closes, hence $\models \exists x(P(x) \to \forall y P(y))$.

(c)

$$\forall x \exists y Q(x,y) \rightarrow \forall y \exists x Q(x,y) : F$$

$$|$$

$$\forall x \exists y Q(x,y) : T, \forall y \exists x Q(x,y) : F$$

$$|$$

$$\exists x Q(x,c_1) : F$$

$$|$$

$$Q(c_1,c_1) : F$$

$$|$$

$$\exists y Q(c_1,y) : T$$

$$|$$

$$Q(c_1,c_2) : T$$

$$|$$

$$Q(c_2,c_1) : F$$

$$|$$

$$\exists y Q(c_2,y) : T$$

$$|$$

$$Q(c_2,c_3) : T$$

$$\vdots$$

$\nvDash \forall x \exists y Q(x,y) \rightarrow \forall y \exists x Q(x,y)$ since the tableau does not terminate. Counter-model: consider the structure $\mathcal{N}$ with Q taken as the predicate $<$.

(e) $\nvDash \forall x \exists y Q(x,y) \rightarrow \exists y \forall x Q(x,y)$ since the tableau does not terminate. Counter-model: consider the structure $\mathcal{N}$ with Q taken as the predicate $<$.

(g) The tableau closes, hence the formula is valid.

(i) The tableau does not terminate, hence
$\nvDash \forall x \exists y P(x,y) \land \forall x \forall y (P(x,y) \rightarrow P(y,x)) \rightarrow \exists x P(x,x)$.

4.2.4 (b)

$$\forall x (A(x) \land B(x)) : T, \forall x A(x) \land \forall x B(x) : F$$

$\forall x A(x) : F$	$\forall x B(x) : F$
$\|$	$\|$
$A(c_1) : F$	$B(c_2) : F$
$\|$	$\|$
$A(c_1) \land B(c_1) : T$	$A(c_1) \land B(c_2) : T$
$\|$	$\|$
$A(c_1) : T, B(c_1) : T$	$A(c_2) : T, B(c_2) : T$
$\times$	$\times$

The tableau closes, hence $\forall x (A(x) \land B(x)) \models \forall x A(x) \land \forall x B(x)$.

(d)

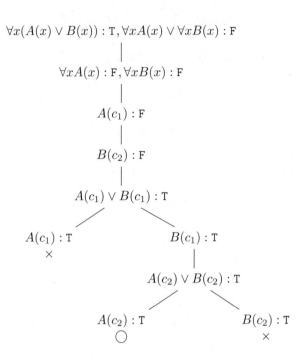

The tableau does not close, hence $\forall x(A(x) \vee B(x)) \nvDash \forall x A(x) \vee \forall x B(x)$.
A counter-model extracted from the tableau: a structure with domain $\{a_1, a_2\}$
where c_1 is interpreted as a_1, c_2 is interpreted as a_2, A is interpreted as $\{a_1\}$,
and B is interpreted as $\{a_2\}$.

(f)

$$\forall x(A(x) \rightarrow B(x)) : \text{T}, \forall x A(x) \rightarrow \forall x B(x) : \text{F}$$
$$|$$
$$\forall x A(x) : \text{T}, \forall x B(x) : \text{F}$$
$$|$$
$$B(c) : \text{F}$$
$$|$$
$$A(c) : \text{T}$$
$$|$$
$$A(c) \rightarrow B(c) : \text{T}$$

$A(c) : \text{F}$ $B(c) : \text{T}$
×   ×

The tableau closes, hence $\forall x(A(x) \rightarrow B(x)) \models \forall x A(x) \rightarrow \forall x B(x)$.

(h)

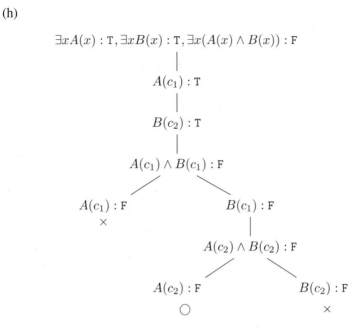

$$\exists x A(x) : \text{T}, \exists x B(x) : \text{T}, \exists x (A(x) \wedge B(x)) : \text{F}$$

$$A(c_1) : \text{T}$$

$$B(c_2) : \text{T}$$

$$A(c_1) \wedge B(c_1) : \text{F}$$

$$A(c_1) : \text{F} \qquad\qquad B(c_1) : \text{F}$$
$$\times$$

$$A(c_2) \wedge B(c_2) : \text{F}$$

$$A(c_2) : \text{F} \qquad\qquad B(c_2) : \text{F}$$
$$\bigcirc \qquad\qquad\qquad \times$$

The tableau does not close, hence $\exists x A(x), \exists x B(x) \nvDash \exists x (A(x) \wedge B(x))$. A counter-model extracted from the tableau: a structure with domain $\{a_1, a_2\}$ where c_1 is interpreted as a_1, c_2 is interpreted as a_2, A is interpreted as $\{a_1\}$, and B is interpreted as $\{a_2\}$.

(j)

$$\exists x (A(x) \vee B(x)) : \text{T}, \exists x A(x) \vee \exists x B(x) : \text{F}$$

$$A(c) \vee B(c) : \text{T}$$

$$A(c) : \text{T} \qquad\qquad B(c) : \text{T}$$

$$\exists x A(x) : \text{F}, \exists x B(x) : \text{F} \qquad \exists x A(x) : \text{F}, \exists x B(x) : \text{F}$$

$$A(c) : \text{F} \qquad\qquad B(c) : \text{F}$$
$$\times \qquad\qquad\qquad \times$$

The tableau closes, hence $\exists x (A(x) \vee B(x)) \vDash \exists x A(x) \vee \exists x B(x)$.

(1)

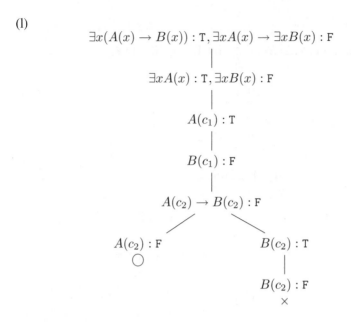

$$\exists x(A(x) \to B(x)) : \text{T}, \exists x A(x) \to \exists x B(x) : \text{F}$$

$$\exists x A(x) : \text{T}, \exists x B(x) : \text{F}$$

$$A(c_1) : \text{T}$$

$$B(c_1) : \text{F}$$

$$A(c_2) \to B(c_2) : \text{F}$$

$A(c_2) : \text{F}$ $B(c_2) : \text{T}$
$\circ$ $B(c_2) : \text{F}$
                          $\times$

The tableau does not close, hence $\exists x(A(x) \to B(x)) \nvDash \exists x A(x) \to \exists x B(x)$.
A counter-model extracted from the tableau: a structure with domain $\{a_1, a_2\}$
where c_1 is interpreted as a_1, c_2 is interpreted as a_2, A is interpreted as $\{a_1\}$,
and B is interpreted as $\emptyset$.

4.2.5 (a) Checking the validity of both consequences:

$$\forall x(P(x) \wedge Q(x)) \models \forall x P(x) \wedge \forall x Q(x):$$

$$\forall x(P(x) \wedge Q(x)) : \text{T}, \forall x P(x) \wedge \forall x Q(x) : \text{F}$$

$\forall x P(x) : \text{F}$ $\forall x Q(x) : \text{F}$

$P(c_1) : \text{F}$ $Q(c_2) : \text{F}$

$P(c_1) \wedge Q(c_1) : \text{T}$ $P(c_2) \wedge Q(c_2) : \text{T}$

$P(c_1) : \text{T}, Q(c_1) : \text{T}$ $P(c_2) : \text{T}, Q(c_2) : \text{T}$
         $\times$ $\times$

$\forall x P(x) \land \forall x Q(x) \models \forall x (P(x) \land Q(x))$:

$$\forall x P(x) \land \forall x Q(x) : \mathrm{T}, \forall x (P(x) \land Q(x)) : \mathrm{F}$$

$$P(c) \land Q(c) : \mathrm{F}$$

$P(c) : \mathrm{F}$ $Q(c) : \mathrm{F}$

$\forall x P(x) : \mathrm{T}, \forall x Q(x) : \mathrm{T}$ $\forall x P(x) : \mathrm{T}, \forall x Q(x) : \mathrm{T}$

$P(c) : \mathrm{T}$ $Q(c) : \mathrm{T}$
$\times$ $\times$

Hence, $\forall x (P(x) \land Q(x)) \equiv \forall x P(x) \land \forall x Q(x)$.

(c) $\forall x P(x) \to \forall x Q(x) \not\equiv \forall x (P(x) \to Q(x))$ because $\forall x P(x) \to \forall x Q(x) \not\models \forall x (P(x) \to Q(x))$:

$$\forall x P(x) \to \forall x Q(x) : \mathrm{T}, \forall x (P(x) \to Q(x)) : \mathrm{F}$$

$$P(c_1) \to Q(c_1) : \mathrm{F}$$

$$P(c_1) : \mathrm{T}, Q(c_1) : \mathrm{F}$$

$\forall x P(x) : \mathrm{F}$ $\forall x Q(x) : \mathrm{T}$

$P(c_2) : \mathrm{F}$ $Q(c_1) : \mathrm{T}$
$\bigcirc$ $\times$

A counter-model extracted from the tableau: a structure with domain $\{a_1, a_2\}$ where c_1 is interpreted as a_1, c_2 is interpreted as a_2, P is interpreted as $\{a_1\}$ and Q is interpreted as $\emptyset$.

(d) $\exists x P(x) \land \exists x Q(x) \not\equiv \exists x (P(x) \land Q(x))$ because $\exists x P(x) \land \exists x Q(x) \not\models \exists x (P(x) \land Q(x))$:

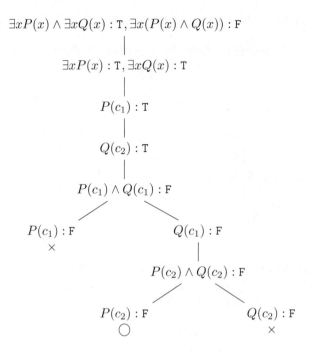

$$\exists x P(x) \land \exists x Q(x) : \text{T}, \exists x (P(x) \land Q(x)) : \text{F}$$

$$\exists x P(x) : \text{T}, \exists x Q(x) : \text{T}$$

$$P(c_1) : \text{T}$$

$$Q(c_2) : \text{T}$$

$$P(c_1) \land Q(c_1) : \text{F}$$

$P(c_1) : \text{F}$ $Q(c_1) : \text{F}$
$\times$

$$P(c_2) \land Q(c_2) : \text{F}$$

$P(c_2) : \text{F}$ $Q(c_2) : \text{F}$
$\bigcirc$ $\times$

A counter-model extracted from the tableau: a structure with domain $\{a_1, a_2\}$ where c_1 is interpreted as a_1, c_2 is interpreted as a_2, P is interpreted as $\{a_1\}$ and Q is interpreted as $\{a_2\}$.

(f) Checking the validity of both consequences:

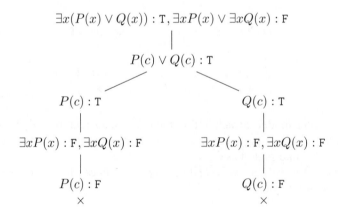

$$\exists x (P(x) \lor Q(x)) : \text{T}, \exists x P(x) \lor \exists x Q(x) : \text{F}$$

$$P(c) \lor Q(c) : \text{T}$$

$P(c) : \text{T}$ $Q(c) : \text{T}$

$\exists x P(x) : \text{F}, \exists x Q(x) : \text{F}$ $\exists x P(x) : \text{F}, \exists x Q(x) : \text{F}$

$P(c) : \text{F}$ $Q(c) : \text{F}$
$\times$ $\times$

Therefore, $\exists x P(x) \vee \exists x Q(x) \models \exists x (P(x) \vee Q(x))$:

$$\exists x P(x) \vee \exists x Q(x) : \text{T}, \exists x (P(x) \vee Q(x)) : \text{F}$$

$\exists x P(x) : \text{T}$ $\exists x Q(x) : \text{T}$

$P(c_1) : \text{T}$ $Q(c_2) : \text{T}$

$P(c_1) \vee Q(c_1) : \text{F}$ $P(c_2) \vee Q(c_2) : \text{F}$

$P(c_1) : \text{F}, Q(c_1) : \text{F}$ $P(c_2) : \text{F}, Q(c_2) : \text{F}$
        $\times$ $\times$

Therefore, $\exists x (P(x) \vee Q(x)) \equiv \exists x P(x) \vee \exists x Q(x)$.

(h) $\exists x \forall y Q(x, y) \not\equiv \exists y Q(y, y)$ since $\exists y Q(y, y) \not\models \exists x \forall y Q(x, y)$:

$$\exists y Q(y, y) : \text{T}, \exists x \forall y Q(x, y) : \text{F}$$

$$Q(c_1, c_1) : \text{T}$$

$$\forall y Q(c_1, y) : \text{F}$$

$$Q(c_1, c_2) : \text{F}$$

$$\exists y Q(c_2, y) : \text{F}$$

$$Q(c_2, c_3) : \text{F}$$

$$\vdots$$

A counter-model: consider the structure $\mathcal{N}$ with Q taken as $=$.

(j) Checking the validity of $\forall x \forall y Q(x, y) \leftrightarrow \forall x \forall y Q(y, x)$:

$$\forall x \forall y Q(x, y) \leftrightarrow \forall x \forall y Q(y, x) : \text{F}$$

|

$$(\forall x \forall y Q(x, y) \rightarrow \forall x \forall y Q(y, x)) \wedge (\forall x \forall y Q(y, x) \rightarrow \forall x \forall y Q(x, y)) : \text{F}$$

$$\forall x \forall y Q(x, y) \rightarrow \forall x \forall y Q(y, x) : \text{F} \qquad\qquad \forall x \forall y Q(y, x) \rightarrow \forall x \forall y Q(x, y) : \text{F}$$

|

$$\forall x \forall y Q(x, y) : \text{T}, \forall x \forall y Q(y, x) : \text{F} \qquad\qquad \forall x \forall y Q(y, x) : \text{T}, \forall x \forall y Q(x, y) : \text{F}$$

|

$$\forall y Q(y, c_1) : \text{F} \qquad\qquad\qquad\qquad\qquad \forall y Q(c_3, y) : \text{F}$$

|

$$Q(c_2, c_1) : \text{F} \qquad\qquad\qquad\qquad\qquad\quad Q(c_3, c_4) : \text{F}$$

|

$$\forall y Q(c_2, y) : \text{T} \qquad\qquad\qquad\qquad\qquad \forall y Q(y, c_4) : \text{T}$$

|

$$Q(c_2, c_1) : \text{T} \qquad\qquad\qquad\qquad\qquad\quad Q(c_3, c_4) : \text{T}$$

$$\times \qquad\qquad\qquad\qquad\qquad\qquad\qquad\qquad \times$$

The tableau closes, hence $\forall x \forall y Q(x, y) \equiv \forall x \forall y Q(y, x)$.

4.2.6 (a) Using predicates $L(x)$ for "x is a lawyer" and $G(x)$ for "x is greedy," we can formalize A and B as follows:

$$A : \neg \forall x (L(x) \rightarrow G(x)), \text{and}$$

$$B : \exists x L(x) \wedge \neg \forall x G(x).$$

Checking the validity of both consequences:
The tableau for $\neg \forall x (L(x) \rightarrow G(x)) \models \exists x L(x) \wedge \neg \forall x G(x)$ closes, hence it is valid.
However, $\exists x L(x) \wedge \neg \forall x G(x) \nvDash \neg \forall x (L(x) \rightarrow G(x))$:

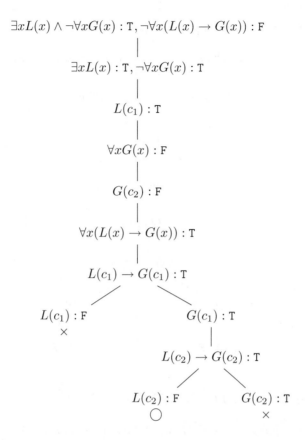

A counter-model extracted from the tableau: a structure with domain $\{a_1, a_2\}$ where c_1 is interpreted as a_1, c_2 is interpreted as a_2, and both L and G are interpreted as $\{a_1\}$.

Therefore, $A \not\equiv B$.

(c) Using predicates $T(x)$ for "x is talking" and $L(x)$ for "x is listening," we can formalize A and B as follows:

$$A : \exists x T(x) \rightarrow \forall x L(x), \text{ and}$$
$$B : \exists x \neg L(x) \rightarrow \neg \exists x T(x).$$

Checking the validity of both logical consequences:

$$\exists x T(x) \rightarrow \forall x L(x) \models \exists x \neg L(x) \rightarrow \neg \exists x T(x):$$

$$\exists x T(x) \rightarrow \forall x L(x) : \text{T}, \exists x \neg L(x) \rightarrow \neg \exists x T(x) : \text{F}$$
|
$$\exists x \neg L(x) : \text{T}, \neg \exists x T(x) : \text{F}$$
|
$$\neg L(c_1) : \text{T}$$
|
$$L(c_1) : \text{F}$$
|
$$\exists x T(x) : \text{T}$$
|
$$T(c_2) : \text{T}$$

$$\exists x T(x) : \text{F} \qquad \qquad \forall x L(x) : \text{T}$$
| |
$$T(c_2) : \text{F} \qquad \qquad L(c_1) : \text{T}$$
$$\times \qquad \qquad \times$$

$$\exists x \neg L(x) \rightarrow \neg \exists x T(x) \models \exists x T(x) \rightarrow \forall x L(x):$$

$$\exists x \neg L(x) \rightarrow \neg \exists x T(x) : \text{T}, \exists x T(x) \rightarrow \forall x L(x) : \text{F}$$
|
$$\exists x T(x) : \text{T}, \forall x L(x) : \text{F}$$
|
$$T(c_1) : \text{T}$$
|
$$L(c_2) : \text{F}$$

$$\exists x \neg L(x) : \text{F} \qquad \qquad \neg \exists x T(x) : \text{T}$$
| |
$$\neg L(c_2) : \text{F} \qquad \qquad \exists x T(x) : \text{F}$$
| |
$$L(c_2) : \text{T} \qquad \qquad T(c_1) : \text{F}$$
$$\times \qquad \qquad \times$$

Both tableaux close, hence $A \equiv B$.

4.2.7   (b)   Using predicates $N(x)$ for "x is a natural number," $I(x)$ for "x is an integer," and $O(x)$ for "x is odd," we can formalize the argument as follows:

$$\frac{\forall x(N(x) \to I(x)), \exists x(I(x) \land O(x))}{\exists x(N(x) \land O(x))}.$$

The argument is not valid:

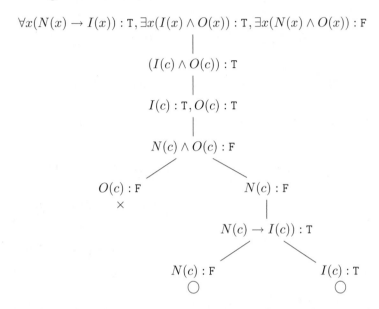

$\forall x(N(x) \to I(x)) : \mathrm{T}, \exists x(I(x) \land O(x)) : \mathrm{T}, \exists x(N(x) \land O(x)) : \mathrm{F}$

$(I(c) \land O(c)) : \mathrm{T}$

$I(c) : \mathrm{T}, O(c) : \mathrm{T}$

$N(c) \land O(c) : \mathrm{F}$

$O(c) : \mathrm{F}$ $N(c) : \mathrm{F}$

$\times$

$N(c) \to I(c)) : \mathrm{T}$

$N(c) : \mathrm{F}$ $I(c) : \mathrm{T}$

$\bigcirc$ $\bigcirc$

        A counter-model extracted from the tableau: a structure with domain $\{a\}$ where c is interpreted as a, both I and O are interpreted as $\{a\}$, and N is interpreted as the empty set. Another counter-model: keep the original interpretation of N and I and change the interpretation of O to mean "x is negative."

  (d)   Notation: $S(x)$ means "x is selling," $B(x)$ means "x is buying," and $N(x)$ means "x is negotiating."

        Formalization:

        $\forall x S(x) \to \exists y(B(y) \lor N(y)), \neg \exists x(S(x) \land B(x)) \models \exists x(S(x) \to N(x))$.

        The tableau closes, hence the argument is logically correct:

$$\forall x S(x) \rightarrow \exists y(B(y) \lor N(y)) : \text{T}, \ \neg \exists x(S(x) \land B(x)) : \text{T}, \ \exists(S(z) \rightarrow N(z)) : \text{F}$$

$$\downarrow$$

$$\exists x(S(x) \land B(x)) : \text{F}$$

$\forall x S(x) : \text{F}$ $\exists y(B(y) \lor N(y)) : \text{T}$

$\downarrow$ $\downarrow$

$S(c_1) : \text{F}$ $B(c_2) \lor N(c_2) : \text{T}$

$\downarrow$ $\downarrow$

$S(c_1) \rightarrow N(c_1) : \text{F}$ $S(c_2) \rightarrow N(c_2) : \text{F}$

$\downarrow$ $\downarrow$

$S(c_1) : \text{T}, N(c_1) : \text{F}$ $S(c_2) : \text{T}, N(c_2) : \text{F}$

$\times$ $\downarrow$

$S(c_2) \land B(c_2) : \text{F}$

$\times$ $B(c_2) : \text{F}$

$B(c_2) : \text{T}$ $N(c_2) : \text{T}$

$\times$ $\times$

(f) Using predicates $P(x)$ for "x is a penguin," $B(x)$ for "x is a bird," $W(x)$ for "x is white," and $E(x)$ for "x eats butterflies," we can formalize the argument as follows:

$$\frac{\forall x(P(x) \rightarrow B(x)), \exists x(P(x) \land W(x)), \neg \exists x(W(x) \land E(x))}{\neg \forall x(P(x) \rightarrow E(x))}.$$

The argument is valid:

$$\forall x(P(x) \rightarrow B(x)) : \text{T}, \exists x(P(x) \wedge W(x)) : \text{T}$$
$$\neg \exists x(W(x) \wedge E(x)) : \text{T}, \neg \forall x(P(x) \rightarrow E(x)) : \text{F}$$

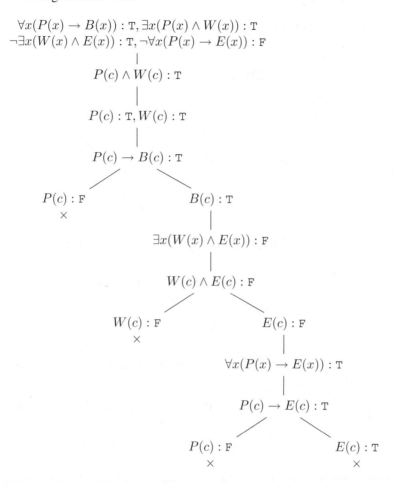

(h) Using predicates $B(x)$ for "x is a bachelor," $M(x)$ for "x is a man," $W(x)$ for "x is a woman," $K(x, y)$ for "x knows y," and $L(x, y)$ for "x loves y," we can formalize the argument as follows:

$$\exists x(B(x) \wedge \forall y(W(y) \wedge K(x, y) \rightarrow L(x, y)))$$
$$\forall x(B(x) \rightarrow M(x))$$
$$\underline{W(Eve) \wedge \forall x(M(x) \rightarrow K(x, Eve))}$$
$$\exists x(B(x) \wedge \exists y(W(y) \wedge L(x, y))).$$

The argument is logically valid:

$$\exists x(B(x) \land \forall y(W(y) \land K(x,y) \to L(x,y))) : \text{T}, \forall x(B(x) \to M(x)) : \text{T}$$
$$W(Eve) \land \forall x(M(x) \to K(x,Eve)) : \text{T}, \exists x(B(x) \land \exists y(W(y) \land L(x,y))) : \text{F}$$

|

$$B(c) \land \forall y(W(y) \land K(c,y) \to L(c,y)) : \text{T}$$

|

$$B(c) : \text{T}, \forall y(W(y) \land K(c,y) \to L(c,y)) : \text{T}$$

|

$$W(Eve) : \text{T}, \forall x(M(x) \to K(x,Eve)) : \text{T}$$

|

$$B(c) \land \exists y(W(y) \land L(c,y)) : \text{F}$$

$B(c) : \text{F}$ $\exists y(W(y) \land L(c,y)) : \text{F}$
  ×

$$W(Eve) \land L(c,Eve) : \text{F}$$

$W(Eve)) : \text{F}$ $L(c,Eve) : \text{F}$
    ×

$$W(Eve) \land K(c,Eve) \to L(c,Eve) : \text{T}$$

$W(Eve) \land K(c,Eve) : \text{F}$ $L(c,Eve) : \text{T}$
                        ×

$W(Eve) : \text{F}$ $K(c,Eve) : \text{F}$
    ×

$$M(c) \to K(c,Eve) : \text{T}$$

$M(c) : \text{F}$ $K(c,Eve) : \text{T}$
                    ×

$$B(c) \to M(c) : \text{T}$$

$B(c) : \text{F}$ $M(c) : \text{T}$
    ×   ×

(i) In this case, the argument is formalized as follows and is *not* valid.

$$\exists x(B(x) \wedge \forall y(W(y) \wedge K(x,y) \to L(x,y)))$$
$$\forall x(B(x) \to M(x))$$
$$\underline{W(Eve) \wedge \forall x(M(x) \to K(x,Eve))}$$
$$\forall x(B(x) \to \exists y(W(y) \wedge L(x,y)))$$

Section 4.3

4.3.1 (b) $\vdash_{\mathbf{ND}} \models \exists x A(x) \to \exists y A(y)$:

$$\cfrac{\cfrac{[\exists x A(x)]^2 \quad \cfrac{[A(c)]^1}{\exists y A(y)}}{\exists y A(y)} \; 1}{\exists x A(x) \to \exists y A(y)} \; 2$$

(c) $\forall x A(x) \vdash_{\mathbf{ND}} \; \neg \exists x \neg A(x)$:

$$\cfrac{\cfrac{[\exists x \neg A(x)]^2 \quad \cfrac{[\neg A(c)]^1 \quad \cfrac{\forall x A(x)}{A(c)}}{\bot} \; 1}{\bot}}{\neg \exists x \neg A(x)} \; 2$$

4.3.2 (a) $\forall x(P(x) \vee Q) \vdash_{\mathbf{ND}} \; \forall x P(x) \vee Q$:

$$\cfrac{\neg Q \vee Q \;^{(*)} \quad \cfrac{\cfrac{[\neg Q]^1 \quad \cfrac{\cfrac{\forall x(P(x) \vee Q)}{P(c) \vee Q} \;^{(**)}}{\neg Q \to P(c)}}{\cfrac{P(c)}{\cfrac{\forall x P(x)}{\forall x P(x) \vee Q}}} \quad \cfrac{[Q]^1}{\forall x P(x) \vee Q}}{\forall x P(x) \vee Q} \; 1}{\forall x P(x) \vee Q}$$

(*) Here we use the derivation $\vdash_{\mathbf{ND}} \; \neg Q \vee Q$ from Section 2.4.2.
(**) Here we use the derivation $A \vee B \vdash_{\mathbf{ND}} \neg B \to A$ from Section 2.4.2.

(c)

$$\cfrac{\exists x(P(x) \wedge Q) \quad \cfrac{\cfrac{\cfrac{[P(c) \wedge Q]^1}{P(c)}}{\exists x P(x)} \quad \cfrac{[P(c) \wedge Q]^1}{Q}}{\exists x P(x) \wedge Q}}{\exists x P(x) \wedge Q} {\scriptstyle 1}$$

(e) $\exists x P(x) \to Q \vdash_{\mathbf{ND}} \ \forall x(P(x) \to Q)$:

$$\cfrac{\cfrac{\cfrac{\cfrac{[P(c)]^1}{\exists x P(x)} \quad \exists x P(x) \to Q}{Q}}{P(c) \to Q} {\scriptstyle 1}}{\forall x(P(x) \to Q)}$$

(h) $\exists x(P(x) \to Q), \forall x P(x) \vdash_{\mathbf{ND}} \ Q$:

$$\cfrac{\cfrac{\cfrac{\forall x P(x)}{P(c)} \quad [P(c) \to Q]^1}{Q} \quad \exists x(P(x) \to Q)}{Q} {\scriptstyle 1}$$

(j) $\exists x(P(x) \vee \neg Q) \vdash_{\mathbf{ND}} \ Q \to \exists x P(x)$:

$$\cfrac{\cfrac{\exists x(P(x) \vee \neg Q) \quad \cfrac{\cfrac{[P(c) \vee \neg Q]^1 \quad Q}{P(c)} {\scriptstyle (*)}}{\exists x P(x)}}{\exists x P(x)} {\scriptstyle 1}}{Q \to \exists x P(x)} {\scriptstyle 2}$$

(*) Here we use the propositional derivation $P \vee \neg Q, Q \vdash_{\mathbf{ND}} P$, which is left as an exercise.

4.3.3 (b) $\forall x A(x) \vee \forall x B(x) \vdash_{\mathbf{ND}} \ \forall x(A(x) \vee B(x))$:

$$\cfrac{\forall x A(x) \vee \forall x B(x) \quad \cfrac{\cfrac{\cfrac{[\forall x A(x)]^1}{A(c)}}{A(c) \vee B(c)}}{\forall x(A(x) \vee B(x))} \quad \cfrac{\cfrac{\cfrac{[\forall x B(x)]^1}{B(c)}}{A(c) \vee B(c)}}{\forall x(A(x) \vee B(x))}}{\forall x(A(x) \vee B(x))} {\scriptstyle 1}$$

(d) No. Consider for example the structure $\mathcal{H}$ with A and B interpreted as M and W, respectively.

(f) $\forall x(A(x) \to B(x)) \vdash_{\textbf{ND}} \forall x A(x) \to \forall x B(x)$:

$$
\dfrac{\dfrac{\dfrac{[\forall x A(x)]^1}{A(x)} \quad \dfrac{\forall x(A(x) \to B(x))}{A(x) \to B(x)}}{\dfrac{B(x)}{\dfrac{\forall x B(x)}{\forall x A(x) \to \forall x B(x)}\,{}^1}}}{}
$$

(h) $\exists x(A(x) \wedge B(x)) \vdash_{\textbf{ND}} \exists x A(x) \wedge \exists x B(x)$:

$$
\dfrac{\exists x(A(x) \wedge B(x)) \qquad \dfrac{\dfrac{[A(c) \wedge B(c)]^1}{A(c)}}{\exists x A(x)} \quad \dfrac{\dfrac{[A(c) \wedge B(c)]^1}{B(c)}}{\exists x B(x)}}{\dfrac{\exists x A(x) \wedge \exists x B(x)}{\exists x A(x) \wedge \exists x B(x)}\,{}^1}
$$

(j) $\exists x A(x) \vee \exists x B(x) \vdash_{\textbf{ND}} \exists x(A(x) \vee B(x))$:

$$
\dfrac{\exists x A(x) \vee \exists x B(x) \qquad \dfrac{[\exists x A(x)]^3 \quad \dfrac{\dfrac{[A(c)]^1}{A(c) \vee B(c)}}{\exists x(A(x) \vee B(x))}}{\exists x(A(x) \vee B(x))}\,{}^1 \qquad \dfrac{[\exists x B(x)]^3 \quad \dfrac{\dfrac{[B(k)]^2}{A(k) \vee B(k)}}{\exists x(A(x) \vee B(x))}}{\exists x(A(x) \vee B(x))}\,{}^2}{\exists x(A(x) \vee B(x))}\,{}^3
$$

(l) No. Consider for example the structure $\mathcal{N}$ with $A(x)$ interpreted as "x is prime" and $B(x)$ to mean "$x < x$."

(n) $\forall y \exists x(P(x) \to Q(y)) \vdash_{\textbf{ND}} \forall x P(x) \to \forall z Q(z)$:

$$
\dfrac{\dfrac{\forall y \exists x(P(x) \to Q(y))}{\exists x(P(x) \to Q(d))} \qquad \dfrac{\dfrac{[\forall x P(x)]^2}{P(d)} \quad [P(d) \to Q(c)]^1}{Q(c)}}{\dfrac{\dfrac{Q(c)}{\forall z Q(z)}}{\forall x P(x) \to \forall z Q(z)}\,{}^2}\,{}^1
$$

4.3.4 (b) Using the predicates $B(x)$ for "x is a bird," $P(x)$ for "x is a penguin," $F(x)$ for "x can fly," and $W(x)$ for "x is white," the argument is formalized as follows.

$$
\dfrac{\forall x(P(x) \to B(x)), \neg \exists x(P(x) \wedge F(x)), \forall x((B(x) \wedge W(x)) \to F(x))}{\neg \exists x(P(x) \wedge W(x))}.
$$

This argument is logically correct because the conclusion is derivable from the premises in **ND**.

(c) Using the predicates above, but where $F(x)$ means "*x is a fish*" plus $E(x, y)$ for "*x eats y*," the argument is formalized as follows.

$$\frac{\forall x(P(x) \rightarrow B(x)), \exists x(P(x) \wedge W(x)), \forall x(P(x) \rightarrow \exists y(E(x, y) \wedge F(y)))}{\exists x(B(x) \wedge W(x) \wedge \exists y(E(x, y) \wedge F(y)))}.$$

This argument is logically correct because the conclusion is derivable from the premises in **ND**.

Section 4.4

4.4.2 Transformation into a prenex DNF and a prenex CNF:

$\exists z(\exists x Q(x, z) \rightarrow (\exists x P(x) \vee \neg \exists z P(z)))$
$\equiv \exists z(\neg \exists x Q(x, z) \vee (\exists x P(x) \vee \forall z \neg P(z)))$
$\equiv \exists z(\forall x \neg Q(x, z) \vee \exists x P(x) \vee \forall z \neg P(z))$
$\equiv \exists z(\forall x \neg Q(x, z) \vee \exists y P(y) \vee \forall w \neg P(w))$
$\equiv \exists z \exists y \forall x \forall w(\neg Q(x, z) \vee P(y) \vee \neg P(w))$ (PCNF and PDNF)
Skolemization: $\forall x \forall w(\neg Q(x, c) \vee P(d) \vee \neg P(w))$
Clausal form: $\{\neg Q(x, c), P(d), \neg P(w)\}$

4.4.4 Transformation into a prenex DNF and a prenex CNF:

$\forall z(\exists y P(y) \leftrightarrow Q(z, y))$
$\equiv \forall z((\exists y P(y) \rightarrow Q(z, y)) \wedge (Q(z, y) \rightarrow \exists y P(y)))$
$\equiv \forall z((\neg \exists y P(y) \vee Q(z, y)) \wedge (\neg Q(z, y) \vee \exists y P(y)))$
$\equiv \forall z((\forall y \neg P(y) \vee Q(z, y)) \wedge (\neg Q(z, y) \vee \exists y P(y)))$
$\equiv \forall z((\forall u \neg P(u) \vee Q(z, y)) \wedge (\neg Q(z, y) \vee \exists v P(v)))$
$\equiv \forall z \exists v \forall u((\neg P(u) \vee Q(z, y)) \wedge (\neg Q(z, y) \vee P(v)))$ (PCNF)
$\equiv \forall z \exists v \forall u((\neg P(u) \wedge \neg Q(z, y)) \vee (\neg P(u) \wedge P(v)) \vee (Q(z, y) \wedge \neg Q(z, y)) \vee$
$\quad (Q(z, y) \wedge P(v)))$
$\equiv \forall z \exists v \forall u((\neg P(u) \wedge \neg Q(z, y)) \vee (\neg P(u) \wedge P(v)) \vee (Q(z, y) \wedge P(v)))$ (PDNF)
Skolemization: $\forall z \forall u((\neg P(u) \vee Q(z, y)) \wedge (\neg Q(z, y) \vee P(f(z))))$
Clausal form: $\{\{\neg P(u), Q(z, y)\}, \{\neg Q(z, y), P(f(z))\}\}$

4.4.6 Transformation into a prenex DNF and a prenex CNF:

$\forall x(\neg \forall y Q(x, y) \wedge P(z)) \rightarrow \exists z(\forall y Q(z, y) \wedge \neg P(x))$
$\equiv \neg \forall x(\neg \forall y Q(x, y) \wedge P(z)) \vee \exists z(\forall y Q(z, y) \wedge \neg P(x))$
$\equiv \exists x(\forall y Q(x, y) \vee \neg P(z)) \vee \exists z(\forall y Q(z, y) \wedge \neg P(x))$
$\equiv \exists u(\forall y Q(u, y) \vee \neg P(z)) \vee \exists w(\forall v Q(w, v) \wedge \neg P(x))$
$\equiv \exists u \exists w \forall y \forall v((Q(u, y) \vee \neg P(z)) \vee (Q(w, v) \wedge \neg P(x)))$ (PDNF)
$\equiv \exists u \exists w \forall y \forall v((Q(u, y) \vee \neg P(z) \vee Q(w, v)) \wedge (Q(u, y) \vee \neg P(z) \vee \neg P(x)))$
$\qquad\qquad\qquad\qquad\qquad\qquad\qquad\qquad\qquad\qquad\qquad\qquad\qquad$ (PCNF)

Skolemization:
$\forall y \forall u((Q(c_1, y) \vee \neg P(z) \vee Q(c_2, v)) \wedge (Q(c_1, y) \vee \neg P(z) \vee \neg P(x)))$
Clausal form: $\{\{Q(c_1, y), \neg P(z), Q(c_2, v)\}, \{Q(c_1, y), \neg P(z), \neg P(x)\}\}$

4.4.7   Transformation into a prenex CNF:

$\neg(\forall y(\forall z Q(y,z) \to P(z)) \to \exists z(P(z) \wedge \forall x(Q(z,y) \to Q(x,z))))$

$\equiv \forall y(\forall z Q(y,z) \to P(z)) \wedge \neg\exists z(P(z) \wedge \forall x(Q(z,y) \to Q(x,z)))$

$\equiv \forall y(\neg\forall z Q(y,z) \vee P(z)) \wedge \forall z(\neg P(z) \vee \neg\forall x(\neg Q(z,y) \vee Q(x,z)))$

$\equiv \forall y(\exists z \neg Q(y,z) \vee P(z)) \wedge \forall z(\neg P(z) \vee \exists x(Q(z,y) \wedge \neg Q(x,z)))$

$\equiv \forall y_1(\exists z_1 \neg Q(y_1,z_1) \vee P(z)) \wedge \forall z_2(\neg P(z_2) \vee \exists x(Q(z_2,y) \wedge \neg Q(x,z_2)))$

$\equiv \forall y_1 \exists z_1(\neg Q(y_1,z_1) \vee P(z)) \wedge \forall z_2 \exists x(\neg P(z_2) \vee (Q(z_2,y) \wedge \neg Q(x,z_2)))$

$\equiv \forall y_1 \exists z_1 \forall z_2 \exists x((\neg Q(y_1,z_1) \vee P(z)) \wedge (\neg P(z_2) \vee Q(z_2,y)) \wedge (\neg P(z_2) \vee \neg Q(x,z_2)))$

Skolemization:

$\forall y_1 \forall z_2((\neg Q(y_1,f(y_1)) \vee P(z)) \wedge (\neg P(z_2) \vee Q(z_2,y)) \wedge (\neg P(z_2) \vee \neg Q(g(y_1,z_2),z_2)))$

Clausification:

$C_1 = \{\neg Q(y_1, f(y_1)), P(z)\},$
$C_2 = \{\neg P(z_2), Q(z_2,y)\},$
$C_3 = \{\neg P(z_2), \neg Q(g(y_1,z_2),z_2)\}.$

Clausal form: $\{C_1, C_2, C_3\}$.

4.4.9   Transformation into a prenex DNF and a prenex CNF:

$\neg(\forall y(\neg\exists z Q(y,z) \to P(z)) \to \exists z((P(z) \to Q(z,y)) \wedge \neg\exists x R(x,y,z)))$

$\equiv \forall y(\neg\exists z Q(y,z) \to P(z)) \wedge \forall z((P(z) \wedge \neg Q(z,y)) \vee \exists x R(x,y,z))$

$\equiv \forall y(\exists z Q(y,z) \vee P(z)) \wedge \forall z((P(z) \wedge \neg Q(z,y)) \vee \exists x R(x,y,z))$

$\equiv \forall v(\exists w Q(v,w) \vee P(z)) \wedge \forall u((P(u) \wedge \neg Q(u,y)) \vee \exists x R(x,y,u))$

$\equiv \forall v \exists w \forall u \exists x((Q(v,w) \vee P(z)) \wedge ((P(u) \wedge \neg Q(u,y)) \vee R(x,y,u)))$ ($\star$)

$\equiv \forall v \exists w \forall u \exists x(((Q(v,w) \vee P(z)) \wedge (P(u) \wedge \neg Q(u,y))) \vee ((Q(v,w) \vee P(z)) \wedge R(x,y,u)))$

$\equiv \forall v \exists w \forall u \exists x((Q(v,w) \wedge P(u) \wedge \neg Q(u,y)) \vee (P(z) \wedge P(u) \wedge \neg Q(u,y)) \vee (Q(v,w) \wedge R(x,y,u)) \vee (P(z) \wedge R(x,y,u)))$ **(PDNF)**

$\equiv \forall v \exists w \forall u \exists x((Q(v,w) \vee P(z)) \wedge (P(u) \vee R(x,y,u)) \wedge (\neg Q(u,y) \vee R(x,y,u)))$**(PCNF from ($\star$))**

Skolemization:

$\forall v \forall u((Q(v,f(v)) \vee P(z)) \wedge (P(u) \vee R(g(v,u),y,u)) \wedge (\neg Q(u,y) \vee R(g(v,u),y,u)))$

Clausal form:

$\{\{Q(v,f(v)), P(z)\}, \{P(u), R(g(v,u),y,u)\}, \{\neg Q(u,y), R(g(v,u),y,u)\}\}$

Section 4.5

4.5.2   (a) Transformation of $\neg(\forall x P(x) \to \forall y P(y))$ into a clausal form:

$$C_1 = \{P(x)\} \text{ and } C_2 = \{\neg P(c)\}$$

for some Skolem constant c.

Applying Resolution: unify $P(c)$ and $P(x)$ with $\mathsf{MGU}[c/x]$, and then resolve C_1 with C_2 to obtain $\{\}$.

(c) Transformation of $\neg\exists x(P(x) \to \forall y P(y))$ into a clausal form:

$$C_1 = \{P(x)\} \text{ and } C_2 = \{\neg P(f(y))\}$$

for some Skolem function f.

Applying Resolution: unify $P(x)$ and $P(f(y))$ with $\mathsf{MGU}[f(y)/x]$, and then resolve C_1 with C_2 to obtain $\{\}$.

(e) Transformation of $\neg\exists x\neg A(x)$ and $\neg\forall x A(x)$ into a clausal form:

$$C_1 = \{\neg A(c)\} \text{ and } C_2 = \{A(x)\}$$

for some Skolem constant c.

Applying Resolution: unify $A(x)$ and $A(c)$ with $\mathsf{MGU}[c/x]$, and then resolve C_1 with C_2 to obtain $\{\}$.

(h) Transformation of $\exists x\exists y A(x, y)$ and $\neg\exists y\exists x A(x, y)$ into a clausal form:

$$C_1 = \{A(c_1, c_2)\} \text{ and } C_2 = \{\neg A(x, y)\}$$

for Skolem constants c_1 and c_2.

Applying Resolution: unify $A(c_1, c_2)$ and $A(x, y)$ with $\mathsf{MGU}[c_1/x, c_2/y]$, and then resolve C_1 with C_2 to obtain $\{\}$.

4.5.3 (a) Transforming each of $\forall x(P(x) \lor Q)$ and $\neg(\forall x P(x) \lor Q)$ to a clausal form:

$$C_1 = \{P(x), Q\}, C_2 = \{\neg P(c)\}, \text{ and } C_3 = \{\neg Q\}$$

for some Skolem constant c.

Now, applying the Resolution rule successively, we get

$$C_4 = Res(C_1, C_2) = \{Q\} \qquad\qquad \mathsf{MGU}[c/x]$$
$$C_5 = Res(C_3, C_4) = \{\}.$$

The empty clause is derived, hence the logical consequence holds.

(c) Transforming each of $\exists x(P(x) \land Q)$ and $\neg(\exists x P(x) \land Q)$ to a clausal form:

$$C_1 = \{P(c)\}, C_2 = \{Q\}, \text{ and } C_3 = \{\neg P(x), \neg Q\}$$

for some Skolem constant c.

Now, applying the Resolution rule successively:

$$C_4 = Res(C_1, C_3) = \{\neg Q\} \qquad\qquad \mathsf{MGU}[c/x]$$
$$C_5 = Res(C_2, C_4) = \{\}.$$

The empty clause is derived, hence the logical consequence holds.

(e) Transforming each of $\forall x(Q \to P(x))$, Q and $\neg\forall x P(x)$ to a clausal form:

$$C_1 = \{\neg Q, P(x)\}, C_2 = \{Q\}, \text{ and } C_3 = \{\neg P(c)\}$$

for some Skolem constant c.

Now, applying the Resolution rule successively:

$$C_4 = Res(C_1, C_3) = \{\neg Q\} \qquad \text{MGU}[c/x]$$
$$C_5 = Res(C_2, C_4) = \{\}.$$

The empty clause is derived, hence the logical consequence holds.

(h) Transforming each of $\exists x(Q \rightarrow P(x))$, Q and $\neg \exists x P(x)$ to a clausal form:

$$C_1 = \{\neg Q, P(c)\}, C_2 = \{Q\}, \text{ and } C_3 = \{\neg P(x)\}$$

for some Skolem constant c.

Now, applying the Resolution rule successively:

$$C_4 = Res(C_1, C_3) = \{\neg Q\} \qquad \text{MGU}[c/x]$$
$$C_5 = Res(C_2, C_4) = \{\}.$$

The empty clause is derived, hence the logical consequence holds.

4.5.4 (a) First, we transform each of $\forall x A(x)$, $\forall x B(x)$ and $\neg \forall x(A(x) \wedge B(x))$ to clausal form:

$$C_1 = \{A(x)\}, C_2 = \{B(x)\}, \text{ and } C_3 = \{\neg A(c), \neg B(c)\}$$

for some Skolem constant c.

Now, applying the Resolution rule successively, we get

$$C_4 = Res(C_1, C_2) = \{\neg B(c)\} \qquad \text{MGU}[c/x]$$
$$C_5 = Res(C_2, C_4) = \{\} \qquad \text{MGU}[c/x].$$

The empty clause is derived, hence the logical consequence holds.

(c) Transforming each of $\forall x(A(x) \vee B(x))$ and $\neg(\forall x A(x) \vee \forall x B(x))$ to clausal form:

$$C_1 = \{A(x), B(x)\}, C_2 = \{\neg A(c_1)\}, \text{ and } C_3 = \{\neg B(c_2)\}$$

for Skolem constants c_1 and c_2. The only possible resolutions on this set of clauses are as follows. Unify $A(x)$ and $A(c_1)$ and resolve C_1 and C_2 to obtain $C_4 = \{B(c_1)\}$. Likewise, unify $B(x)$ and $B(c_2)$ and resolve C_1 and C_3 to obtain $C_5 = \{A(c_2)\}$. No more clauses are derivable. The empty clause therefore cannot be derived. Hence $\forall x(A(x) \vee B(x)) \nvDash \forall x A(x) \vee \forall x B(x)$. A counter-model: for instance, consider the structure $\mathcal{H}$ with A and B taken as the predicates M and W, respectively.

(e) The empty clause cannot be derived, hence the logical consequence does not hold. A counter-model: consider for example the structure $\mathcal{H}$ with A and B taken as the predicates M and W, respectively.

(g) The empty clause cannot be derived, hence the logical consequence does not hold. A counter-model: for instance, consider the structure $\mathcal{H}$ with A and B taken as the predicates M and W, respectively.

(i) We transform each of $\neg(\exists x A(x) \vee \exists x B(x))$ and $\exists x (A(x) \vee B(x))$ to clausal form:

$$C_1 = \{A(c), B(c)\}, C_2 = \{\neg A(x)\}, \text{ and } C_3 = \{\neg B(y)\}$$

for some Skolem constant c. Now, applying the Resolution rule successively, we get

$$C_4 = Res(C_1, C_2) = \{B(c)\} \qquad \text{MGU}[c/x]$$
$$C_5 = Res(C_3, C_4) = \{\} \qquad \text{MGU}[c/y].$$

The empty clause is derived, hence the logical consequence holds.

(k) Transforming each of $\exists x A(x) \rightarrow \exists x B(x)$ and $\neg \exists x (A(x) \rightarrow B(x))$ to clausal form:

$$C_1 = \{\neg A(x), B(f(x))\}, C_2 = \{A(z)\}, \text{ and } C_3 = \{\neg B(z)\}$$

for some Skolem function f. Now, applying the Resolution rule successively, we get

$$C_4 = Res(C_1, C_2) = \{B(f(x))\} \qquad \text{MGU}[x/z]$$
$$C_5 = Res(C_3, C_4) = \{\} \qquad \text{MGU}[f(x)/z].$$

The empty clause is derived, hence the logical consequence holds.

4.5.5 (b) Using predicates $N(x)$ for "x is a natural number," $I(x)$ for "x is an integer," and $O(x)$ for "x is odd," we can formalize the argument as follows:

$$\frac{\forall x (N(x) \rightarrow I(x)), \exists x (I(x) \wedge O(x))}{\exists x (N(x) \wedge O(x))}.$$

Then we transform the formulas $\forall x (N(x) \rightarrow I(x))$, $\exists x (I(x) \wedge O(x))$, and $\neg \exists x (N(x) \wedge O(x))$ to clausal form:

$$C_1 = \{\neg N(x), I(x)\}, C_2 = \{I(c)\}, C_3 = \{O(c)\}, \text{ and}$$
$$C_4 = \{\neg N(y), \neg O(y)\},$$

for some Skolem constant c. Applying the Resolution rule we get

$$C_5 = Res(C_1, C_2) = \{\neg N(c)\} \qquad \text{MGU}[c/x]$$
$$C_6 = Res(C_3, C_4) = \{\neg N(c)\} \qquad \text{MGU}[c/y].$$

No more resolvents can be obtained, hence the empty clause cannot be derived and the argument is not valid.

(d) Using predicates $P(x)$ for "x is a penguin," $B(x)$ for "x is a bird," $W(x)$ for "x is white," and $E(x)$ for "x eats some fish," we can formalize the argument as follows:

$$\frac{\forall x (P(x) \rightarrow B(x)), \exists x (P(x) \wedge W(x)), \forall x (P(x) \rightarrow E(x))}{\exists x (W(x) \wedge B(x) \wedge E(x))}.$$

Now we transform the formulas $\forall x(P(x) \rightarrow B(x)), \exists x(P(x) \wedge W(x))$, $\forall x(P(x) \rightarrow E(x))$, and $\neg \exists x(W(x) \wedge B(x) \wedge E(x))$ to clausal form:

$$C_1 = \{\neg P(x), B(x)\} \qquad\qquad C_2 = \{P(c)\}$$
$$C_3 = \{W(c)\} \qquad\qquad C_4 = \{\neg P(x), E(x)\}$$
$$C_5 = \{\neg W(x), \neg B(x), \neg E(x)\}$$

for some Skolem constant c. Now, applying the Resolution rule successively, we get

$$C_6 = Res(C_3, C_5) = \{\neg B(c), \neg E(c)\} \qquad \text{MGU}[c/x]$$
$$C_7 = Res(C_4, C_6) = \{\neg P(c), \neg B(c)\} \qquad \text{MGU}[c/x]$$
$$C_8 = Res(C_1, C_7) = \{\neg P(c)\} \qquad \text{MGU}[c/x]$$
$$C_9 = Res(C_2, C_8) = \{\}.$$

The empty clause is derived, hence the argument is valid.

(f) Using predicates $Y(x)$ for "x is yellow," $P(x)$ for "x is a plonk," and $Q(x)$ for "x is a qlink," we can formalize the argument as follows:

$$\frac{\neg \exists x(Y(x) \wedge P(x) \wedge Q(x)), \exists x(P(x) \vee \neg Q(x))}{\exists x \neg (Y(x) \wedge Q(x))}.$$

We next transform the formulas $\neg \exists x(Y(x) \wedge P(x) \wedge Q(x))$, $\exists x(P(x) \vee \neg Q(x))$, and $\neg \exists x \neg (Y(x) \wedge Q(x))$ to clausal form:

$$C_1 = \{\neg Y(x), \neg P(x), \neg Q(x)\} \qquad C_2 = \{P(c), \neg Q(c)\}$$
$$C_3 = \{Y(x)\} \qquad\qquad C_4 = \{Q(x)\}$$

for some Skolem constant c.
Now, applying the Resolution rule successively, we get

$$C_5 = Res(C_1, C_2) = \{\neg Y(c), \neg Q(c)\} \qquad \text{MGU}[c/x]$$
$$C_6 = Res(C_3, C_5) = \{\neg Q(c)\} \qquad \text{MGU}[c/x]$$
$$C_7 = Res(C_4, C_6) = \{\} \qquad \text{MGU}[c/x].$$

The empty clause is derived, hence the argument is valid.

4.5.6 We formalize A and B in the domain of all men, using $P(x)$ for "x is happy" and $Q(x)$ for "x is drunk" as follows:
$A := \neg \exists x(P(x) \rightarrow Q(x))$
$B := \exists y P(y) \wedge \neg \exists z Q(z)$.
(a) First we check if A implies B.
Clausification of A and $\neg B$:
$C_1 := \{P(x), \neg Q(x)\}, C_2 := \{\neg P(y)\}, C_3 := \{Q(s_1)\}$,
where s_1 is a new Skolem constant.

Application of Resolution:
$C_4 := Res(C_1, C_2) = \{\neg Q(x)\}, MGU : [x/y]\ C_5 := Res(C_3, C_4) = \{\},$
$MGU : [s_1/x].$
The empty clause is derived and therefore the implication holds.

(b) We now check if B implies A.

Clausification of B and $\neg A$:
$C_1 := \{\neg P(s_1), Q(s_1)\}, C_2 := \{P(s_2)\}, C_3 := \{\neg Q(z)\},$
where s_1, s_2 are new Skolem constants.

Application of Resolution:
$C_4 := Res(C_1, C_3) = \{Q(s_2)\}, MGU : [s_3/z]$
No new clauses can be obtained and therefore the consequence does not hold.
A counter-model: $\mathfrak{J} = \langle X, P^{\mathfrak{J}}, Q^{\mathfrak{J}} \rangle,$
where $X = \{a_1, a_2\}, P^{\mathfrak{J}} = \{a_1\}, Q^{\mathfrak{J}} = \emptyset.$
Now, B is true if we choose y to be a_1 as $\neg \exists Q(z)$ is always true in this model. For A, we can choose x to be a_2. Then $P(x)$ is false and it follows that $\exists x(P(x) \rightarrow Q(x))$ is true and A is false. The consequence therefore fails.

4.5.7 (a) The argument can be written as follows:

$$\forall x(F(x) \rightarrow E(x))$$
$$\forall x(H(x) \rightarrow \neg \exists y(E(y) \wedge L(x,y)))$$
$$\exists x((G(x) \vee F(x)) \wedge L(\text{Kermit}, x))$$
$$\frac{\forall x(\exists y(G(y) \wedge L(x,y)) \rightarrow \neg \exists z(T(z) \wedge L(x,z)))}{\exists x(T(x) \wedge L(\text{Kermit}, x)) \rightarrow \neg H(\text{Kermit})}$$

From all premises and the negated conclusion we obtain the following clauses:

$C_1 = \{\neg F(x), E(x)\}$ $\qquad\qquad\qquad$ $C_2 = \{\neg H(x), \neg E(x),$
$\qquad\qquad\qquad\qquad\qquad\qquad\qquad\qquad\qquad \neg L(x, y)\}$

$C_3 = \{G(c_1), F(c_1)\}$ $\qquad\qquad\qquad$ $C_4 = \{L(\text{Kermit}, c_1)\}$

$C_5 = \{\neg G(y), \neg L(x, y), \neg T(z), \neg L(x,z)\}$ $\quad$ $C_6 = \{T(c_2)\}$

$C_7 = \{L(\text{Kermit}, c_2)\}$ $\qquad\qquad\qquad$ $C_8 = \{H(\text{Kermit})\}$

for Skolem constants c_1 and c_2.

Now, applying Resolution successively, we get

$C_9 = Res(C_5, C_7) = \{\neg G(y), \neg L(\text{Kermit}, y), \neg T(c_2)\}$ $\qquad$ $MGU[\text{Kermit}/x, c_2/z]$
$C_{10} = Res(C_6, C_9) = \{\neg G(y), \neg L(\text{Kermit}, y)\}$
$C_{11} = Res(C_3, C_{10}) = \{F(c_1), \neg L(\text{Kermit}, c_1)\}$ $\qquad$ $MGU[c_1/y]$
$C_{12} = Res(C_1, C_{11}) = \{E(c_1), \neg L(\text{Kermit}, c_1)\}$ $\qquad$ $MGU[c_1/x]$
$C_{13} = Res(C_2, C_{12}) = \{\neg H(x), \neg L(x, c_1), \neg L(\text{Kermit}, c_1)\}$ $\quad$ $MGU[c_1, y]$
$C_{14} = Res(C_8, C_{13}) = \{\neg L(\text{Kermit}, c_1)\}$ $\qquad$ $MGU[\text{Kermit}/x]$
$C_{15} = Res(C_4, C_{14}) = \{\}.$

The empty clause is derived, hence the argument is valid.

(c) We can formalize the argument as follows:

$$\forall x(T(x) \rightarrow ((E(x) \wedge \neg G(x)) \vee (G(x) \wedge \neg E(x))))$$
$$\forall y(T(y) \wedge Fl(y) \rightarrow E(y))$$
$$\forall x(D(x) \rightarrow Fl(x))$$
$$\forall x(G(x) \wedge U(x) \rightarrow T(x))$$
$$\exists x(G(x) \wedge U(x))$$
$$\overline{\exists x(U(x) \wedge \neg D(x))}$$

From all premises and the negated conclusion we obtain the following clauses:

$$C_1 = \{\neg T(x), E(x), G(x)\} \qquad C_2 = \{\neg T(x), \neg E(x), \neg G(x)\}$$
$$C_3 = \{\neg T(y), \neg Fl(y), E(y)\} \qquad C_4 = \{\neg D(u), Fl(u)\}$$
$$C_5 = \{\neg G(v), \neg U(v), T(v)\} \qquad C_6 = \{G(c)\}$$
$$C_7 = \{U(c)\} \qquad\qquad\qquad C_8 = \{\neg U(w), D(w)\}$$

for some Skolem constant c.

Now, applying the Resolution rule successively, we get

$$C_9 = Res(C_3, C_4) = \{\neg T(y), E(y), \neg D(y)\} \qquad \mathsf{MGU}[y/u]$$
$$C_{10} = Res(C_8, C_9) = \{\neg T(y), E(y), \neg U(y)\} \qquad \mathsf{MGU}[y/w]$$
$$C_{11} = Res(C_2, C_{10}) = \{\neg T(y), \neg U(y), \neg G(y)\} \qquad \mathsf{MGU}[y/x]$$
$$C_{12} = Res(C_5, C_{11}) = \{\neg G(y), \neg U(y)\} \qquad \mathsf{MGU}[y/v]$$
$$C_{13} = Res(C_6, C_{12}) = \{\neg U(c)\} \qquad\qquad \mathsf{MGU}[c/y]$$
$$C_{14} = Res(C_7, C_{13}) = \{\}.$$

The empty clause is derived, hence the argument is valid.

4.5.8 (a) Using predicates $B(x)$ for "x is a bachelor," $M(x)$ for "x is a man," $W(x)$ for "x is a woman," $K(x, y)$ for "x knows y," and $L(x, y)$ for "x loves y," we can formalize the argument as follows:

$$\exists x(B(x) \wedge \forall y(W(y) \wedge K(x, y) \rightarrow L(x, y)))$$
$$\forall x(B(x) \rightarrow M(x))$$
$$W(Eve) \wedge \forall x(M(x) \rightarrow K(x, Eve))$$
$$\overline{\exists x(B(x) \wedge \exists y(W(y) \wedge L(x, y)))}$$

We then obtain the following clausal form:

$$C_1 = \{B(c)\} \qquad\qquad\qquad C_2 = \{\neg W(y), \neg K(c, y), L(c, y)\}$$
$$C_3 = \{\neg B(z), M(z)\} \qquad\qquad C_4 = \{W(Eve)\}$$
$$C_5 = \{\neg M(w), K(w, Eve)\} \qquad C_6 = \{\neg B(u), \neg W(v), \neg L(u, v)\}$$

for some Skolem constant c. Now, applying Resolution successively, we get

$$C_7 = Res(C_2, C_6) = \{\neg B(c), \neg W(y), \neg K(c,y)\} \qquad \text{MGU}[c/u, y/v]$$
$$C_8 = Res(C_5, C_7) = \{\neg B(c), \neg M(c), \neg W(Eve)\} \quad \text{MGU}[c/w, Eve/y]$$
$$C_9 = Res(C_3, C_8) = \{\neg B(c), \neg W(Eve)\} \qquad\qquad \text{MGU}[c/z]$$
$$C_{10} = Res(C_1, C_9) = \{\neg W(Eve)\}$$
$$C_{11} = Res(C_4, C_{10}) = \{\}.$$

The empty clause is derived, hence the argument is valid.

(c) We formalize the argument using the following non-logical symbols: J to mean "Juan;" L to mean "Lara;" $W(x)$ to mean "x is a woman;" $M(x)$ to mean "x is a man;" $B(x)$ to mean "x is brave;" $P(x)$ to mean "x is pretty;" and $A(x, y)$ to mean "x admires y."
Here is the formalization of the argument:

"Every man admires some brave woman":
$$P_1 \equiv \forall x(M(x) \rightarrow \exists y(W(y) \wedge B(y) \wedge A(x,y))).$$

"Juan is a man who admires every pretty woman":
$$P_2 \equiv M(J) \wedge \forall y(W(y) \wedge P(y) \rightarrow A(J,y)).$$

"Lara is a brave or pretty woman": $P_3 \equiv W(L) \wedge (B(L) \vee P(L)).$

"Some man admires Lara": $Q \equiv \exists x(M(x) \wedge A(x,L)).$

Clausification:
$$P_1: C_1 = \{\neg M(x), W(f(x))\}, C_2 = \{\neg M(x), B(f(x))\},$$
$$C_3 = \{\neg M(x), A(x, f(x))\},$$
$$P_2: C_4 = \{M(J)\}, C_5 = \{\neg W(y), \neg P(y), A(J,y)\},$$
$$P_3: C_6 = \{W(L)\}, C_7 = \{B(L), P(L)\},$$
$$\neg Q: C_8 = \{\neg M(z), \neg A(z, L)\}.$$

Running Resolution:
$$C_9 = (C_4, C_8) = \{\neg A(J, L)\}$$
$$C_{10} = Res(C_1, C_4) = \{W(f(J))\},$$
$$C_{11} = Res(C_2, C_4) = \{B(f(J))\},$$
$$C_{12} = Res(C_3, C_4) = \{A(J, f(J))\},$$
$$C_{13} = (C_5, C_6) = \{\neg P(L), A(J, L)\}$$
$$C_{14} = (C_9, C_{13}) = \{\neg P(L)\}$$
$$C_{15} = (C_7, C_{14}) = \{B(L)\}$$
$$C_{16} = (C_5, C_9) = \{\neg W(L), \neg P(L)\}$$
$$C_{17} = (C_1, C_5) = \{\neg M(x), \neg P(f(x)), A(J, f(x))\},$$
$$C_{18} = (C_4, C_{17}) = \{\neg P(f(J)), A(J, f(J))\}.$$

No more new clauses can be derived. The empty clause has not been derived. Therefore, the argument is not logically correct.
A counter-model can be extracted from the execution above with universe $U = \{\text{Juan, Lara, X}\}$, where:
$M = \{\text{Juan}\}$, $W = B = \{\text{Lara, X}\}$, $P = \emptyset$, $f(u) = \text{X}$ for every $u \in U$, $A = \{(\text{Juan,X})\}$.

4.5.9 We use the predicates:

$T(x)$ for "x can talk,"

$W(x)$ for "x can walk,"

and $C(x, y)$ for "x can construct y."

(a) Formalizing the premises:

$P_1 = \forall x(T(x) \rightarrow W(x))$

$P_2 = \exists x(W(x) \land \forall y(T(y) \rightarrow C(x, y)))$

Formalizing the conclusion:

$Q = \exists x(T(x) \land \exists z(W(z) \land C(x, z)) \land \forall y(T(y) \rightarrow C(x, y)))$

Prenex normal form, Skolemization, and clausification (using different variables in the different clauses):

$P_1 \equiv \forall x(\neg T(x) \lor W(x))$. Clause: $C_1 = \{\neg T(x), W(x)\}$.

$P_2 \equiv \exists x \forall y(W(x) \land (\neg T(y) \lor C(x, y)))$.

Skolemized: $W(c) \land (\neg T(y) \lor C(c, y))$.

Clauses: $C_2 = \{W(c)\}$, $C_3 = \{\neg T(y), C(c, y)\}$.

Negated conclusion:

$\neg Q = \neg \exists x(T(x) \land \exists z(W(z) \land C(x, z)) \land \forall y(T(y) \rightarrow C(x, y)))$

$\equiv \forall x(\neg T(x) \lor \forall z(\neg W(z) \lor \neg C(x, z)) \lor \exists y(T(y) \land \neg C(x, y)))$

$\equiv \forall x \exists y \forall z(\neg T(x) \lor \neg W(z) \lor \neg C(x, z) \lor (T(y) \land \neg C(x, y)))$.

Skolemization produces:

$\forall x \forall z(\neg T(x) \lor \neg W(z) \lor \neg C(x, z) \lor (T(f(x)) \land \neg C(x, f(x))))$.

Clauses:

$C_4 = \{\neg T(x_1), \neg W(z_1), \neg C(x_1, z_1), T(f(x_1))\}$;

$C_5 = \{\neg T(x_2), \neg W(z_2), \neg C(x_2, z_2), \neg C(x_2, f(x_2))\}$.

Note that the clause C_5 can be equivalently simplified to the subsuming one

$C_5' = \{\neg T(x_2), \neg W(z_2), \neg C(x_2, f(x_2))\}$.

Running Resolution (most unifications are obvious):

$Res(C_1, C_4)$ (unifying $W(x)$ with $W(z_1)$) :=

$C_6 = \{\neg T(z_1), \neg T(x_1), \neg C(x_1, z_1), T(f(x_1))\}$;

$Res(C_1, C_5')$ (unifying $W(x)$ with $W(z_2)$) :=

$C_7 = \{\neg T(z_2), \neg T(x_2), \neg C(x_2, f(x_2))\}$;

$Res(C_2, C_4) := C_8 = \{\neg T(x_1), \neg C(x_1, c), T(f(x_1))\}$;

$Res(C_2, C_5') := C_9 = \{\neg T(x_2), \neg C(x_2, f(x_2))\}$;

(At this stage, the clause C_2 is 'used up'.)

$Res(C_1, C_8) := C_{10} = \{\neg T(x_1), \neg C(x_1, c), W(f(x_1))\}$;

. . .

It is easy to see that infinitely many clauses can be derived. However, the empty clause cannot be derived, hence the argument above is not logically correct. A counter-model: consider a universe with two robots $\{R_1 \text{ and } R_2\}$ such that:

R_1 can talk and walk, but cannot construct any robot;

R_2 cannot talk, but can walk and can construct R_1.

(b) Using the predicates defined in the previous exercise, as well as the following: $R(x)$ for "x is a robot" and $L(x)$ for "x can reason logically," we can formalize the argument as follows.

    i. Some devices are robots: $\exists x R(x)$.

    ii. All robots can talk or walk or reason logically:
       $\forall x(R(x) \to (T(x) \lor W(x) \lor L(x)))$.

    iii. No non-talking robot can reason logically:
       $\neg \exists x(R(x) \land \neg T(x) \land L(x))$.

    iv. All talking robots can construct every logically reasoning device:
       $\forall x(R(x) \land T(x) \to \forall y(L(y) \to C(x,y)))$.

    v. A device can construct a robot only if it is itself a robot:
       $\forall x(\exists y(R(y) \land C(x,y)) \to R(x))$.

    vi. Some logically reasoning devices can construct every non-walking robot:
       $\exists x(L(x) \land \forall y((R(y) \land \neg W(y)) \to C(x,y)))$.

    vii. Some talking robot can construct itself:
       $\exists x(R(x) \land T(x) \land C(x,x))$.

The empty clause is not derivable, hence the argument is not logically correct.

(c) The modified last premise formalizes as

$$\forall x(L(x) \to \forall y((R(y) \land \neg W(y)) \to C(x,y))).$$

The empty clause is still not derivable, hence the resulting argument is not logically correct.

(d) The added premise formalizes as

$$R(\text{Creepy}) \land \neg W(\text{Creepy}).$$

The empty clause is now derivable, hence the resulting argument is logically correct.

4.5.10 Indeed, the unit clause $\sigma(s) = \sigma(t)$ can be added as an instance of the reflexivity axiom $u = u$ and then, by unifying and resolving with $\neg s = t \lor P(s)$, used to derive $\sigma(P(t))$.

4.5.11 See algebraic derivations of all but the last exercises in many standard textbooks on group theory. For the last derivation, see the appendix in Wos and Robinson (1969).

Section 4.6

No solutions are provided for these exercises, but the interested reader can find detailed proofs of completeness for each of the deductive systems studied here in the references provided at the end of the section.

Section 5.1

5.1.2 (b) Here we combine direct reasoning with reasoning by contradiction. Let $a, b, x \in \mathbb{R}$, and assume $a < b$ and $(x - a)(x - b) < 0$. From $(x - a)(x - b) < 0$ we have that either $x - a < 0$ and $x - b > 0$ or $x - a > 0$ and $x - b < 0$, hence either $x < a$ and $b < x$ or $a < x$ and $x < b$. The first case leads to a contradiction with $a < b$ because $x < a$ and $b < x$ imply $b < a$ by transitivity of $<$, and $b < a$ implies not $a < b$ by irreflexivity of $<$. The second case implies the claim $a < x < b$.

(d) We use a proof by contraposition. Let $m, n \in \mathbb{Z}$, and suppose mn is even, yet it is not the case that m is even or n is even. So, both m and n are not even, hence they are odd. Then there are integers a and b such that $m = 2a + 1$ and $n = 2b + 1$. Therefore,

$$mn = (2a + 1)(2b + 1) = 4ab + 2a + 2b + 1 = 2(2ab + a + b) + 1,$$

which means that mn is odd, hence not even: a contradiction.

(f) We use a direct proof here. Let $a, b \in \mathbb{Z}$, and assume ab is odd. This is only possible if both a and b are odd. Hence, $a = 2m + 1$ and $b = 2n + 1$ for some integers m and n. Then we obtain $a + b = 2m + 1 + 2n + 1 = 2m + 2n + 2 = 2(m + n + 1)$; $a + b$ is therefore even.

(h) We give a proof by contradiction. Suppose there are positive integers m and n such that $\frac{1}{m} + \frac{1}{n} > 2$. But, since m and n are positive integers we have that $m \geq 1$ and $n \geq 1$, therefore $\frac{1}{m} \leq 1$ and $\frac{1}{n} \leq 1$, hence $\frac{1}{m} + \frac{1}{n} \leq 2$: a contradiction.

Section 5.2

Exercises on sets and operations on sets

I only sketch a few sample proofs here, presented as semi-formal mathematical arguments. I leave it to the interested reader to formalize them completely, in **ND** or another deductive system.

5.2.4 (a) We proceed with proof by contradiction: suppose $A \cap \emptyset$ is non-empty and take any element of x of $A \cap \emptyset$. Then, in particular, $x \in \emptyset$, which is impossible. $A \cap \emptyset$ is therefore the empty set.

(g) Suppose $X \subseteq A$ and $X \subseteq B$, we need to show that $X \subseteq A \cap B$. To that end, suppose that $x \in X$. Since $X \subseteq A$ we have $x \in A$, and since $X \subseteq B$ we also have $x \in B$. By the definition of intersection, we therefore have $x \in A \cap B$.

5.2.5 (a) To prove that $A \cup \emptyset \subseteq A$, suppose that $x \in A \cup \emptyset$. Hence $x \in A$ or $x \in \emptyset$. But since it cannot be the case that $x \in \emptyset$, it must be the case that $x \in A$. For the inclusion $A \subseteq A \cup \emptyset$, suppose that $x \in A$. Then $x \in A$ or $x \in \emptyset$, so $x \in A \cup \emptyset$.

5.2.6   (a) Suppose that $x \in A - A$. That means $x \in A$ and $x \notin A$, which is impossible. There can therefore be no element $x \in A - A$, that is, $A - A = \emptyset$.

      (e) Suppose $x \in (A - B) \cap B$. This means $x \in (A - B)$ and $x \in B$. But $x \in (A - B)$ means $x \in A$ and $x \notin B$. So it cannot be the case that $x \in B$, which is a contradiction. We conclude that $(A - B) \cap B$ has no elements, that is, $(A - B) \cap B = \emptyset$.

      (k) We will prove both inclusions. To prove that $A - (B \cup C) \subseteq (A - B) \cap (A - C)$, suppose that $x \in A - (B \cup C)$. Then, $x \in A$ and $x \notin (B \cup C)$, that is, $x \in A$ but $x \notin B$ and $x \notin C$. Then $x \in A - B$ and $x \in A - C$, that is, $x \in (A - B) \cap (A - C)$.
For the inclusion $(A - B) \cap (A - C) \subseteq A - (B \cup C)$, suppose that $x \in (A - B) \cap (A - C)$. This means that $x \in (A - B)$ and $x \in (A - C)$, that is, $x \in A$ but $x \notin B$ and $x \notin C$. Since $x \notin B$ and $x \notin C$ we have $x \notin (B \cup C)$. Combining these with $x \in A$, we have $x \in A - (B \cup C)$.

5.2.7   (a) The proof goes through a chain of "if and only if"s:
$x \in A \cap (B \cup C)$
iff $x \in A$ and $x \in (B \cup C)$
iff $x \in A$, and $x \in B$ or $x \in C$
iff $x \in A$ and $x \in B$, or $x \in A$ and $x \in C$
iff $x \in A \cap B$ or $x \in A \cap C$
iff $x \in (A \cap B) \cup (A \cap C)$.
Note that for the third "iff" we made use of the propositional tautology $p \wedge (q \vee r) \equiv (p \wedge q) \vee (p \wedge r)$, which looks very much like the set-theoretic property we are proving. This is not accidental, as there is a close relationship between set theory and logic. Think about it!

5.2.8   (g) In order to construct a subset A of X, for each element of X we must decide whether it will be in A or not and these choices completely determine A. For each element of X there are therefore two choices: "in" or "out." The number of ways to form the subset A is therefore $\underbrace{2 \times 2 \times \times \cdots \times 2}_{n\text{times}} = 2^n$.

5.2.9   (a) For the inclusion $A \times (B \cup C) \subseteq (A \times B) \cup (A \times C)$, suppose that $x \in A \times (B \cup C)$. Then $x = (y, z)$ for some $y \in A$ and $z \in (B \cup C)$. Since $z \in (B \cup C)$, we have $z \in B$ or $z \in C$. So $(y, z) \in A \times B$ or $(y, z) \in A \times C$, and hence $x = (y, z) \in (A \times B) \cup (A \times C)$.
For the inclusion $(A \times B) \cup (A \times C) \subseteq A \times (B \cup C)$, suppose that $x = (y, z) \in (A \times B) \cup (A \times C)$. Then $(y, z) \in (A \times B)$ or $(y, z) \in (A \times C)$. Thus $y \in A$, while $z \in B$ or $z \in C$. We therefore have $y \in A$ while $z \in B \cup C$, and hence $x = (y, z) \in A \times (B \cup C)$.

Exercises on functions

5.2.10 Since f is surjective, we have $\mathsf{rng}(f) = B = \mathsf{dom}(f^{-1})$. To prove that f^{-1} is injective, suppose that $b_1, b_2 \in B$ and that $f^{-1}(b_1) = f^{-1}(b_2)$. Since $\mathsf{rng}(f) = B$, there are $a_1, a_2 \in A$ such that $f(a_1) = b_1$ and $f(a_2) = b_2$. Thus $a_1 = f^{-1}(f(a_1)) = f^{-1}(b_1) = f^{-1}(b_2) = f^{-1}(f(a_2)) = a_2$. Since we now have $a_1 = a_2$, the injectivity of f allows us to conclude that $f(a_1) = f(a_2)$ and therefore that $b_1 = b_2$, as desired.

To show that f^{-1} is surjective, suppose that $a \in A$. We must show that there is a $b \in B$ such that $f^{-1}(b) = a$. Take $f(a) \in B$: we have that $f^{-1}(f(a)) = a$.

5.2.12 (a) Suppose f and g are injective, and that $a_1, a_2 \in A$ such that $gf(a_1) = gf(a_2)$. We must show that $a_1 = a_2$. Thus $g(f(a_1)) = g(f(a_2))$. But since g is injective this means that $f(a_1) = f(a_2)$, which in turn means that $a_1 = a_2$, since f is injective.

(c) Follows immediately from Proposition 194, (a) and (b).

(e) Suppose that gf is surjective and that $c \in C$. We need to find an $b \in B$ such that $g(b) = c$. By the surjectivity of gf there is an $a \in A$ such that $gf(a) = c$, that is, $g(f(a)) = c$. Then $f(a) \in B$ is the desired element of B.

5.2.13 By Proposition 194(3) gf is bijective and therefore we know that it has an inverse $(gf)^{-1}$. Clearly $\mathsf{dom}(f^{-1}g^{-1}) = C = \mathsf{dom}((gf)^{-1})$ and $\mathsf{rng}(f^{-1}g^{-1}) = A = \mathsf{rng}((gf)^{-1})$. We need to show that $f^{-1}g^{-1}(c) = (gf)^{-1}(c)$ for all $c \in C$. To this end, let $c \in C$ arbitrarily. Then, because gf is surjective, there is an $a \in A$ such that $gf(a) = c$. By definition of inverses $(gf)^{-1}(c) = (gf)^{-1}(gf(a)) = a$. Now $f^{-1}g^{-1}(c) = f^{-1}g^{-1}(gf(a)) = f^{-1}g^{-1}(g(f(a))) = f^{-1}(g^{-1}(g(f(a))))$. By the associativity of function composition (Proposition 193) this last expression is equal to $f^{-1}((g^{-1}g)(f(a))) = f^{-1}(f(a)) = a$.

5.2.14 (a) For the sake of definiteness, suppose that $f : A \to B$. To prove the left-to-right implication suppose that f is injective and take any mappings $g_1 : C \to A$ and $g_2 : C \to A$. Further suppose that $g_1 f = g_2 f$. We must show that $g_1 = g_2$. So let $c \in C$ arbitrarily. Then $fg_1(c) = fg_2(c)$, that is, $f(g_1(c)) = f(g_2(c))$. Since f is injective, it follows that $g_1(c) = g_2(c)$.

To prove the converse we will proceed by contraposition, so suppose that f is not injective. Then there are $a_1, a_2 \in A$ such that $a_1 \neq a_2$ but $f(a_1) = f(a_2)$. We need to show that the left cancellation rule does not hold. For this it is sufficient to find two mappings g_1 and g_2 such that $fg_1 = fg_2$ but $g_1 \neq g_2$. Let $g_1 : A \to A$ and $g_2 : A \to A$ be maps such that g_1 is the identity map on A, that is, $g_1(a) = a$ for all $a \in A$, and $g_2(a) = a$ for all $a \in A - \{a_1, a_2\}$ but $g(a_1) = a_2$ and $g(a_2) = a_1$. Then it is easy to check that $fg_1 = fg_2$, but $g_1 \neq g_2$.

Exercises on binary relations

5.2.15 (a) For the inclusion $\mathsf{dom}(R \cup S) \subseteq \mathsf{dom}(R) \cup \mathsf{dom}(S)$, suppose that $a \in \mathsf{dom}(R \cup S)$, that is, there exists $b \in B$ such that $(a, b) \in R \cup S$. But then $(a, b) \in R$ or $(a, b) \in S$, that is, $a \in \mathsf{dom}(R)$ or $a \in \mathsf{dom}(S)$, that is, $a \in \mathsf{dom}(R) \cup \mathsf{dom}(S)$.

For the converse inclusion, suppose that $a \in \mathsf{dom}(R) \cup \mathsf{dom}(S)$, that is, $a \in \mathsf{dom}(R)$ or $a \in \mathsf{dom}(S)$, that is, there is a $b \in B$ such that $(a, b) \in R$ or $(a, b) \in S$. But then $(a, b) \in R \cup S$, and hence $a \in \mathsf{dom}(R \cup S)$.

5.2.16 (a) $b \in \mathsf{dom}(R^{-1})$ iff $\exists a \in A$ such that $(b, a) \in R^{-1}$ iff $\exists a \in A$ such that $(a, b) \in R$ iff $b \in \mathsf{rng}(R)$.

(d) $(b, a) \in (R \cup S)^{-1}$ iff $(a, b) \in (R \cup S)$ iff $(a, b) \in R$ or $(a, b) \in S$ iff $(b, a) \in R^{-1}$ or $(b, a) \in S^{-1}$ iff $(b, a) \in R^{-1} \cup S^{-1}$.

5.2.17 To be definite, let $R \subseteq A \times B$, $S \subseteq B \times C$, and $T \subseteq C \times D$. Then $(a, d) \in (R \circ S) \circ T$, iff there exists a $c \in C$ such that $(a, c) \in (R \circ S)$ and $(c, d) \in T$, iff there exists $c \in C$ and $b \in B$ such that $(a, b) \in R$, $(b, c) \in S$, and $(c, d) \in T$, iff there exists $b \in B$ such that $(a, b) \in R$ and $(b, d) \in S \circ T$, iff $(a, d) \in R \circ (S \circ T)$.

5.2.19 (c) Suppose that R is asymmetric, that is, $\forall x \forall y (xRy \rightarrow \neg yRx)$. Now suppose that $(x, y) \in R^{-1} \cap R$, that is, $(x, y) \in R^{-1}$ and $(x, y) \in R$, that is, $(y, x) \in R$ and $(x, y) \in R$, which is impossible by the asymmetry of R. Hence, $R^{-1} \cap R = \emptyset$.

For the converse we argue contrapositively. Suppose that R is not asymmetric, that is $\neg \forall x \forall y (xRy \rightarrow \neg yRx)$, that is, there exists $x, y \in X$ such that $(x, y) \in R$ and $(y, x) \in R$, that is, $(x, y) \in R$ and $(x, y) \in R^{-1}$, that is, $(x, y) \in R \cap R^{-1}$. So we conclude that $R^{-1} \cap R \neq \emptyset$.

(e) Suppose R is connected. To show that $R \cup R^{-1} \cup E_X = X^2$ we only need to show that $X^2 \subseteq R \cup R^{-1} \cup E_X$ since the converse containment holds for any relation. To that end, take an arbitrary $(x, y) \in X^2$. If $x = y$, then $(x, y) = (x, x) \in E_X$ and we are done. If $x \neq y$, then since R is connected we know that xRy or yRx, that is, xRy or $xR^{-1}y$, that is, $(x, y) \in R \cup R^{-1}$. Conversely, suppose that $X^2 = R \cup R^{-1} \cup E_X$. For all $x, y \in X$, if $x \neq y$, we therefore have $(x, y) \in R$ or $(x, y) \in R^{-1}$. That is, for all $x, y \in X$, we have $(x, y) \in R$ or $(y, x) \in R$, which precisely means that R is connected.

5.2.21 Hint: consider the intersection of all reflexive/symmetric/transitive relations containing the given relation.

5.2.22 (c) Let $\mathcal{X} = \{S \mid R \subseteq S \subseteq X^2$, where S is transitive$\}$, that is, $\mathcal{X}$ is the set consisting of all transitive binary relations on X which contain R. We must show that $R^{\mathrm{tran}} = \bigcap \mathcal{X}$. We begin by noting that $\mathcal{X} \neq \emptyset$, since X^2 is transitive and $R \subseteq X^2$. By definition of $\mathcal{X}$ we have that $R \subseteq S$ for all $S \in \mathcal{X}$, so $R \subseteq \bigcap \mathcal{X}$ by Proposition 186(7). Also, $\bigcap \mathcal{X}$ is transitive: suppose that $(x, y), (y, z) \in \bigcap \mathcal{X}$. Then $(x, y), (y, z) \in S$ for all $S \in \mathcal{X}$ and, since each

$S \in \mathcal{X}$ is transitive, $(x, z) \in S$ for all $S \in \mathcal{X}$. Hence $(x, z) \in \bigcap \mathcal{X}$. We have therefore established that $\bigcap \mathcal{X}$ is a transitive binary relation containing R. All that remains is to show that $\bigcap \mathcal{X}$ is the *smallest* such relation. To that end, suppose that $R' \subseteq X^2$ is transitive and $R \subseteq R'$. Then, by definition of $\mathcal{X}$ we have $R' \in \mathcal{X}$, and hence $\bigcap \mathcal{X} \subseteq R'$.

5.2.24   The proofs of (1), (2), and (4) are left as an exercise. We prove (3) by proving the contrapositive. Suppose that $[x]_R \cap [y]_R \neq \emptyset$. Then there is at least one element $z \in [x]_R \cap [y]_R$, and so xRz and yRz. By symmetry we also have zRx, and then by transitivity from yRz and zRx we get yRx, that is, $x \in [y]_R$.

5.2.25   In order to show that $\{[x]_R \mid x \in X\}$ satisfies the conditions of the definition of partition, we need to show that $\bigcup \{[x]_R \mid x \in X\} = X$ and that for any two different equivalence classes $[x]_R$ and $[y]_R$ it holds that $[x]_R \cap [y]_R = \emptyset$.

Since $[x]_R \subseteq X$ for each $x \in X$ we immediately have that $\bigcup \{[x]_R \mid x \in X\} \subseteq X$. For the sake of the inclusion in the other direction, let $y \in X$ arbitrarily. By Proposition 204(1), $y \in [y]_R$ and therefore $y \in \bigcup \{[x]_R \mid x \in X\}$. We have therefore established that $\bigcup \{[x]_R \mid x \in X\} = X$.

Next suppose that $[x]_R \neq [y]_R$. By Proposition 204(4) this implies that $[x]_R \cap [y]_R = \emptyset$. This completes the proof.

5.2.27   (b) To show that $\tilde{R}$ is well defined means, as usual, to show that the definition of $\tilde{R}$ does not depend on the particular choice of representatives of equivalence classes. In particular, we need to show that if $[x]_\sim = [x']_\sim$ and $[y]_\sim = [y']_\sim$, then $[x]_\sim \tilde{R}[y]_\sim$ iff $[x']_\sim \tilde{R}[y']_\sim$. To that purpose, suppose that $[x]_\sim = [x']_\sim$ and $[y]_\sim = [y']_\sim$ and that $[x]_\sim \tilde{R}[y]_\sim$. We have to show that $[x']_\sim \tilde{R}[y']_\sim$. Since $[x]_\sim \tilde{R}[y]_\sim$ we have xRy. From the facts $[x]_\sim = [x']_\sim$ and $[y]_\sim = [y']_\sim$ it follows by Proposition 204 that $x \sim x'$ and $y \sim y'$. Now applying the definition of $\sim$ to these last two facts we have xRx', $x'Rx$, yRy', and $y'Ry$. Next applying the fact that R is transitive to the facts $x'Rx$ and xRy yields $x'Ry$, and then again applying the transitivity of R to the facts $x'Ry$ and yRy' yields $x'Ry'$. But then, by the definition of $\tilde{R}$, we have $[x']_\sim \tilde{R}[y']_\sim$ as we wanted. A symmetric argument proves that $[x]_\sim \tilde{R}[y]_\sim$ if $[x']_\sim \tilde{R}[y']_\sim$.

Now that we have established that $\tilde{R}$ is well defined, we proceed to proving that it is a partial order $X/_\sim$.

Reflexivity:
Since R is reflexive we have xRx for all $x \in X$ and hence that $[x]_\sim \tilde{R}[x]_\sim$ for all $[x]_\sim \in X/_\sim$.

Transitivity:
Suppose $[x]_\sim \tilde{R}[y]_\sim$ and $[y]_\sim \tilde{R}[z]_\sim$. By the definition of $\tilde{R}$ we have xRy and yRz and hence by the transitivity of R that xRz. Again, by applying the definition of $\tilde{R}$ we get $[x]_\sim \tilde{R}[z]_\sim$.

Anti-symmetry:
Suppose that $[x]_\sim \tilde{R}[y]_\sim$ and $[y]_\sim \tilde{R}[x]_\sim$. By the definition of $\tilde{R}$ we have xRy and yRx and consequently, by the definition of $\sim$, that $x \sim y$. But then, by Proposition 204, $[x]_\sim = [y]_\sim$.

Exercises on ordered sets

5.2.33 (a) We prove (1). Suppose that x and x' are infima of Y. Then, since x is a lower bound of Y we must have $x \leq x'$. Similarly, since x' is a lower bound of Y we must have $x' \leq x$. Then $x = x'$, by the anti-symmetry of $\leq$.

5.2.34 It is sufficient to exhibit an infinite strictly descending chain of elements of $\mathcal{P}(X)$. The key idea is that what remains after removing a single element from an infinite set is again an infinite set.

5.2.35 (a) ($\Rightarrow$) Suppose that $(X, \leq)$ is well-founded but, for the sake of contradiction, that there is a subset $\emptyset \neq Y \subseteq X$ such that Y has no minimal element. Take an arbitrary element of Y, and call it y_1. Since y_1 cannot be a minimal element of Y there must be another element, y_2 say, such that $y_2 \in Y$ and $y_1 > y_2$. Again, since y_2 cannot be a minimal element of Y, there must be another element, y_3 say, such that $y_3 \in Y$ and $y_1 > y_2 > y_3$. Continuing in this way we can construct an infinite descending chain, contradicting our assumption that $(X, \leq)$ is well-founded. (This step of the proof requires a special axiom of set theory know as the **Axiom of Choice**.)

($\Leftarrow$) We prove the contrapositive. Suppose that $(X, \leq)$ is not well-founded. If $(X, \leq)$ is not well-founded it must contain an infinite descending chain, say $x_1 > x_2 > x_3 > \cdots$. Now let $Y = \{x_1, x_2, x_3, \dots \}$. Clearly, Y is a non-empty subset of X which does not have a least element.

5.2.36 Assume the contrary, that is, $X \setminus P \neq \emptyset$. Then $X \setminus P$ has a minimal element x. Then all elements of X less than x belong to P, hence x must belong to P: a contradiction.

Section 5.3

Exercises on Mathematical Induction

5.3.1 1.1 For $n = 0$, $0^2 + 0 = 0$, which is even. Our inductive hypothesis is that $k^2 + k$ is even. Now, for $n = k + 1$,

$$(k + 1)^2 + k + 1 = k^2 + 2k + 1 + k + 1 = k^2 + k + 2(k + 1).$$

We know that $k^2 + k$ is even from the inductive hypothesis. Besides, $2(k + 1)$ is also even, by definition. The sum of two even numbers is even, so $k^2 + k$ is also even.

1.3 First, the set $\{1\}$ has two subsets, namely $\{1\}$ and $\emptyset$, so clearly, the powerset of $\{1\}$ has 2^1 elements. Our inductive hypothesis is that the powerset of $\{1, 2, 3, \dots, k\}$ has 2^k elements for some $k \geq 1$. Now, let $A = \{1, 2, \dots, k, k + 1\}$. Choose an element $a \in A$ and set $A' = A - \{a\}$. Note that $\mathcal{P}(A) = \{X \subseteq A \mid a \notin X\} \cup \{X \subseteq A \mid a \in X\}$. It is clear that these sets are disjoint, so to find the number of elements in $\mathcal{P}(A)$, we need only find the number of the elements in each of these sets and add them together. First, clearly $\mathcal{P}(A') = \{X \subseteq A \mid a \notin X\}$ and so, since A' has k elements, $\{X \subseteq A \mid a \notin X\}$ has 2^k elements by the inductive hypothesis. Next, note that $X = Y \cup \{a\}$ for all $X \in \{X \subseteq A \mid a \in X\}$, where $Y \in \mathcal{P}(A')$. Since there are k elements in the set A', there are 2^k such Y by the inductive hypothesis. Hence, the set $\{X \subseteq A \mid a \in X\}$ has 2^k elements. In total, $\mathcal{P}(A)$ therefore has $2^k + 2^k = 2 \cdot 2^k = 2^{k+1}$ elements.

1.5 For $n = 1$ both sides equal 1. The inductive hypothesis is that

$$1 + 3 + 5 + \cdots + 2k - 1 = k^2$$

for some $k \geq 1$. Then, for $n = k + 1$ we have

$$1 + 3 + 5 + \cdots + 2k - 1 + 2(k + 1) - 1 = k^2 + 2k + 2 - 1$$
$$= k^2 + 2k + 1$$
$$= (k + 1)^2.$$

1.7 For $n = 1$ we have $1^3 = 1$ and $\frac{1^2(1+1)^2}{4} = \frac{4}{4} = 1$.
The inductive hypothesis is that

$$1^3 + 2^3 + 3^3 + \cdots + k^3 = \frac{k^2(k + 1)^2}{4}$$

for some $k \geq 1$. Then, for $n = k + 1$ we have

$$1^3 + 2^3 + 3^3 + \cdots + k^3 + (k + 1)^3 = \frac{k^2(k + 1)^2}{4} + (k + 1)^3$$
$$= \frac{k^2(k + 1)^2 + 4(k + 1)^3}{4} = \frac{(k + 1)^2(k^2 + 4(k + 1))}{4}$$
$$= \frac{(k + 1)^2(k^2 + 4k + 4)}{4} = \frac{(k + 1)^2(k + 2)^2}{4}$$

1.9 For $n = 1$ we have $\frac{1}{1 \times 3} = \frac{1}{3}$ and $\frac{3+5}{4(1+1)(1+2)} = \frac{8}{24} = \frac{1}{3}$.
The inductive hypothesis is

$$\frac{1}{1 \times 3} + \frac{1}{2 \times 4} + \cdots + \frac{1}{k \times (k + 2)} = \frac{k(3k + 5)}{4(k + 1)(k + 2)}$$

for some $k \geq 1$. Then, for $n = k + 1$ we have

$$\frac{1}{1 \times 3} + \frac{1}{2 \times 4} + \cdots + \frac{1}{k \times (k+2)} + \frac{1}{(k+1)(k+3)}$$

$$= \frac{k(3k+5)}{4(k+1)(k+2)} + \frac{1}{(k+1)(k+3)} = \frac{k(3k+5)(k+3) + 4(k+2)}{4(k+1)(k+2)(k+3)}$$

$$= \frac{3k^3 + 14k^2 + 19k + 8}{4(k+1)(k+2)(k+3)} = \frac{(k+1)(3k^2 + 11k + 8)}{4(k+1)(k+2)(k+3)}$$

$$= \frac{(k+1)(3k+8)(k+1)}{4(k+1)(k+2)(k+3)} = \frac{(k+1)(3(k+1)+5)}{4((k+1)+1)((k+1)+2)}.$$

5.3.2 Suppose there are natural numbers greater than or equal to n_0, for which P does not hold. The set of such numbers has a least element m. First, m must be greater than n_0, since P does hold for n_0. Since m is the least natural number for which P does not hold, P must hold for all natural numbers less than m. Hence, by assumption, P must holds for m, which is a contradiction.

5.3.3 First, assume the PCMI holds. Now, suppose that some property P holds for some natural number $n_0 \geq 0$, and whenever P holds for some natural number $m \geq n_0$, it holds for $m + 1$. Then if P holds for all natural numbers less that $m + 1$ and greater than or equal n_0, then it also holds for $m + 1$. By PMCI, P therefore holds for every natural number greater than or equal to n_0. We have therefore proved the PMI.

Conversely, assume the PMI holds. We will prove the PCMI as follows. Suppose that the property P holds for some natural number n_0, and for any natural number $k \geq n_0$, whenever P holds for all natural numbers greater than or equal to n_0 and less than k, then P holds for k. We consider the property P^* where $P^*(n)$ means that $P(n)$ holds for all natural numbers greater than or equal to n_0 and less than n. We will prove using the PMI that $P^*(n)$ holds for all natural numbers greater than or equal to n_0 which, by the assumption above, will imply that $P(n)$ holds for all natural numbers greater than or equal to n_0. Indeed, $P^*(n_0)$ holds vacuously, because there are no natural numbers greater than or equal to n_0 and less than n_0. Now, suppose that $P^*(k)$ holds for some $k \geq n_0$. To show that $P^*(k + 1)$ also holds, it is sufficient to note first that $P(k + 1)$ holds by the starting assumption and the assumption that $P^*(k)$ holds, and second that the natural numbers greater than or equal to n_0 and less than $k + 1$ are precisely k and all those that are greater than or equal to n_0 and less than k. P therefore holds for all of them, hence $P^*(k + 1)$ holds. This completes the proof.

5.3.4 Suppose not. Then there must be a least non-interesting natural number. But then, this makes it quite interesting, no?

Exercises on Peano Arithmetic

Here are just two sample proofs, using the Induction Principle formalized by the Induction Scheme and previous exercises from the list.

5.3.7.17 Proof by induction on y.

When $y = \mathbf{0}$ the claim holds because:

(a) $PA \vdash x \leq \mathbf{0} \leftrightarrow x = \mathbf{0}$, by definition of $\leq$, (11), and (12);

(b) $PA \vdash \neg(x < \mathbf{0})$ because of the above.

Now, we are to prove $PA \vdash (x \leq y \leftrightarrow (x < y \vee x = y)) \rightarrow (x \leq s(y) \leftrightarrow (x < s(y) \vee x = s(y)))$. The strategy will now be to prove the consequent (without essential use of the antecedent). To simplify the argument we reason partly semantically (skipping some propositional steps) by showing the following.

(a) $PA \vdash x \leq s(y) \leftrightarrow (x \leq y \vee x = s(y))$ (see Exercise 7.16).

(b) $PA \vdash x < s(y) \leftrightarrow (x \leq s(y) \wedge x \neq s(y))$ by definition of $<$.

(c) Therefore $PA \vdash (x < s(y) \vee x = s(y)) \leftrightarrow ((x \leq s(y) \wedge x \neq s(y)) \vee x = s(y))$.

(d) $PA \vdash ((x \leq s(y) \wedge x \neq s(y)) \vee x = s(y)) \leftrightarrow (x \leq y \vee x = s(y))$
because $((A \wedge \neg B) \vee B) \leftrightarrow (A \vee B)$ is a tautology.

This completes the induction.

5.3.7.20 Proof by induction on y.

$PA \vdash \mathbf{0} \neq \mathbf{0} \rightarrow x \leq x \times \mathbf{0}$ holds because of the false antecedent.

To prove $PA \vdash (y \neq \mathbf{0} \rightarrow x \leq x \times y) \rightarrow (s(y) \neq \mathbf{0} \rightarrow x \leq x \times s(y))$ it is sufficient to prove $PA \vdash (y \neq \mathbf{0} \rightarrow x \leq x \times y) \rightarrow (x \leq x \times s(y))$. This holds because $PA \vdash x \times s(y) = x \times y + x$ by (PA6) and $PA \vdash x \leq x \times y + x$ by definition of $\leq$ and commutativity of $+$.

This completes the induction.

Section 5.4

5.4.3 (a) (Q1) :- ancestor(Elisabeth, Harry).
　　　Answer: "No".

(b) (Q2) :- ancestor(Diana, Y).
　　　Answer: {William, Harry, George}.

(c) (Q3) :- ancestor(X, Harry).
　　　Answer: {Diana}.

(d) (Q4) :- ancestor(X, George).
　　　Answer: {William, Diana, Charles, Elisabeth}.

References

D'Agostino, M., Gabbay, D.M., Hähnle, R., and Posegga, J. (eds.) 1999. *Handbook of Tableau Methods*. Kluwer Academic Publishing, Dordrecht.

Barwise, J. and Echemendy, J. 1999. *Language, Proof and Logic*. CSLI Publication, Stanford.

Ben-Ari, M. 2012. *Mathematical Logic for Computer Science*, third edition. Springer, New York.

van Benthem, J., van Ditmarsch, H., van Eijck, J., and Jaspars, J. 2014. *Logic in Action*. Open Course Project, University of Amsterdam, http://www.logicinaction.org/docs/lia.pdf (accessed 7 April 2016).

Boole, G. 2005. *An Investigation of the Laws of Thought*. Project Gutenberg, eBook [#15114].

Boolos, G., Burgess, J., and Jeffrey, R. 2007. *Computability and Logic*, fifth edition. Cambridge University Press, Cambridge.

Bornat, R. 2005. *Proof and Disproof in Formal Logic*. Oxford Texts in Logic, Oxford University Press, Oxford.

Carroll, L. 1886. *The Game of Logic*. MacMillan and Co., London. Available from the Project Gutenberg as Ebook at http://www.gutenberg.org/ebooks/4763 (Accessed 6 April 2016).

Carroll, L. 1897. *Symbolic Logic*. MacMillan and Co., London. Available from the Project Gutenberg as Ebook at http://www.gutenberg.org/files/28696/28696-h/28696-h.htm (accessed 6 April 2016).

Chang, C.-L. and Lee, R. C.-T. 1997. *Symbolic Logic and Mechanical Theorem Proving*. Academic Press, Cambride, MA.

Chiswell, I. and Hodges, W. 2007. *Mathematical Logic*. Oxford University Press, Oxford.

Conradie, W. and Goranko, V. 2015. *Logic and Discrete Mathematics: A Concise Introduction*. John Wiley & Sons, Chichester.

Copi, I.M., Cohen, C., and McMahon, K. 2010. *Introduction to Logic*, 14th edition. Pearson, New York.

Curry, H. 1977. *Foundations of Mathematical Logic*. Dover Publications, Mineola.

van Dalen, D. 1983. *Logic and Structure*, second edition. Universitext, Springer, New York.

Day, M. 2012. *An Introduction to Proofs and the Mathematical Vernacular*. Virginia Polytechnic Institute and State University, Blacksburg. Available at http://www.math.vt.edu/people/day/ProofsBook (accessed 6 April 2016).

Devlin, K. 2004. *Sets, Functions, and Logic: An Introduction to Abstract Mathematics*, third edition. Chapman Hall/CRC, Boca Raton.

Devlin, K. 2012. *Introduction to Mathematical Thinking*. Keith Devlin Publishing, Standford.

Ebbinghaus, H.-D., Flum, J., and Thomas, W. 1996. *Mathematical Logic*, second edition. Springer, New York.

Enderton, H. 2001. *A Mathematical Introduction to Logic*, second edition. Academic Press, New York.

Fitting, M. 1996. *First-Order Logic and Automated Theorem Proving*, second edition. Springer, New York.

Gallier, J. 1986. *Logic for Computer Science: Foundations of Automatic Theorem Proving*, revised 2003/ Dover Publications, Mineola.

Gamut, L. T. F. (a collective pseudonym for Johan van Benthem, Jeroen Groenendijk, Dick de Jongh, Martin Stokhof, and Henk Verkuyl) 1991. *Logic, Language and Meaning, Volume I: Introduction to Logic*. University of Chicago Press, Chicago.

Garnier, R. and Taylor, J. 1996. *100 % Mathematical Proof*. John Wiley & Sons, New York.

Halbach, V. 2010. *The Logic Manual*. Oxford University Press, Oxford.

Hamilton, A.G. 1988. *Logic for Mathematicians*, revised edition. Cambridge University Press, Cambridge.

Hedman, S. 2004. *A First Course in Logic. An Introduction to Model Theory, Proof Theory, Computability, and Complexity*. Oxford University Press, Oxford.

Hodges, W. 2001. Elementary predicate logic. In: *Handbook of Philosophical Logic* (eds D. M. Gabbay and F. Guenthner), second edition. Kluwer Academic Publishers, Dordrecht, pp. 1–129.

Hodges, W. 2005. *Logic*, second edition. Penguin, London.

Huth, M. and Ryan, M. 2004. *Logic in Computer Science Modelling and Reasoning about Systems*, second edition. Cambridge University Press, Cambridge.

Jeffrey, R. 1994. *Formal Logic: Its Scope and Limits*, third edition. McGraw-Hill, New York.

Kalish, D. and Montague, R. 1980. *Logic: Techniques of Formal Reasoning*, second edition. Harcourt Brace Jovanovich, San Diego.

Kleene, S. C. 2002. *Mathematical Logic*, Dover Publications, USA. (Originally published by North-Holland in 1962.)

Makinson, D. 2008. *Sets, Logic and Maths for Computing*. Springer, New York.

Mendelson, E. 1997. *Introduction to Mathematical Logic*, fourth edition. CRC Press, New York.

Nederpelt, R. and Kamareddine, F. 2004. *Logical Reasoning: A First Course*. Kings College, London.

Nerode, A. and Shore, R. 1993. *Logic for Applications*. Springer, New York.

Nilsson, U. and Małuszyński, J. 1995. *Logic, Programming and Prolog*, second edition. John Wiley & Sons Ltd. Also available at http://www.ida.liu.se/~ulfni53/lpp/ (accessed 6 April 2016).

van Oosten, J. 1999. *Introduction to Peano Arithmetic*. Utrecht University. Also available at http://www.staff.science.uu.nl/~ooste110/syllabi/peanomoeder.pdf} (accessed 6 April 2016).

Prawitz, D. 2006. *Natural Deduction: A Proof-Theoretical Study*. Dover Publications, USA. (Originally published at Stockholm University in 1965.)

Robinson, A. and Voronkov, A. (eds.) 2001. *Handbook of Automated Reasoning*, volumes 1 and 2. Elsevier.

Russell, S. and Norvig, P. 2003. *Artificial Intelligence: A Modern Approach*. Prentice-Hall, New York.

Shoenfield, J. 1967. *Mathematical Logic*. Addison-Wesley, Boston.

Smith, P. 2003. *An Introduction to Formal Logic*. Cambridge University Press, Cambridge.

Smith, P. 2013. *An Introduction to Gödel's Theorems*, second edition. Cambridge University Press, Cambridge.

Smullyan, R. 1995. *First-Order Logic*. Dover Publications, USA.

Smullyan, R. 1998. *The Riddle of Scheherazade and Other Amazing Puzzles*. Mariner Books, Boston, MA.

Smullyan, R. 2009a. *The Lady or the Tiger? and Other Logic Puzzles*. Dover Publications, USA.

Smullyan, R. 2009b. *Satan, Cantor and Infinity: Mind-Boggling Puzzles*. Dover Publications, USA.

Smullyan, R. 2011. *What Is the Name of This Book? The Riddle of Dracula and Other Logical Puzzles*. Dover Publications, USA.

Smullyan, R. 2013. *The Gödelian Puzzle Book: Puzzles, Paradoxes and Proofs*. Publications, USA.

Smullyan, R. 2014. *A Beginner's Guide to Mathematical Logic*. Dover Publications, USA, Books on Mathematics.

Solow, D. 1990. *How to Read and Do Proofs: An Introduction to Mathematical Thought Processes*. John Wiley & Sons, New York.

Spivey, M. 1995. *An Introduction to Logic Programming Through Prolog*. Prentice-Hall International, New York. Also available at http://spivey.oriel.ox.ac.uk/wiki/files/lp/logic.pdf (accessed 10 April 2016).

Tarski, A. 1965. *Introduction to Logic and to the Methodology of Deductive Sciences*. Oxford University Press, Oxford.

Velleman, D. 2006. *How To Prove It. (A Structured Approach)*, second edition. Cambridge University Press, Cambridge.

Wos, L. T. and Robinson, G. A. 1969. *Paramodulation and theorem proving in first-order theories with equality*. Elsevier, New York, Machine Intelligence Vol 4, pp. 135–150.

Index

Logic as a Tool: A Guide to Formal Logical Reasoning, First Edition. Valentin Goranko.
© 2016 John Wiley & Sons, Ltd. Published 2016 by John Wiley & Sons, Ltd.